VIRUSES AND CANCER

SYMPOSIA OF THE
SOCIETY FOR GENERAL MICROBIOLOGY*

* Published by the Cambridge University Press, except for the first Symposium, which was
published by Blackwell's Scientific Publications Limited

VIRUSES AND CANCER

EDITED BY

P. W. J. RIGBY AND N. M. WILKIE

THIRTY-SEVENTH SYMPOSIUM OF
THE SOCIETY FOR GENERAL MICROBIOLOGY
HELD AT
THE UNIVERSITY OF WARWICK
APRIL 1985

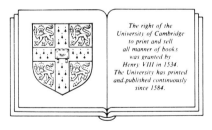

The right of the
University of Cambridge
to print and sell
all manner of books
was granted by
Henry VIII in 1534.
The University has printed
and published continuously
since 1584.

Published for the Society for General Microbiology

CAMBRIDGE UNIVERSITY PRESS

CAMBRIDGE

LONDON   NEW YORK   NEW ROCHELLE

MELBOURNE   SYDNEY

Published by the Press Syndicate of the University of Cambridge
The Pitt Building, Trumpington Street, Cambridge CB2 1RP
32 East 57th Street, New York, NY 10022, USA
10 Stamford Road, Oakleigh, Melbourne 3166, Australia

First published 1985

Printed in Great Britain at The Pitman Press, Bath

Library of Congress catalogue card number: 84–23136

British Library Cataloguing in Publication Data

Society for General Microbiology. *Symposium*
(*37th: 1985: University of Warwick*)
Viruses and cancer. – (Symposia of the Society
for General Microbiology; 37)
1. Viral carcinogenesis   2. Oncogenic viruses
I. Title   II. Rigby, P.W.J.
III. Wilkie, N. M.   IV. Series
616.99′40194   RC268.57
ISBN 0 521 26867 2

CONTRIBUTORS

BARBACID, M. Developmental Oncology Section, NCI-Frederick Cancer Research Facility, Frederick, Maryland 21701, USA

BRUCK, C. Department of Molecular Biology, University of Brussels, Belgium

BURNY, A. Faculty of Agronomy, Gembloux, and Department of Molecular Biology, University of Brussels, Belgium

BYRD, P. J. Department of Cancer Studies, University of Birmingham, The Medical School, Birmingham B15 2TJ, UK

CHANG, S. E. Marie Curie Memorial Foundation Research Institute, The Chart, Oxted, Surrey RH8 0TL, UK

CLEUTER, Y. Department of Molecular Biology, University of Brussels, Belgium

COUEZ, D. Department of Molecular Biology, University of Brussels, Belgium

DESCHAMPS, J. Department of Molecular Biology, University of Brussels, Belgium

DICKSON, C. Imperial Cancer Research Fund, PO Box 123, Lincoln's Inn Fields, London WC2A 3PX, UK

EPSTEIN, M. A. Department of Pathology, University of Bristol Medical School, University Walk, Bristol BS8 1TD, UK

GALLIMORE, P. H. Department of Cancer Studies, University of Birmingham, The Medical School, Birmingham B15 2TJ, UK

GALLO, R. C. Laboratory of Tumor Cell Biology, Developmental Therapeutic Program, Division of Cancer Treatment, National Cancer Institute, Bethesda, Maryland 20205, USA

GHYSDAEL, J. Department of Molecular Biology, University of Brussels, Belgium

GRAND, R. J. A. Department of Cancer Studies, University of Birmingham, The Medical School, Birmingham B15 2TJ, UK

GREGOIRE, D. Department of Molecular Biology, University of Brussels, and National Institute for Veterinary Research, Uccle-Brussels, Belgium

GRIFFIN, B. E. Imperial Cancer Research Fund, PO Box 123, Lincoln's Inn Fields, London WC2A 3PX, UK

HALPERN, M. S. The Wistar Institute, Philadelphia, Pennsylvania 19104, USA

HATTORI, S. Department of Viral Oncology, Cancer Institute, Kami-Ikebukuro, Toshima-ku, Tokyo 170, Japan

HOWLEY, P. M. Laboratory of Pathology, National Cancer Institute, National Institutes of Health, Bethesda, Maryland 20205, USA

KARRAN, L. Imperial Cancer Research Fund, PO Box 123, Lincoln's Inn Fields, London WC2A 3PX, UK

KETTMANN, R. Faculty of Agronomy, Gembloux, and Department of Molecular Biology, University of Brussels, Belgium

KING, D. Imperial Cancer Research Fund, PO Box 123, Lincoln's Inn Fields, London WC2A 3PX, UK

KIYOKAWA, T. Department of Viral Oncology, Cancer Institute, Kami-Ikebukuro, Toshima-ku, Tokyo 170, Japan

MAMMERICKX, M. Department of Molecular Biology, University of Brussels, and National Institute for Veterinary Research, Uccle-Brussels, Belgium

MARBAIX, G. Department of Molecular Biology, University of Brussels, Belgium

MARTIN-ZANCA, D. Developmental Oncology Section, NCI-Frederick Cancer Research Facility, Frederick, Maryland 21701, USA

MASON, W. S. Institute for Cancer Research, Philadelphia, Pennsylvania 19111, USA

MOLNAR-KIMBER, K. L. Institute for Cancer Research, Philadelphia, Pennsylvania, 19111, USA

NEIL, J. C. Beatson Institute for Cancer Research, Wolfson Laboratory for Molecular Pathology, Bearsden, Glasgow G61 1BD, UK

CONTRIBUTORS

NEWBOLD, J. Department of Microbiology and Immunology, University of North Carolina, Chapel Hill, North Carolina 27514, USA

PETERS, G. Imperial Cancer Research Fund, PO Box 123, Lincoln's Inn Fields, London WC2A 3PX, UK

PORTETELLE, D. Faculty of Agronomy, Gembloux, and Department of Molecular Biology, University of Brussels, Belgium

RABSON, M. S. Laboratory of Pathology, National Cancer Institute, National Institutes of Health, Bethesda, Maryland 20205, USA

RATNER, L. Laboratory of Tumor Cell Biology, Developmental Therapeutic Program, Division of Cancer Treatment, National Cancer Institute, Bethesda, Maryland 20205, USA

RIGBY, P. W. J. Cancer Research Campaign, Eukaryotic Molecular Genetics Research Group, Department of Biochemistry, Imperial College of Science and Technology, London SW7 2AZ, UK

ROGLER, C. Liver Research Center, Albert Einstein College of Medicine, Bronx, New York 10461, USA

SANTOS, E. Developmental Oncology Section, NCI-Frederick Cancer Research Facility, Frederick, Maryland 21701, USA

SARIN, P. S. Laboratory of Tumor Cell Biology, Developmental Therapeutic Program, Division of Cancer Treatment, National Cancer Institute, Bethesda, Maryland 20205, USA

SEIKI, M. Department of Viral Oncology, Cancer Institute, Kami-Ikebukuro, Toshima-ku, Tokyo 170, Japan

SUKUMAR, G. Developmental Oncology Section, NCI-Frederick Cancer Research Facility, Frederick, Maryland 21701, USA

SUMMERS, J. Institute for Cancer Research, Philadelphia, Pennsylvania 19111, USA

WATANABE, T. Department of Viral Oncology, Cancer Institute, Kami-Ikebukuro, Toshima-ku, Tokyo 170, Japan

WEISS, R. A. Institute of Cancer Research, Chester Beatty Laboratories, Fulham Road, London SW3 6JB, UK

WILKIE, N. M. Beatson Institute for Cancer Research, Wolfson

Laboratory for Molecular Pathology, Bearsden, Glasgow G61 1BD, UK

WONG-STAAL, F. Laboratory of Tumor Cell Biology, Developmental Therapeutic Program, Division of Cancer Treatment, National Cancer Institute, Bethesda, Maryland 20205, USA

YOSHIDA, M. Department of Viral Oncology, Cancer Institute, Kami-Ikebukuro, Toshima-ku, Tokyo 170, Japan

YU-CHANG YANG Laboratory of Pathology, National Cancer Institute, National Institutes of Health, Bethesda, Maryland 20205, USA

ZARBL, H. Developmental Oncology Section, NCI-Frederick Cancer Research Facility, Frederick, Maryland 21701, USA

ZUCKERMAN, A. J. Department of Medical Microbiology and WHO Collaborating Centre for Reference and Research on Viral Hepatitis, London School of Hygiene and Tropical Medicine (University of London), London WC1E 7HT, UK

ZUR HAUSEN, H. Deutsches Krebsforschungszentrum, Im Neuenheimer Feld 280, D-6900 Heidelberg 1, Federal Republic of Germany

CONTENTS

EDITORS' PREFACE

The field of tumour virology could be said to have its origins in the studies of Ellerman and Bang and of Peyton Rous in the period between 1908 and 1911. These early pioneers reported the successful transmission of avian tumours by cell- and bacteria-free extracts of tumour cells. As we now know, the transmissible agents in these experiments were retroviruses. Subsequently, Shope (in 1933) and Gross (in 1953) identified other viruses which induce benign warts in rabbits and parotid tumours in laboratory mice. These agents proved to have small DNA genomes and to belong to the group of viruses we now know as papovaviruses. The rapid expansion of virus research during the late 1950s and the 1960s soon led to the realisation that other virus groups, including adenoviruses, herpesviruses and poxviruses, have oncogenic potential or at least the potential for altering the control of cell proliferation.

Despite the abundant evidence from laboratory studies for the oncogenic potential of these viruses, many workers remained sceptical about their role in naturally occurring cancer. The experiments were considered to be interesting *in vitro* assays for virus functions but with no real relevance to the genesis of tumours under natural conditions. However, with the advent of improved methods of diagnosis and classification of tumours and the impact of the new molecular technologies on the detection and analysis of tumour-associated viruses, this view has altered dramatically. Of crucial importance to this changing perception has been the analysis of virus-related cancers in animals, where the role of the virus in tumourigenesis can be tested directly. It is now abundantly clear that horizontally transmitted viruses play important causative roles in a number of major cancers of animals and humans. Thus, retroviruses directly induce lymphoid malignancies in cats, cattle and almost certainly in man. Interestingly, they can also cause immune deficiency syndromes in animals, which may be relevant to the reported association of a human T-Cell leukaemia virus with Acquired Immune Deficiency Syndrome (AIDS). Hepatitis B viruses, which have some characteristics in common with retroviruses although they contain a small DNA genome, are closely associated with primary hepatocellular carcinoma in animals and in man. Papovaviruses are now known to play a major role in the genesis of carcinomas in cattle and in urogenital cancer in humans. Although the role of *Herpes simplex* virus in human urogenital carcinoma is still unclear, another human herpesvirus, Epstein–Barr virus, is an important co-factor in the genesis of Burkitt's lymphoma and the more numerically important undifferentiated nasopharyngeal carcinoma.

The emerging realisation of the relevance of viruses to major naturally occurring cancers underlines the importance of continued support for research. Despite recent major advances we still know remarkably little about the mechanisms of transmission, replication, persistence and malignant transformation. There is a need to refine epidemiological and diagnostic methods, and to search for as yet unrecognised virus–tumour associations and ways of limiting the horizontal spread of tumour viruses. In this Symposium volume, we attempt to summarise recent progress in these areas and present new information from major laboratories engaged in relevant original research.

We are indebted to all the authors who willingly spared time from their busy schedules to prepare their manuscripts, to the patient and long-suffering staff of Cambridge University Press, and to Roger Berkeley, Meetings Secretary to the Society for General Microbiology.

Cancer Research Campaign,   Peter Rigby
Eukaryotic Molecular Genetics Research Group,
Department of Biochemistry,
Imperial College of Science and Technology

Beatson Institute for Cancer Research,   Neil Wilkie
Wolfson Laboratory for Molecular Pathology,
Bearsden, Glasgow

UNRAVELLING THE COMPLEXITIES
OF CARCINOGENESIS

ROBIN A. WEISS

Institute of Cancer Research, Chester Beatty Laboratories, Fulham Road, London SW3 6JB, UK

'Cancer stands at what might be the meeting-place, but what is in fact the no-man's land, between the three disciplines of heredity, development, and infection', declared Darlington (1948) at a symposium on the genetics of cancer nearly 40 years ago. Since then, cancer has indeed become the meeting-place and the terrain has been mapped and developed largely by studies of viral carcinogenesis. Oncogenic viruses have been more amenable to molecular analysis than the host genome, and thus the way they effect changes in the regulation of host cells during malignant transformation can be examined in great detail following infection. With integrating viruses, the intimate association between viral and host genomes has blurred the distinction between heredity and infection. Retroviruses, for example, may on the one hand capture and transduce host oncogenes and on the other may themselves become sequestered in the germ line of their hosts as Mendelian elements (Weiss, 1982).

Viruses are important causes of naturally occurring malignant disease. Although they are not generally cited as environmental carcinogens, viruses probably play as great a role in human cancer mortality as any other agents. Table 1 lists eight viruses which are known from laboratory experimentation to have oncogenic properties, or which are implicated by epidemiological studies for an aetiological role in human tumours. These oncogenic viruses represent all the major families of DNA viruses except the pox group (some of which can be oncogenic in animals), and the one family of RNA viruses, the retroviruses, which form DNA proviruses intracellularly. Members of each family are reviewed by other contributors to this volume.

Oncogenic viruses also serve as model carcinogens giving us insight into malignant transformation generally. Table 2 lists a number of ways by which viruses may induce or promote tumour formation. For the indirect mechanisms, the virus itself need not have transforming properties; for example, the development of Kaposi's sarcoma in

Table 1. *Human viruses with oncogenic properties*

Virus family	Type	Human tumour	Co-factors
Adeno	Types 2, 5, 12	None	
Hepadna	Hepatitis B (HBV)	Hepatocellular carcinoma	Aflatoxin, alcohol, smoking
Herpes	Epstein–Barr (EBV)	Burkitt's lymphoma Immunoblastic lymphoma	Malaria Immunodeficiency
		Nasopharyngeal carcinoma	Nitrosamines, HLA genotype
	Simplex (HSV-II)	Cervical neoplasia?	
	Cytomegalo (CMV)	Kaposi's sarcoma?	Immunodeficiency, HLA genotype
Papova	Papilloma (HPV)	Warts Cervical neoplasia Skin cancer	Smoking, HSV? Genetic disorders, sunlight
	Polyoma (BK, JC)	Neural tumours?	
Retro	Human T-cell leukaemia (HTLV-1)	Adult T-cell leukaemia– lymphoma	Uncertain

Table 2. *Possible mechanisms of tumour induction by viruses*

A. Indirect – neither the tumour cells nor their precursors need to have been infected by the virus.

   1. Cell proliferation following immune stimulation or regeneration of damaged tissue may allow tumours to arise from initiated cells among the expanding cell population.

   2. Immune suppression induced by one virus may allow cells infected by another, oncogenic virus to develop into neoplasia.

B. Direct – the tumour cells or their precursors have been infected by the virus.

   1. The virus is no longer present in the tumour cells, but a transient, 'hit-and-run' infection has caused a heritable neoplastic change in the tumour cell lineage.

   2. All or part of the viral genome persists in the tumour cells:

      a. Integrated viral DNA alters cellular gene arrangement and expression (insertional mutagenesis).

      b. Integrated or free viral DNA expresses a gene product which initiates or maintains cell transformation (viral oncogenes).

30% of cases of acquired immune deficiency syndrome (AIDS) implicates the AIDS retrovirus not as a transforming virus for this tumour but rather as a predisposing co-factor for the tumour which may be primarily induced by another co-factor (perhaps cytomegalo virus). The viruses selected for discussion in this symposium all show evidence of direct oncogenesis through the persistence of some viral sequences in the tumour cells. Molecular analysis of the viral genome and, where integration occurs, of the insertion sites in the host genome, has thrown light upon a set of genes playing a role in the malignant phenotype. The study of 'oncogenes', first identified in viruses, has brought together previously disparate fields of research in chemical carcinogenesis, karyotypic analysis, human tumour cell biology and viral transformation. Thus the recent lessons learned in molecular virology have been at the forefront of modern carcinogenesis studies.

It is the identification of specific genes involved in oncogenesis that has yielded greatest dividends and will continue to do so as the properties and behaviour of the proteins they encode are elucidated. However, it would be oversimplistic to assume that alterations in the expression or coding sequence of one oncogene in a single somatic cell can explain the whole of cancer. The changes from a normal to a fully malignant phenotype take place by many steps, and are not necessarily linear in their progress. Neither do the 'unit characters' (Foulds, 1958) that constitute the malignant phenotype always comprise the same set, nor is there yet a one-to-one correlation between expression of a particular oncogene and a unit character of malignancy. But it is now timely to look back at the pioneering work of Rous, Berenblum and Foulds, who conceived the stepwise pro-gression of carcinogenesis, in the light of our present knowledge of molecular and viral oncology.

It is also appropriate in a symposium on viruses and cancer to remind ourselves that carcinogenesis is multifactorial in its aetiology. Not all infected individuals, let alone all infected cells, are destined to develop malignancy. Various factors contribute to virus-associated cancer, such as genetic predisposition and environmental or dietary carcinogens other than the oncogenic virus itself. It is not easy to make quantitative calculations of the 'risk factor' to be apportioned to each of the multiple agents, including the virus. The most detailed data have accrued from epidemiological studies of human cancer.

There may be more opportunities for prevention and early detec-tion in cancers with a viral causative factor than in other types of cancer. Strategies for the prevention of primary infection are discussed by Epstein and by Zuckerman (this volume) and successful early monitoring of nasopharyngeal carcinoma based on screening for IgA antibodies for Epstein–Barr virus (EBV) has been pursued by Zeng (1984) in China. Immunodeficiency, whether inherited, acquired by infection or by immunosuppressive treatment, acts as a most important risk factor in viral carcinogenesis but not generally in cancers of non-viral aetiology (Kinlen, 1982; Weiss, 1984). One may therefore speculate on the possibilities for intervention when natural immunological controls appear to falter. Those cancers in which the viral genome persists and in which viral antigens are expressed may be the most appropriate for new attempts at immunotherapy.

The genetic basis of cancer, and the concepts of multistage and multifactorial carcinogenesis will now be reviewed briefly to set the scene for the more specialised papers on oncogenic viruses that follow.

SOMATIC MUTATION IN ONCOGENESIS

The notion that carcinogenesis involves somatic genetic change accompanied the rediscovery of Mendelian genetics at the turn of the century, when Boveri identified chromosomes as the repository of genetic material and speculated on chromosome imbalance in cancer (see Boveri, 1929). The cytogenetics of cancer has been assiduously studied since that time (Levan, 1956; Klein, 1983), and with Nowell and Hungerford's (1960) discovery of the Philadelphia chromosome, now known to be a translocation at the site of the *abl* oncogene (Heisterkamp *et al.*, 1983), the search for specific chromosomal deletions and translocations was initiated.

The link between mutation and cancer was first drawn by Muller (1927), who demonstrated the mutagenicity of ionizing radiation in *Drosophila* when X-rays were already known to be carcinogenic, and Auerbach and Robson (1946) first showed that a chemical carcinogen, mustard gas, also induced mutations in *Drosophila*. It is now well known that most carcinogens or their active metabolites are mutagens (McCann *et al.*, 1975), and that even between stereoisomers of benzo[a]pyrene-diolepoxide there is a precise correlation between chemical reaction with DNA, mutagenicity in animal cells, and carcinogenicity *in vivo* (Brookes and Osborne,

1982). Inherited syndromes in man caused by lesions in DNA repair pathways frequently result in a predisposition to certain cancers, though the relationship is complex and would not appear to implicate point mutations (Cairns, 1981). Oncogenic viruses cause genetic change by introducing new genetic material into cells, by insertional mutagenesis on integration, and by inducing chromosome breakage.

The somatic mutation hypothesis of cancer presupposes the clonal origin of the tumour cell population, and where it has been amenable to analysis, clonality has been upheld (Fialkow, 1976; Greaves, 1982). Mendelian inheritance of cancer predisposition also supports the mutational hypothesis, of which heritable retinoblastoma in man is a good example (Murphree and Benedict, 1984).

That cancer involves irreversible genetic change has, however, been challenged by those who hold that cancer is wholly epigenetic in character, requiring no more genetic change than normal processes of differentiation (in which only lymphoid cells have been shown to undergo gene rearrangement). The malignant state may be no more permanent than the commitment to particular pathways of differentiation. In this view, cancer may be envisaged as a population of cells with an essentially normal phenotype, which are partially arrested in maturation, and in which the processes of proliferation and differentiation have become uncoupled (Greaves, 1982). While such derangement is usually found to involve genetic change, mutation might not invariably be required for the development of neoplasia.

The epigenetic nature of cancer has been advocated by Smithers (1962a) based on observations on the spontaneous regression of human tumours undergoing further maturation, e.g. neuroblastoma. Smithers (1962b) extended this view to attack what he called 'cytologism' or the individual cellular basis of malignancy, and more recently Rubin (1980) has attacked what one might call 'geneticism' in oncology. 'Cancer is no more a disease of cells than a traffic jam is a disease of cars,' wrote Smithers (1962b), continuing 'A lifetime of study of the internal-combustion engine would not help anyone to understand our traffic problems. A traffic jam is due to a failure of the normal relationship between driven cars and their environment and can occur whether they themselves are running normally or not.' Yet it is the reductionist analysis of the molecular genetics of tumour cells that has given us recent insight into cancer.

The interaction of tumour cells with normal tissues and their dependence on normal regulatory phenomena has been the subject of much study, from hormone dependency to the social behaviour of cells in culture (Abercrombie, 1979). Most tumour cells are not so anarchic that they can grow independently of all host factors (the cancer 'seed' requires the right 'soil'), but the interaction of tumour cells with the host environment is an elective process of stimulus and response in which, contrary to Smithers' view, the essential lesion almost always exists in the behaviour of the neoplastic cell, and is inherited within the clonal tumour cell population.

Experiments to determine the reversibility of the oncogenic phenotype while retaining proliferative potential have yielded evidence suggesting that some malignancies may be almost entirely epigenetic. McKinnel, Deggins and Labat (1969) introduced nuclei derived from Lucké carcinoma cells (a renal carcinoma induced by a herpes virus) into activated enucleated eggs of *Rana pipiens*. Following transplantation, a small proportion of the tumour nuclei supported the development of tadpoles with normal, differentiated tissues, but it has not been unequivocally demonstrated that these nuclei came from tumour cells rather than the stroma. Mouse teratocarcinomas containing pluripotential embryonal carcinoma (EC) cells have been the subject of extensive study as these cancer cells can differentiate into many types of apparently normal cell. The introduction of EC cells into the embryonic blastocyst allows incorporation into the inner cell mass, and hence the embryo. Using genetic markers it has been shown that EC cells contribute to most, if not all, the normal tissues of the mouse, although such mice have a higher frequency of teratocarcinoma (Papaioannou *et al.*, 1975). It has also been claimed that the descendents of the tumour cells can form a normal germ-line (Mintz and Illmensee, 1975). Similarly, crown-gall tumours in tobacco plants, which are induced by transfer of a bacterial plasmid, can give rise to normal tissues and plants (Braun, 1959). Thus certain pluripotent cancer cells can be restored to heritable normal behaviour by transplantation into, or cultivation in, a suitable environment. It will in future be of interest to determine whether activated oncogenes introduced into the germ-line of mice cause a genetic predisposition to cancer.

These experiments show that the cancer phenotype is not inexorably irreversible but do not wholly refute the somatic mutation hypothesis. The evidence of karyotypic changes in tumour cells, of the mutagenic properties of most carcinogens, and of the alteration of specific genes in oncogenesis, strongly suggests that cancer usually involves genetic change.

Table 3. *Specific genes involved in oncogenesis*

1. Viral genes essential for replication which also contribute to cell transformation: e.g. transforming genes of most DNA tumour viruses.
2. Viral genes superfluous for replication, derived from host cellular genes: e.g. most oncogenes of retroviruses.
3. Cellular genes identified by homology to retrovirus oncogenes: e.g. *myc, ras, abl.*
4. Cellular genes at sites of viral integration: e.g. *myc, erb*-B, *int*-1 and *int*-2.
5. Cellular genes at sites of chromosome translocation: e.g. *myc, abl.*
6. Cellular genes identified by neoplastic transformation following DNA transfer: e.g. *ras*, B-*lym.*
7. Cellular genes activated or repressed in virus-transformed and other neoplastic cells: e.g. MHC genes.
8. Cellular genes whose products interact with those of viral oncogenes: e.g. p53 with SV40 large T.

SPECIFIC GENES IN ONCOGENESIS

The wealth of information that accrued from studies of chemical and physical carcinogens showing that somatic mutation occurs in oncogenesis gave no indication of whether neoplastic transformation was caused by genetic deletion of essential functions (Van Potter, 1962) or the mutational 'activation' of a limited set of specific genes. The latter concept of specific oncogenes, even if it requires deletion of dominant, normal alleles, is now ascendant. It arose from studies of transforming viruses and has been developed more recently from direct molecular genetic analysis of tumour cells.

Table 3 lists different classes of genes involved in oncogenesis. It is remarkable how the same families of genes have been repeatedly identified by different methods of isolation. In particular, retrovirus oncogenes were correctly postulated to have arisen from cellular genes (Weiss, 1973; Fischinger and Haapala, 1974; Stehelin *et al.*, 1976; Bishop, 1984). Homologues of viral oncogenes have been found in host DNA at sites of virus integration and of chromosome translocation, and in more than one independently derived retrovirus, as well as by transfection of DNA derived from naturally occurring or chemically induced tumours (Duesberg, 1983; Klein, 1983; Marshall, 1984; Santos *et al.*, this volume). These findings suggest that the same limited set of genes is involved in tumours induced by disparate carcinogens. Analysis of these genes, and their products, can tell us much about the molecular changes in neoplastic cells (Bishop, 1984; Cooper, 1984; Hunter, 1984; Marshall, 1984).

Table 4. *Retrovirus oncogenes*

Oncogene	Product
src	Tyrosine phosphokinases
fes, fps	or structurally related
fgr	proteins
yes	
abl	
ros	
mos	
mil, raf	
erb-B	EGF receptor
sis	PDGF
ras (*Ha-, Ki-*)	GTP-ase
myc	DNA-binding
myb	
fms	Uncertain
erb-A	
fos	
ets	
ski	
rel	

Table 4 summarises retrovirus oncogenes and their products. Some 20 viral oncogenes of cellular origin have been defined, and encode proteins with different functions, located at different sites within the infected cell (Hunter, 1984). Some are tissue-specific in their effects but most are not. Figure 1 depicts the sites of action of oncogene products, ranging from growth factors and receptors to nuclear proteins. The viral oncogenes appear to be mutated from their cellular 'proto-oncogenes' (Duesberg, 1983), and this may give them more neoplastic properties. Moreover, ectopic expression, often in excess amounts due to viral transcriptional control, may also promote cell transformation. Similarly, cellular oncogenes may be 'activated' to a neoplastic function by point mutation of a coding sequence, e.g. *ras* genes (Santos *et al.*, this volume), or by gene amplification and/or ectopic expression, e.g. *myc*, in which the necessity for mutation in the coding sequence remains uncertain (Klein, 1983; Marshall, 1984). So either qualitative or quantitative changes can cause otherwise normal genes to act as oncogenes.

Transforming retroviruses generally carry only one oncogene, though some have captured two, derived from unlinked genes in the host genome. However, most naturally occurring, endemic retro-

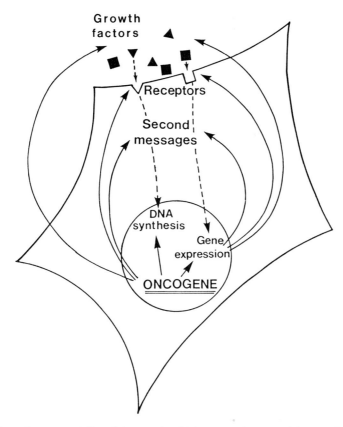

Fig. 1. Schematic representation of the ways in which untoward c-*onc* activity may disrupt the normal regulation of cell growth. The figure depicts a cell, with nucleus, that bears on its surface receptors (open square and triangle) for external stimuli such as growth factors (solid squares and triangles). Arrows on the left of the figure indicate that an oncogene in the cell DNA may directly encode molecules acting as external factors, receptors for these factors or components of second messages that lead to specific gene expression or DNA synthesis. Arrows within the nucleus indicate that DNA synthesis or specific gene expression may be mediated directly by the action of an oncogene product located in the nucleus. Arrows on the right of the figure suggest that oncogene-induced gene expression may have an effect on reception and processing of environmental factors. (From Wyke and Weiss, 1984.)

viruses have no oncogenes, and the epidemic transmission of on-cogene-bearing retroviruses appears to be limited (Weiss, 1982; Neil, this volume). The tumours induced by naturally occurring retro-viruses frequently develop from those cells in which the provirus has integrated adjacent to a particular cellular oncogene, thereby ac-tivating it (Hayward *et al.*, 1981; Dickson and Peters, this volume). Alternatively, naturally occurring cancer can also arise from a cell infected with an oncogene-containing retrovirus (e.g. spontaneous feline T-lymphomas; see Neil, this volume). In this case it is not

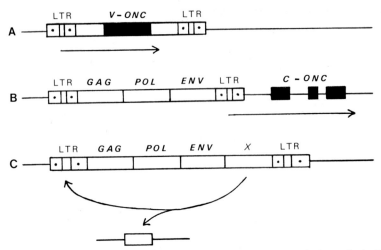

Fig. 2. Molecular models of retrovirus oncogenesis. The proviral DNA is shown flanked by 5′ and 3′ long terminal repeats (LTRs) integrated into host cellular DNA. (A) LTR-initiated transcription of a transduced oncogene (v-*onc*) within a defective viral genome. (B) Integration of a non-defective provirus in the vicinity of a cellular proto-oncogene (c-*onc*) leads to its activation via enhancer or promoter sequences in the adjacent LTR. (C) The HTLV genome carries an *X* gene region whose product, acting in positive feedback, enhances transcription of its own LTR and also other transcription units. The transcription of v-*onc* and c-*onc* genes in A and B is *cis*-controlled, whereas activation of the LTR or cellular genes by the *X* product in C is *trans*. Model B requires site-specific integration of the provirus, while models A and C are independent of the host chromosomal locus.

known whether the oncogene-containing virus is generated (by recombination with cellular sequences) in the same animal or transmitted by horizontal infection from another animal. However, with lymphomas caused by bovine leukosis virus (Burny *et al.*, this volume) and by human T-cell leukaemia virus (HTLV) (Ratner *et al.*, Yoshida *et al.*, this volume) there is neither evidence of common integration sites in different lymphomas, nor of an oncogene of cellular origin. These viruses have a viral gene region, *X*, extra to the three structural genes (Ratner *et al.*, Yoshida *et al.*, this volume) which encodes a protein thought to bind to the long terminal repeats (LTRs) of the viral genome and which might also bind to tissue-specific enhancer sequences, activating cellular genes and leading to neoplastic transformation (Sodroski, Rosen and Haseltine, 1984). Figure 2 depicts the three models of cell transformation by retroviruses just discussed.

The oncogenes of DNA viruses encode proteins necessary for early stages of viral DNA replication. These genes also activate cellular DNA synthesis and may lead to immortalisation and transformation of the host cell when viral replication is incomplete (see Howley *et*

Table 5. *Synergism between oncogenes*

---

1. Viruses with two or more oncogenes,
   e.g. *a*. Polyoma virus (large, middle and small T-antigens)
       b. Avian erythroblastosis virus (*erb*-A, *erb*-B)
       c. Mill Hill 2 virus (*myc*, *mil*)
2. Human tumours with two or more 'activated' oncogenes,
   e.g. *a*. Burkitt's lymphoma (EBV genes, *myc*, and *ras* or B-*lym*)
       b. HL60 (mutated *ras*, amplified *myc*)
3. Enhanced transformation of cells by two oncogenes,
   e.g. *ras* with *myc* or E1A

---

al., Griffin *et al.*, Mason *et al.*, Gallimore *et al.*, this volume; Lane, Gannon and Winchester, 1982; Rassoulzadegan *et al.*, 1982).

Different oncogenes may act synergistically in inducing neoplastic transformation (Table 5). The presence of more than one oncogene in retroviruses has already been mentioned. DNA viruses such as polyoma virus and adenovirus also carry more than one transforming gene. Some human tumour cells show alteration in more than one cellular oncogene. Furthermore, transformation of cells in culture by cloned, activated *ras* oncogenes can be achieved more efficiently if the cells are already immortalised by a chemical or physical carcinogen (Newbold and Overell, 1983) or are stimulated by co-transformation with an oncogene with quasi-immortalising functions in the nucleus (Land, Parada and Weinberg, 1983; Ruley, 1983). However, the mutated *ras* gene alone can induce transformation of primary cells if overexpressed (Spandidos and Wilkie, 1984).

MULTISTAGE ONCOGENESIS AND TUMOUR PROGRESSION

Rous and Beard (1935) described the stepwise process whereby 'tumors go from bad to worse' in their study of 'progression to carcinoma of virus induced papillomas'. Friedewald and Rous (1944) subsequently coined the terms *initiation* and *promotion* for discrete steps in carcinogenesis. Berenblum greatly extended the two-stage model by his studies of initiating and promoting agents (polycyclic hydrocarbons and croton oil extracts) in chemical carcinogenesis (Berenblum and Shubik, 1947; Berenblum, 1974). Initiating chemicals cause irreversible, heritable but often phenotypically in-apparent mutations in cells. Tumour promotion is at first reversible, but further, irreversible, genetic changes also take place as the

Table 6. *Major steps in carcinogenesis*

INITIATION →	PROMOTION →	PROGRESSION →
Irreversible	Initially reversible	Independent evolution
No threshold	Threshold	Heterogeneity
Dose dependent	Not dose dependent	Increasing autonomy
	Clonal expansion	

tumours progress towards a more malignant state (Table 6). Epidemiological studies of human cancers and their age distribution also led to multistep models of carcinogenesis (Armitage and Doll, 1954; Peto, 1977).

Foulds (1958, 1969) proposed a number of rules for tumour progression, based on his observations of the pathology of experimental and human cancer. He pointed out that the development of neoplasia is discontinuous in time and site and that the stages of carcinogenesis are qualitatively different. The cancer phenotype comprises many unit characters such as growth factor or hormone dependence, invasiveness, or ability to metastasise. A certain number of these need to be acquired for full malignancy to be expressed. Even then, further changes continue to appear with progression to more aggressive forms, and, for example, drug resistance (Skipper, 1983). Foulds realised that such progression was not ordered but resulted from heritable change followed by selection of the more malignant lineages of cells. Regression of tumours might also occur, as already discussed, or there might be persistence of a set of characters for a long period. He wrote: 'The behaviour of tumours is determined by numerous characters that, within wide limits, are independently variable, capable of different combinations and assortments and liable to independent progression' (Foulds, 1969).

Foulds' concepts are as pertinent as ever and are amenable to analysis by molecular approaches (Klein and Klein, 1984). We can now begin to ask whether particular unit characters of malignancy are determined by individual oncogenes, as Rassoulzadegan *et al.* (1982) have shown for polyoma virus transformation, and as Land *et al.* (1983) suggest in categorising oncogenes into 'complementation groups' for the abnormal neoplastic phenotype. Do certain changes

NORMAL   NEOPLASTIC

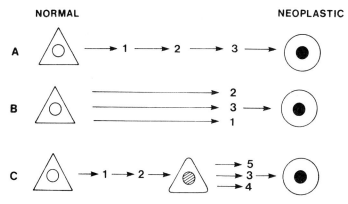

Fig. 3. Multiple steps in neoplasia. The numbers denote arbitrary changes in gene function or expression that lead to neoplasia, and three modes are shown by which a normal cell may be converted into a tumour cell. (A) A sequential model in which the stages are in a fixed order and events important at any particular stage are only effective if the changes required for earlier stages have already occurred. (B) A cumulative model in which changes can occur in any order, neoplasia resulting when a number of changes have accumulated (this number may be less than the total number of possible events). (C) Sequential events convert a normal into a preneoplastic cell which then progresses by cumulative changes in any order to a fully neoplastic condition. (From Wyke and Weiss, 1984.)

have to be expressed before selection for others can occur? Figure 3 outlines possible models of tumour progression, as a series of linear steps, of parallel steps, or of a combination. From Foulds' quotation above, he appeared to favour the possibility of a number of independent, collateral steps, though many students of carcinogenesis have assumed that progression is strictly linear with a set order. A note of caution, however, is that the 'staging' classification of tumours by pathologists in some cases reflects different kinds of tumour rather than serial stages, e.g. cervical intraepithelial neoplasia (zur Hausen, this volume).

In tumour progression, it probably makes no difference whether a nuclear function such as the *myc* gene product is activated before *ras* or vice versa. Moreover, there is evidence that *ras* genes may be activated early or late in the progression of different tumours. Balmain *et al.* (1984) and Santos *et al.* (this volume) detect Ha-*ras* mutation at early stages of chemical carcinogenesis, whereas Albino *et al.* (1984) and Vousden and Marshall (1984) have observed *ras* activation late in progression to metastasis. The mutation of *ras* genes at critical sites reducing the GTP-ase activity of its product (McGrath *et al.*, 1984) may cause tumour progression at any stage, and probably in any cell type.

MULTIFACTORIAL ONCOGENESIS

The development of cancer is multifactorial with a number of risk factors. Which one is the 'crucial cause' of one type of cancer is sometimes hotly debated between proponents of different agents or disciplines of study. Such arguments may be useful in determining the most appropriate means of prevention, though the most incriminating risk factor is not necessarily the one most amenable to elimination. In a symposium on viruses and cancer, it is valuable to recall that viral oncogenesis is no exception to the multifactorial model. With oncogenic retroviruses, for example, malignant thymoma in AKR mice is greatly delayed by slightly reducing dietary intake (Saxton, Boon and Furth, 1944), mammary tumours in C3H mice by the type of woodshaving used as bedding (Heston, 1975), and bursal lymphomatosis in chickens by host genetic factors (Neiman *et al.*, 1980).

Three examples of viral neoplasia in man serve to emphasise multifactorial aetiology:

1. *Nasopharyngeal carcinoma* (NPC) of the undifferentiated type is closely associated with EBV (Epstein, this volume). The EBV genome and nuclear antigen (EBNA) is found in the tumour cells of all cases of undifferentiated NPC, whether occurring in the populations where the disease is endemic or sporadically in other groups. EBV genes can transform epithelial cells *in vitro* (Griffin *et al.*, this volume), although NPC cells are notoriously difficult to establish as cell lines *in vitro*. EBV is a highly prevalent infection in all human populations and there are no notable strain differences in pathogenicity. Yet NPC occurs overwhelmingly in Chinese of Cantonese origin, whether living in Southern China, Hong Kong or South-East Asia, and in certain African populations. In searching for other predisposing co-factors that might restrict NPC to one community when the virus is ubiquitous, two contributory factors have come to light. One is genetic, showing a significant risk (three-fold) for Chinese carrying certain HLA haplotypes (Chan *et al.*, 1983). The other is a strong correlation, with approximately four-fold risk, of weaning infants on a salt-fish diet containing nitrosamines and subsequent development of NPC (Ho *et al.*, 1978; Yu *et al.*, 1981). Thus, viral, dietary and genetic factors contribute towards the high prevalence of NPC among people of Cantonese origin (Table 7). We do not know whether immunisation against EBV infection (Epstein, this volume) or changes in diet at weaning will be the more successful means of

Table 7. *Risk factors in undifferentiated nasopharyngeal carcinoma*

1. EBV infection of target cells
2. Nitrosamines in salt fish ingested in infancy
3. HLA haplotypes A2, BW46, B17

reducing the incidence of this carcinoma. In the meantime, monitoring blood IgA titres for EBV antibodies (Zeng, 1984) is proving a useful method of mass screening for early NPC at a stage when surgery and radiotherapy are highly successful.

2. *Primary hepatocellular carcinoma* (HCC) is a major cause of cancer mortality, though rare in the Western world (Zuckerman, this volume). Unlike NPC, the prevalence of HCC correlates with those areas and communities in which HBV is endemic, and a large prospective study of HCC in Taiwan clearly implicated the virus (Beasley *et al.*, 1981). However, the quantitative incidence of HCC is not perfectly matched with HBV infection as a relatively low prevalence of HCC is found among HBV-infected subjects in countries such as Greenland, New Zealand, and Lesotho. Levels of aflatoxin contamination of staple cereals, pulses and groundnuts also correlates with incidence of HCC (Linsell and Peers, 1977). It appears that both HBV infection and aflatoxin ingestion are significant risk factors for HCC, synergistically increasing tumour incidence (Cook-Mozaffari and van Rensburg, 1984). Alcohol consumption and cigarette smoking are additional risk factors in HCC (Yu *et al.*, 1983). Prevention of primary HBV infection (Zuckerman, this volume) may prevent HCC, and immunisation, if successful, will surely reduce morbidity due to acute and chronic hepatitis. One could argue that HBV infection may exert an indirect effect (Table 8) on HCC by causing

Table 8. *Risk factors in hepatocellular carcinoma*

Agent	Possible effect
Hepatitis B Virus	Transformation of target cells Damage to liver tissue
Aflatoxins	Mutagenesis of target cells Damage to liver tissue
Alcohol	Damage to liver tissue
Smoking	?

Table 9. *Human tumours for which immunodeficiency is a risk factor*

Malignancy	Associated virus
Non-Hodgkin's lymphoma	EBV
Kaposi's sarcoma	CMV ?
Hepatocellular carcinoma	HBV
Cervical carcinoma	HPV-16, 18
Squamous skin carcinoma	HPV ?
Melanoma	HPV or polyoma ?

liver damage that induces compensatory regeneration of uninfected cells initiated by aflatoxin. But both epidemiological considerations (Cook-Mozaffari and Van Rensburg, 1984) and the persistence of HBV sequences in the HCC cells (Mason *et al.*, this volume) suggest that the virus is the initiating carcinogen with aflatoxin acting as a promoting co-factor.

3. *Adult T-cell leukaemia-lymphoma* (Ratner *et al.*, Yoshida *et al.*, this volume) is a distinct form of malignancy of OKT4+ T-cells in which the incidence of disease precisely correlates with geographic regions and communities infected with HTLV-I. Nonetheless, as with most oncogenic viruses, the disease is a rare consequence of infection, developing several decades later. The associated co-factors are not accurately known, but in South-West Japan a higher than expected onset has been noted in the summer months and in subjects infected with microfilaria worms that tend to block lymphatic vessels (Tajima, Tominaga and Suchi, 1982).

Immunological risk factors in viral oncogenesis have been reviewed elsewhere (Kinlen, 1982; Weiss, 1984). One may argue that only those cancers with persistent viral antigens in the tumour cells will be subject to immune surveillance. In that case, immunodeficiencies should only reveal cancers of viral aetiology. Table 9 lists the human malignancies with increased incidence in immunodeficient subjects. While the majority have known associated viruses, a harder search should perhaps be made for viruses in the skin cancers presenting in immunosuppressed patients.

Chronic immunological stimulation may also be a risk factor for malignancy, especially of lymphoid cells. This probably explains why holoendemic malaria is the major risk factor of Burkitt's lymphoma (BL) in children (Epstein, this volume). BL exemplifies many of the complexities outlined in this article. EBV would appear to be an initiating carcinogen causing potential immortalisation of infected B-cells; malaria promotes the development of tumour cell clones; teleocidin-related chemical promoters secreted by bacteria at the roots of decaying milk teeth might explain the typical location of the tumour in the jaw during deciduation (Ito, 1983); a chromosome translocation moving the *myc* gene into a hyperactive immuno-globulin gene domain (Klein, 1983) is associated with a later step towards malignancy (presumably not 'immortalisation' as that function is already provided by EBV), and further 'activated' oncogenes, *ras* and B-*lym* have been identified in BL cells by DNA transfection (Cooper, 1984; Marshall, 1984). The greater risk for boys than for girls remains unexplained.

CONCLUSIONS

There is a real sense of excitement as molecular biologists identify a reasonably small set of genes functionally involved in oncogenesis, and begin to discern the role of the gene products in the normal and neoplastic physiology of the cell. The reductionist, genetic approach pioneered by virologists has yielded great insights into the perturbations that render cells maligant. But the complexity apparent in the natural history of cancer should remind us that molecular methods must be combined with other approaches, such as epidemiological analysis, if we are to unravel and attribute different risk factors and events in the carcinogenetic process.

Regarding human cancer, viruses have emerged as major aetiological agents, with the elucidation of the role of EBV in BL and NPC, of HBV in primary liver cancer, of new strains of papilloma virus in cervical neoplasia, and the discovery of oncogenic human retroviruses. The more we understand the causes of different human malignancies, the more we may hope to take measures to prevent their occurrence, to monitor early tumour development in those at special risk, and to attempt novel treatments designed to interefere with the specific functions that have been altered in the tumour cells.

REFERENCES

ABERCROMBIE, M. (1979). Contact inhibition and malignancy. *Nature*, **281**, 259–62.
ALBINO, A. P., LE STRANGE, R., OLIFF, A. I., FURTH, M. E. and OLD, L. J. (1984). Transforming *ras* genes from human melanoma: A manifestation of tumour heterogeneity? *Nature*, **308**, 69–72.

ARMITAGE, P. and DOLL, R. (1954). The age distribution of cancer and a multi-stage theory of carcinogenesis. *British Journal of Cancer*, **8**, 1–12.

AUERBACH, C. and ROBSON, J. M. (1946). Chemical production of mutations. *Nature*, **157**, 302.

BALMAIN, A., RAMSDEN, M., BOWDEN, G. T. and SMITH, J. (1984). Activation of the mouse cellular Harvey-*ras* gene in chemically induced benign skin papillomas. *Nature*, **307**, 658–60.

BEASLEY, R. P., HWANG, L.-Y., LIN, C.-C. and CHIEN, C.-S. (1981). Hepatocellular carcinoma and hepatitis B virus: A prospective study of 22 707 men in Taiwan. *Lancet*, **ii**, 1129–33.

BERENBLUM, I. (1974). *Carcinogenesis as a Biological Problem*. Oxford, North-Holland.

BERENBLUM, I. and SHUBIK, P. (1947). The role of croton oil applications, associated with a single painting of a carcinogen, in tumour induction of the mouse's skin. *British Journal of Cancer*, **1**, 379–82.

BISHOP, J. M. (1984). Exploring carcinogenesis with retroviruses. In *The Microbe 1984: Part I, Viruses*, ed. B. W. J. Mahy and J. R. Pattison, Society for General Microbiology Symposium 36, pp. 121–47. Cambridge, Cambridge University Press.

BOVERI, T. (1929). *The Origin of Malignant Tumours*. Translated by M. Boveri. Baltimore, Williams & Wilkins.

BRAUN, A. C. (1959). A demonstration of the recovery of the crown-gall tumor cell with the use of complex tumors of single cell origin. *Proceedings of the National Academy of Sciences, USA*, **45**, 932–8.

BROOKES, P. and OSBORNE, M. R. (1982). Mutation in mammalian cells by stereoisomers of *anti*-benzo[a]pyrene-diolepoxide in relation to the extent and nature of the DNA reaction products. *Carcinogenesis*, **3**, 1223–6.

CAIRNS, J. (1981). The origin of human cancers. *Nature*, **289**, 353–7.

CHAN, S. H., DAY, N. E., KUNARATNAM, N., CHIA, K. B. and SIMONS, M. J. (1983). HLA and nasopharyngeal carcinoma in Chinese – a further study. *International Journal of Cancer*, **32**, 171–6.

COOK-MOZAFFARI, P. and VAN RENSBURG, S. (1984). Cancer of the liver. *British Medical Bulletin*, **40**, 342–5.

COOPER, G. M. (1984). Activation of transforming genes in neoplasms. *British Journal of Cancer*, **50**, 137–42.

DARLINGTON, C. D. (1948). The plasmagene theory of the origin of cancer. *British Journal of Cancer*, **2**, 118–26.

DUESBERG, P. H. (1983). Retroviral transforming genes in normal cells? *Nature*, **304**, 219–26.

FIALKOW, P. J., (1976). Clonal origin of human tumors. *Biochimica et Biophysica Acta*, **458**, 283–321.

FISCHINGER, P. J. and HAAPALA, D. K. (1974). Oncoduction. A unifying hypothesis of viral carcinogenesis. *Progress in Experimental Tumor Research*, **19**, 1–22.

FOULDS, L. (1958). The natural history of cancer. *Journal of Chronic Diseases*, **8**, 2–37.

FOULDS, L. (1969). *Neoplastic Development*, vol. 1. London, Academic Press.

FRIEDEWALD, W. F. and ROUS, P. (1944). The initiating and promoting elements in tumor production. *Journal of Experimental Medicine*, **80**, 101–26.

GREAVES, M. F. (1982). Leukaemogenesis and differentiation: A commentary on recent progress and ideas. *Cancer Surveys*, **1**, 189–204.

HAYWARD, W. S., NEAL, B. G. and ASTRIN, S. M. (1981). Activation of a cellular *onc* gene by promoter insertion in ALV-induced lymphoid leukosis. *Nature*, **290**, 475–80.

HEISTERKAMP, N., STEPHENSON, J. R., GROFFEN, J., HANSEN, P. F., DE KLEIN, A., BARTRAM, C. R. and GROSVELD, G. (1983). Localization of the c-*abl* oncogene adjacent to a translocation break point in chronic myelocytic leukaemia. *Nature*, **306**, 239–42.

HESTON, W. E. (1975). Testing for possible effects of cedar wood shavings and diet on occurrence of mammary gland tumors and hepatomas in C3H-A^vy and C3H-A^vy fB mice. *Journal of the National Cancer Institute*, **54**, 1011–14.

HO, J. H. C., HUANG, D. P. and FONG, Y. Y. (1978). Salted fish and nasopharyngeal carcinoma in southern Chinese. *Lancet*, **ii**, 626.

HUNTER, T. (1984). The proteins of oncogenes. *Scientific American*, **251**, 60–9.

ITO, Y. (1983). Possible roles of Epstein–Barr virus, normal flora microbes and promoter plant diterpene esters in etiology of Burkitt's lymphoma. *Leukemia Reviews International*, **1**, 9–10.

KINLEN, L. (1982). Immunosuppressive therapy and cancer. *Cancer Surveys*, **1**, 565–83.

KLEIN, G. (1983). Specific chromosomal translocations and the genesis of B-cell-derived tumors in mice and men. *Cell*, **32**, 311–15.

KLEIN, G. and KLEIN, E. (1984). Oncogene activation and tumor progression. *Carcinogenesis*, **5**, 596–602.

LAND, H., PARADA, L. F. and WEINBERG, R. A. (1983). Tumorigenic conversion of primary embryo fibroblasts requires at least two cooperating oncogenes. *Nature*, **304**, 1596–602.

LANE, D. P., GANNON, J. and WINCHESTER, G. (1982). The complex between p53 and SV40 T antigen. *Advances in Viral Oncology*, **2**, 23–39.

LEVAN, A. (1956). Chromosomes in cancer tissue. *Annals of the New York Academy of Sciences*, **63**, 774–92.

LINSELL, C. A. and PEERS, F. G. (1977). Field studies on liver cell cancer. In *Origins of Human Cancer*, ed. H. H. Hiatt, J. D. Watson and J. A. Winsten, pp. 549–56. Cold Spring Harbor, Cold Spring Harbor Laboratories.

McCANN, J., CHOI, E., YAMASAKI, E. and AMES, B. N. (1975). Detection of carcinogens as mutagens in the *Salmonella*/microsome test: Assay of 300 chemicals. *Proceedings of the National Academy of Sciences, USA*, **72**, 5135–9.

McGRATH, J. P., CAPON, D. J., GOEDDEL, D. V. and LEVINSON, A. D. (1984). Comparative biochemical properties of normal and activated human *ras* p21 protein. *Nature*, **310**, 644–9.

McKINNELL, R. G., DEGGINS, B. A. and LABAT, D. D. (1969). Transplantation of pluripotential nuclei from triploid frog tumors. *Science*, **165**, 394–6.

MARSHALL, C. J. (1984). Human oncogenes. In *RNA Tumor Viruses*, vol. 2, ed. R. Weiss, N. Teich, H. Varmus and J. Coffin. Cold Spring Harbor, Cold Spring Harbor Laboratories, in press.

MINTZ, B. and ILLMENSEE, K. (1975). Normal genetically mosaic mice produced from malignant teratocarcinoma cells. *Proceedings of the National Academy of Sciences, USA*, **72**, 3585–9.

MULLER, H. J. (1927). Artificial transmutation of the gene. *Science*, **66**, 84–7.

MURPHREE, A. L and BENEDICT, W. F. (1984). Retinoblastoma: Clues to human oncogenesis. *Science*, **223**, 1028–33.

NEIMAN, P. E., JORDAN, L., WEISS, R. A. and PAYNE, L. N. (1980). Malignant lymphoma of the bursa of Fabricius: Analysis of early transformation. *Cold Spring Harbor Conference on Cell Proliferation*, **7**, 519–28.

NEWBOLD, R. F. and OVERELL, R. W. (1983). Fibroblast immortality is a prerequisite for transformation by EJ c-*Ha-ras* oncogene. *Nature*, **304**, 648–51.

NOWELL, P. C. and HUNGERFORD, D. A. (1960). A minute chromosome in human chronic granulocytic leukemia. *Science*, **132**, 1497.

Papaioannou, V. E., McBurney, M. W., Gardner, R. L. and Evans, M. J. (1975). Fate of teratocarcinoma cells injected into early mouse embryos. *Nature*, **258**, 70–3.

Peto, R. (1977). Epidemiology, multistage models, and short-term mutagenicity tests. In *Origins of Human Cancer*, ed. H. H. Hiatt, J. D. Watson and J. A. Winsten, pp. 1403–28. Cold Spring Harbor, Cold Spring Harbor Laboratories.

Rassoulzadegan, M., Cowie, A., Carr, A., Glaichenhaus, N., Kamen, R. and Cuzin, F. (1982). The roles of individual polyoma virus early proteins in oncogenic transformation. *Nature*, **300**, 713–18.

Rous, P. and Beard, J. W. (1935). The progression to carcinoma of virus-induced rabbit papillomas (Shope). *Journal of Experimental Medicine*, **62**, 523–48.

Rubin, H. (1980). Is somatic mutation the major mechanism of malignant transformation? *Journal of the National Cancer Institute*, **64**, 995–1000.

Ruley, H. E. (1983). Adenovirus early region 1A enables viral and cellular transforming genes to transform primary cells in culture. *Nature*, **304**, 1602–6.

Saxton, J. A., Boon, M. C. and Furth, J. (1944). Observations on the inhibition of development of spontaneous leukemia in mice by underfeeding. *Cancer Research*, **4**, 401–9.

Skipper, H. E. (1983). The forty-year-old mutation theory of Luria and Delbruck and its pertinence to cancer chemotherapy. *Advances in Cancer Research*, **40**, 331–63.

Smithers, D. W. (1962a). Spontaneous regression of tumours. *Clinical Radiology*, **13**, 132–7.

Smithers, D. W. (1962b). Cancer – an attack on cytologism. *Lancet*, **i**, 493–9.

Sodroski, J. G., Rosen, C. A. and Haseltine, W. A. (1984). *Trans*-acting transcriptional activation of the long terminal repeat of human T lymphotropic viruses in infected cells. *Science*, **225**, 381–5.

Spandidos, D. A. and Wilkie, N. M. (1984). Malignant transformation of early passage rodent cells by a single mutated human oncogene. *Nature*, **310**, 469–75.

Stehelin, D., Varmus, H. E., Bishop, J. M. and Vogt, P. K. (1976). DNA related to the transforming gene(s) of avian sarcoma viruses is present in normal avian DNA. *Nature*, **260**, 170–3.

Tajima, K., Tominaga, S. and Suchi, T. (1982). Clinico-epidemiological analysis of adult T-cell leukemia. *Gann Monograph*, **28**, 197–210.

Van Potter, R. (1962). Enzyme studies on the deletion hypothesis of carcinogenesis. In *M. D. Anderson Symposium on The Molecular Basis of Neoplasia*, pp. 367–99. Austin, University of Texas Press.

Vousden, K. M. and Marshall, C. J. (1984). Three different activated *ras* genes in mouse tumours: Evidence for oncogene activation during progression of a mouse lymphoma. *EMBO Journal*, **3**, 913–17.

Weiss, R. A. (1973). Transmission of cellular genetic elements by RNA tumour viruses. In *Possible Episomes in Eukaryotes, IV Lepetit Colloquium*, ed. L. G. Silvestri, pp. 130–41. Amsterdam, North-Holland.

Weiss, R. A. (1982). The persistence of retroviruses. In *Virus Persistence*, ed. B. W. J. Mahy, A. C. Minson and G. K. Darby, Society for General Microbiology Symposium 33, pp. 267–88. Cambridge, Cambridge University Press.

Weiss, R. A. (1984). Viruses and human cancer. In *The Microbe 1984: Part I, Viruses*, ed. B. W. J. Mahy and J. R. Pattison, Society for General Microbiology Symposium 36, pp. 211–40. Cambridge, Cambridge University Press.

Wyke, J. A. and Weiss, R. A. (1984). The contribution of tumour viruses to human and experimental oncology. *Cancer Surveys*, **3**, 1–24.

Yu, M. C., Ho, J. H. C., Ross, R. K. and Henderson, B. E. (1981). Naso-pharyngeal carcinoma in Chinese – salted fish or inhaled smoke? *Preventive Medicine*, **10**, 15–24.

Yu, M. C., Mack, T., Hanisch, R., Peters, R. L., Henderson, B. E. and Pike, M. C. (1983). Hepatitis, alcohol consumption, cigarette smoking, and hepatocellular carcinoma in Los Angeles. *Cancer Research*, **43**, 6077–9.

Zeng, Y. (1984). Serological screening programme for nasopharyngeal carcinoma in China. In *Advances in Viral Oncology*, ed. G. Klein. New York, Raven Press, in press.

MOLECULAR BIOLOGY OF THE REPLICATION OF HEPATITIS B VIRUSES

WILLIAM S. MASON*, MICHAEL S. HALPERN†, JOHN NEWBOLD‡, CHARLES ROGLER§, KATHERINE L. MOLNAR-KIMBER*, and JESSE SUMMERS*

*Institute for Cancer Research, Philadelphia, PA 19111, USA
†The Wistar Institute, Philadelphia, PA 19104, USA
‡Department of Microbiology and Immunology, University of North Carolina, Chapel Hill, NC 27514, USA
§Liver Research Center, Albert Einstein College of Medicine, Bronx, NY 10461, USA

The hepadna virus family is a group of small DNA viruses with biological and biochemical similarities to human hepatitis B virus (HBV). These similarities include a novel mode of genome replication, involving reverse transcription of an RNA intermediate, the apparent capacity to maintain persistent infections at the cellular level, and a narrow host range including, but not limited to, parenchymal cells of the liver. The hepadna viruses have drawn considerable attention because of the impact of HBV on public health (Szmuness, 1975). As with the other members of the family, HBV can cause either acute or chronic infections, the latter often persisting for the lifetime of the individual. Although acute infections may have a range of effects, from asymptomatic to liver failure, chronic infections pose a more serious problem. Persistent infections by HBV are associated not only with chronic liver disease but, after several decades, with development of hepatocellular carcinoma (Beasley, 1982). In many parts of the world, chronic infections result from neonatal transmission from infected mothers, and in areas where HBV infections are endemic, liver cancer is a leading cause of death.

The status of HBV in the generation of human disease is discussed in the article by Zuckerman (this volume). Here we discuss the results of studies on animal models for HBV, with emphasis on the molecular biology of virus reproduction and the evolution of persistent infections. Mention is also made of recent histological and molecular studies, indicating that the tropism of hepadna viruses is not as limited as originally thought, extending as it does beyond the liver to a variety of organs and tissues.

ANIMAL MODELS FOR HUMAN HEPATITIS B VIRUS

Human hepatitis B-like viruses have been identified in woodchucks (*Marmota monax*: Summers, Smolec and Snyder, 1978), ground squirrels (*Spermophilus beecheyi*: Marion *et al.*, 1980) and domestic ducks (Mason, Seal and Summers, 1980; Zhou, 1980). As with the human virus, the animal viruses are all highly hepatotropic and are released into the bloodstream from infected cells, producing a recognizable viremia. Two distinct forms of viral particles are invariably found in infectious sera: the surface antigen particles and virions. The former, which are present in vast excess, are noninfectious, lack viral DNA, and contain viral envelope protein as the only identifiable virus constituent. These surface antigen particles appear in mammals as rods and spheres with diameters of *ca* 22 nm and, in ducks, as pleomorphic, roughly spherical, particles with diameters ranging from 35 to 60 nm. In contrast, virions are spherical particles with a diameter of *ca* 40 nm, are infectious, and possess an internal core structure that contains the viral genome (Dane, Cameron and Briggs, 1970).

The most diagnostic features of the known hepadna viruses, establishing tentative criteria for identification of new members of the family, are the structure and organization of the viral genome. In all cases, the viral genome is a relaxed, circular, DNA with a size of about 3 kb (Fig. 1). One strand is always complete, with unique 3′ and 5′ termini, including a protein covalently bound to the 5′ end (Gerlich and Robinson, 1980; Ganem, Greenbaum and Varmus, 1982a; Molnar-Kimber *et al.*, 1983). The other strand is incomplete, in some or all virions in a population, containing a discrete 5′ terminus and a 3′ terminus that is heterogeneous in location; this results in a genome that is up to 50% single stranded. The circular conformation is maintained by a cohesive overlap of *ca* 70–300 bp between the 5′ ends of the two DNA strands. It is a matter of interest that virions possess an endogenous DNA polymerase that can act, at least *in vitro*, to repair the single-stranded gap in the viral genome. Since it may at least be argued that the single-stranded gap reflects release of immature virions (Mason *et al.*, 1981; Ruiz-Opazo, Chakraborty and Shafritz, 1982), demonstration of the gap and of the endogenous polymerase activity might not be diagnostic for all new members of the hepadna virus family; i.e. hepadna viruses that were not released into the bloodstream until both DNA strands were complete.

The hepadna virus genomes are not only similar in structure but

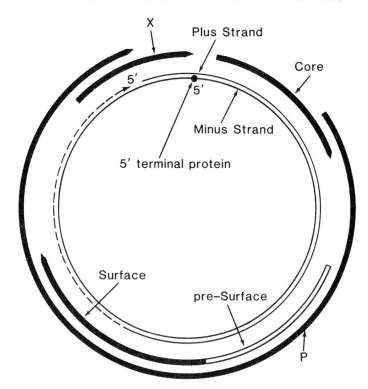

Fig. 1. The genome of human hepatitis B virus (HBV). The open reading frames of HBV (3182 bp), beginning with an AUG initiation codon, have been defined by the work of Galibert *et al.* (1979), Pasek *et al.* (1979), and Valenzuela *et al.* (1979). Relative to the location of the cohesive overlap, a virtually identical pattern of open reading frames has been observed for woodchuck hepatitis virus (Galibert *et al.*, 1982) and ground squirrel hepatitis virus (Seeger *et al.*, 1984*a*). The location of the ends of the cohesive overlap are approximate. Duck hepatitis B virus shows a similar pattern to HBV, except that X and core are fused (Mandart *et al.*, 1984). The X-core fusion gene begins within the duck virus cohesive overlap, which has been precisely mapped (Molnar-Kimber *et al.*, 1984).

also in genetic organization (Fig. 1). From analyses of the complete DNA sequence of all four viruses, it appears that all usable open reading frames have the same polarity, leading to the prediction that the viral mRNAs will be complementary to the complete strand of the viral genome (Galibert *et al.*, 1979; Pasek *et al.*, 1979; Valenzuela *et al.*, 1979; Galibert, Chen and Mandart, 1982; Mandart, Kay and Galibert, 1984; Seeger, Ganem and Varmus, 1984*a*). This has been demonstrated to be the case for the duck hepatitis B virus (DHBV) (Mason *et al.*, 1981) and woodchuck hepatitis virus (WHV) (C. W. Ogston & J. Summers, unpublished). Thus, by convention, the complete strand is the minus-strand. Comparison of the positions of the four open reading frames, beginning with an AUG initiator codon, indicates virtual identity among the three mammalian viruses, as illustrated for HBV in Fig. 1, and only a minor difference between DHBV and the mammalian viruses. Assignment of the major surface antigen and core proteins of HBV to the indicated regions is discussed by Tiollais, Charnay and Vyas (1981). Assignment of the corresponding products of the ground squirrel and woodchuck viruses to these regions derives in part from the high degree of DNA sequence homology between corresponding regions of the three mammalian viruses. Overall, ground squirrel hepatitis virus (GSHV) and HBV have about 50% nucleotide homology (Seeger *et al.*, 1984*a*), WHV and HBV have about 60–70% homology (Galibert *et al.*, 1982), and DHBV and HBV have less than 50% homology (Mandart *et al.*, 1984). The products of the X and P reading frames have not yet been identified; however, DNA sequence analysis suggests that frame P may code for a viral DNA polymerase (Toh, Hayashida and Miyata, 1983). It should be noted that the main difference between the open reading frames of DHBV and of the mammalian viruses is an apparent fusion between the X and core genes (Mandart *et al.*, 1984), resulting in a DHBV core gene product that is almost twice as large as the core polypeptide of the mammalian viruses (Newbold, Mason and Summers, 1984).

Despite these striking similarities, WHV, GSHV, and DHBV differ sufficiently, especially in virus–host interactions, to lend themselves to different sorts of experimental investigations.

Woodchuck hepatitis virus was first identified in a colony of wild-caught woodchucks at the Philadelphia Zoo that exhibited a high incidence of chronic liver disease and hepatocellular carcinoma (Summers *et al.*, 1978; Synder, Tyler and Summers, 1982). In woodchucks, active liver disease is generally associated with chronic WHV infection (Popper *et al.*, 1981), and hepatocellular carcinoma appears to develop in most or all chronically infected animals after 2–4 years (Snyder and Summers, 1980). At present, woodchucks are the only experimental model for analysis of the progression of hepadnavirus infection from chronic liver disease to cancer. Among woodchucks at the Philadelphia Zoo, infection was generally apparent at the time the animals were captured and introduced into the colony. Woodchucks have not yet been extensively bred in captivity, so the mode of natural transmission of WHV is uncertain (Summers, 1981). Perinatal transmission of HBV from infected mothers is thought to be common, although cryptic transplacental infections have not been eliminated as a route of vertical transmission. Whether identical modes of transmission characterize WHV is unknown. Extensive breeding of captive woodchucks should allow a resolution of this issue and may permit an exploration of the role, if any, of transplacental infection in vertical transmission by HBV.

Unlike WHV infection of woodchucks, GSHV infection of ground squirrels produces, at worst, a mild hepatitis detectable upon histological observation (Ganem *et al.*, 1982*b*; Marion *et al.*, 1983). Although both acute and chronic infections can be produced by experimental inoculation, the routes of natural transmission within ground squirrel populations are unknown (Ganem *et al.*, 1982*b*; Marion *et al.*, 1983; Marion and Robinson, 1983). It seems likely, from comparison of experimentally and naturally infected ground squirrels, that the absence of significant liver disease among naturally infected animals is not merely because of immunological tolerance resulting from infection during embryonic development or at birth. Rather, generation of asymptomatic infections appears to involve factors intrinsic to GSHV, ground squirrels, or both. Seeger *et al.* (1984*b*) have pointed out that the high degree of DNA sequence homology (80%) between GSHV and WHV (Seeger *et al.*, 1984*a*), including conservation of some restriction endonuclease sites, as well as the infectivity in ground squirrels of cloned viral DNAs, permits construction of WHV × GSHV recombinant viruses. The interesting possibility of determining the roles of virus and host in pathogenesis by examining the consequences of infecting ground squirrels or woodchucks with appropriate recombinant viruses has been noted, and the infectivity of at least one recombinant virus has been established (Seeger *et al.*, 1984*b*).

DHBV remains the only HBV-like virus in a domestic animal. In the United States, DHBV has been found as a chronic infection in 10% or more of the birds in some commercial flocks of Pekin ducks (*Anas domesticus*: Mason *et al.*, 1980; Marion *et al.*, 1984) and is also present in other varieties of domestic ducks (Mason *et al.*, 1981). The main route for natural transmission in flocks of Pekin ducks is vertical, through the eggs laid by viremic ducks, resulting in a chronic viremia that first appears during embryonic development (Mason *et al.*, 1980; O'Connell, Urban and London, 1983). Significant hepatitis has not been associated with this route of transmission, perhaps owing to an immunologic tolerance; in contrast, significant lesions are observed in at least some ducks inoculated with DHBV *in ovo* or at 1-day post hatch (Marion *et al.*, 1984). Like ground squirrels, chronic infections in domestic ducks have not been correlated with hepatocellular carcinoma (Omata *et al.*, 1983). To date, the DHBV system has been useful for studies on the molecular biology of hepadna virus replication in infected liver (Summers and Mason, 1982*a*), on the tissue specificity of virus infection (Halpern *et al.*, 1983), and on vertical transmission to embryos (O'Connell *et al.*, 1983), work made possible in large part because of the ready availability of infected animals from commercial sources and by the possibility of breeding pure lines of chronically infected and virus-free ducks. Although such research is interesting, the degree to which the latter two lines of investigation will extend to the mammalian systems remains to be determined.

HEPADNAVIRUS REPLICATION

Although a considerable amount is known about viral DNA synthesis and novel viral DNAs that accumulate during chronic infections, apparently as by-products of the replication process, little is known about other details of virus replication, including mechanisms for release of surface antigens and virions from infected cells. For instance, while studies in tissue culture systems suggest that infected hepatocytes continuously shed surface antigen into the bloodstream, direct evidence for persistent shedding of virus is lacking. The possibility of persistent secretion of virus is consistent with immuno-fluorescence microscopy studies of Pekin duck liver that show, in association with a high titre viremia, accumulation of viral surface and core antigens in virtually all cells of the liver (Mason *et al.*, 1984; M . S. Halpern, unpublished). Nevertheless, a rough calculation suggests that the liver contains 50–100 times more copies of viral DNA than the blood, consistent with the formal possibility that virus release is subsequent to cell death that occurs at a low level in the infected tissue. In this case, release of virus might not be correlated chronologically with maximum synthesis of virus in infected cells; i.e. cells might accumulate and store virus for long periods prior to release. Resolutions of this and similar issues pertaining to the dynamics of chronic infections will probably be facilitated by the use of tissue culture systems for hepadnavirus replication, which are still in promising but early stages of development (Pourcel and Summers, 1984). In subsequent discussion of the replication process, we will concentrate on mechanisms of viral DNA synthesis as deduced from analyses of virions and infected liver extracts.

We previously noted (Summers and Mason, 1982b) that the structure of the hepadna virus genome (Fig. 1), as first deduced by Summers, O'Connell and Millman (1975), bears striking similarities to proposed intermediates in reverse transcription of retrovirus RNA (Gilboa *et al.*, 1979), with synthesis of minus-strand DNA preceding synthesis of plus-strand DNA. Evidence in support of this analogy was obtained following the discovery of viral DNA synthesising complexes in the cytoplasm of hepatocytes of chronically infected Pekin ducks. In the electron microscope, these complexes had the structure of viral cores and were able to synthesize both plus- and minus-strand DNA following addition of radioactive deoxynucleoside triphosphates (Summers and Mason, 1982a). Thus, the endogenous reaction of these 'liver cores' was distinct from the virion endogenous reaction, which only synthesizes viral plus-strand DNA. Moreover, the viral core gene product is the structural subunit of virus and liver cores (Newbold *et al.*, 1984; J. E. Newbold, J. Summers and W. S. Mason, unpublished), suggesting that the DNA synthesising complexes are structural intermediates in virion morphogenesis. Analysis of the products and templates of liver cores revealed that minus-strand DNA was synthesized by copying viral RNA, which was rapidly degraded behind the growing point, possibly by a ribonuclease-H activity. In contrast, plus-strand synthesis occurred using the minus-strand as template. Plus-strand DNA is not detected at appreciable levels in association with growing minus-strands (Mason *et al.*, 1982; Summers and Mason, 1982a), implying that extensive plus-strand synthesis requires completion of the minus-strand. If the 5' ends of the cohesive overlap of virion DNA (Fig. 1) represent the origins of synthesis of minus- and plus-strands, this observation would not be surprising. Continuation of plus-strand synthesis past the 5' end of the minus-strand would not be expected until the minus-strand was complete, allowing plus-strand synthesis to jump to the 3' end of the minus-strand template. A similar structural constraint on elongation of plus-strand DNA seems to exist for the murine leukemia viruses, which initiate plus-strand synthesis from a site just downstream from the site of initiation of minus-strand synthesis (Gilboa *et al.*, 1979). Again, extensive plus-strand synthesis requires completion of the minus-strand template. A short plus-strand species (strong-stop plus) extending to the 5' end of the minus-strand template, can be detected as an intermediate in plus-strand elongation.

Observations on the mechanism of DHBV DNA synthesis, described above, raise two somewhat obvious questions. First, do the results with the avian model extend to the mammalian viruses? Secondly, what are the details of DNA synthesis, including the structure and mechanism of synthesis of the RNA template, the sites and primers for initiation of plus- and minus-strand synthesis, and the mechanism of formation of the cohesive overlap of virion DNA?

Evidence for a common mechanism of DNA synthesis comes from Southern (1975) analysis of infected liver DNA using virus-specific hybridization probes. Livers from HBV-infected humans and chimpanzees (Monjardino et al., 1982; Blum et al., 1984b), WHV-infected woodchucks (C. W. Ogston and J. Summers, unpublished), GSHV-infected ground squirrels (Weiser et al., 1983), and DHBV-infected Pekin ducks (Mason et al., 1982) contain similar species of hepadnavirus DNA, including genome-like DNA, supercoiled DNA of unit size, and a heterogeneous population of more rapidly migrating DNAs. Hybridizations with strand-specific probes have shown that the heterogeneous species are predominantly single-stranded minus-strands, shorter than, and equal to, unit length. Minus-strand DNAs with identical electrophoretic mobility are synthesized by reverse transcription in DHBV liver cores, implying that minus-strand DNA is synthesized by reverse transcription for all four hepadnaviruses. Moreover, Miller et al. (1984) have recently isolated cores from human liver that appear to synthesize HBV minus-strand DNA from an RNA template.

The details of the reactions culminating in synthesis of minus-strand and plus-strand DNA are still being investigated. Recent work on the primer and site of initiation of DHBV minus-strand synthesis indicates that the site of initiation corresponds to the 5′ end of the minus-strand of virion DNA and suggests that a protein serves as the primer for reverse transcription (Molnar-Kimber, Summers and Mason, 1984). In particular, a protein is covalently bound to the shortest nascent minus-strands (ca 30 b) labeled in the endogenous DNA polymerase reaction of DHBV liver cores. This is presumably the same protein found attached to the 5′ end of the minus-strand of virion DNA (Gerlich and Robinson, 1980; Ganem et al., 1982a; Molnar-Kimber et al., 1983). The site of initiation and the primer for plus-strand synthesis are unknown. It is tempting to suggest, by analogy with the retroviruses, that an RNA primer for plus-strand synthesis is created by a ribonuclease-H activity associated with a viral DNA polymerase (Smith, Cywinski and Taylor, 1984; Champoux, Gilboa and Baltimore, 1984). This would not explain the puzzling observation of Gerlich and Robinson (1980) that the 5′ end of the plus-strand is blocked to phosphorylation by polynucleotide kinase following treatment with alkaline phosphatase, and alternative plus-strand primers must be considered. No information is available on the site for initiation of plus-strand synthesis, although it seems probable that this is the same as the 5′ end of the plus-strand of virion DNA. In accordance with such a possibility, Molnar-Kimber *et al.* (1984) have reported a 68 base DNA product of DHBV liver cores whose synthesis is inhibited by actinomycin D, implying that the template for this species is DNA rather than RNA. The similarity between the size of this species and the length of the cohesive overlap of DHBV (69 ± 4 bp) (Molnar-Kimber *et al.*, 1984) suggests analogy to the strong-stop plus DNA of retroviruses (Smith *et al.*, 1984; Champoux *et al.*, 1984).

The exact structure of the template for reverse transcription is unknown. Total duck liver extracts contain a substantial amount of polyadenylated RNA of approximately unit length (Mason *et al.*, 1981), and at least some of this RNA is packaged into DHBV liver cores (J. Summers and W. S. Mason, unpublished). Whether this species or a nonpolyadenylated species serves as a template for initiation of minus-strand synthesis remains to be determined. We have been unable to detect nascent minus-strand DNA, radiolabeled in the endogenous reaction of liver cores, in association with polyadenylated RNA (W. S. Mason and J. Summers, unpublished observations). If the polyadenylated RNA serves as a template, this result suggests that the 3′ poly(A) sequences are removed early in minus-strand synthesis. This observation raises the question of the location of the 3′ and 5′ termini of the RNA template relative to the site of initiation of minus-strand synthesis. In the retrovirus system, minus-strand DNA synthesis initiates from a tRNA primer bound 100–200 nucleotides from the 5′ end of the polyadenylated viral RNA template, and poly(A) is presumably removed at an early stage in minus-strand elongation, during or subsequent to the jump from the 5′ to the 3′ end of the template. For hepadna viruses, the locations of the 3′ and 5′ termini of the unit-sized liver core RNA are unknown; however, studies with transcription systems *in vitro* suggest that HBV, at least, contains an RNA polymerase II promotor within the cohesive overlap region, just downstream from the 5′ end of the minus-strand (Chakroborty, Ruiz-Opazo and Shafritz, 1981; Rall *et al.*, 1983).

Apart from the difficulty of characterizing RNA transcripts associated with hepadna virus replication, there is the problem of identifying the template for synthesis of these species. An early step in initiation of infection of the liver by DHBV, occurring within 24 h of virus inoculation, is the appearance of closed circular duplex DNA of unit size. This DNA appears in the liver subsequent to, and may be formed from, relaxed circular virion DNA (Mason *et al.*, 1983). A role for this species as a template for transcription of viral RNAs is suggested by its continued presence as the only major episomal species in the nuclei of chronically infected hepatocytes of ducks and woodchucks (J. Summers, unpublished). As noted earlier, virion DNA synthesis occurs in the cytoplasm (Summers and Mason, 1982*a*), at least for DHBV. Supercoiled DNA in the retrovirus system is a precursor in the formation of integrated DNA, which occurs early in infection, probably before significant synthesis of viral RNA. In this case, integrated DNA serves as a template for transcription and it is uncertain if unintegrated DNA is ever transcriptionally active, although it can be in certain special cases when integration is blocked by deletion of viral sequences around the site of integration on the viral genome (Panganiban and Temin, 1983).

DHBV-infected ducklings provided an opportunity to test whether integration of viral DNA into the DNA of the infected host cell is necessary for production of progeny virus. Virtually all hepatocytes of young infected ducklings stain for DHBV core antigen and thus apparently produce virus. If an integrated form of viral DNA were required for virus production, then every hepatocyte would contain at least one integrated viral genome. To test for this possibility, J.S. and C.R. constructed a library of recombinant phage consisting of *Sac*I fragments of DNA from the liver of a congenitally infected 2-week-old duckling using bacteriophage λgt. Sst 17 (Stringer, 1981). Since *Sac*I does not cleave the DHBV genome, unintegrated forms of viral DNA were not ligated into the vector. Furthermore, all *Sac*I fragments less than 3 kb in length were excluded from the preparation. Approximately 10 cellular genome equivalents of recombinant phage prepared in this way were screened for DHBV sequences. No clones containing integrated viral DNA were detected, suggesting that the frequency of integrated DHBV DNA at that stage of infection was less than one genome per 5–10 cells. If this were so then the majority of virus-producing cells would not have contained integrated DNA, suggesting that integration is not a necessary step for virus production.

Integration of viral DNA, while not essential for replication, may be an inevitable consequence of long-term infections. In man, integration into liver DNA appears clonal, probably owing to focal regeneration subsequent to the integration event (Brechot et al., 1981; Shafritz et al., 1981; Chen et al., 1982; Kam et al., 1982; Hino et al., 1984). A similar observation of clonal integration has been made by one group working with woodchucks (Dejean et al., 1982), whereas another has detected randomly integrated viral DNA after shotgun cloning (Rogler and Summers, 1984). Integration has not yet been described in ground squirrels or ducks. In man, the integrated sequences often remain after all forms of episomal DNA have disappeared. Under this circumstance, virus is no longer synthesized, but surface antigen particles may still be present. Continued synthesis of surface antigen presumably reflects expression from integrated viral sequences.

Clonally integrated DNA is often, though not always (Mitamura et al., 1982; and references in last paragraph), detected in hepatocellular carcinomas developing subsequent to viral infection, suggesting that integration preceded cellular proliferation. Analyses carried out by several groups are consistent with the idea that integration into the host genome is random. Integrated sequences often contain multiple deletions, inversions, and/or duplications, and rearrangements have been detected in flanking cellular sequences (Ogston et al., 1982; Koshy et al., 1983; C. E. Rogler and J. Summers, unpublished observations; C. W. Ogston and J. Summers, unpublished). At the present time, it is unclear if there are preferred sites for integration on the viral genome (see, however, Dejean et al., 1984b; Koch et al., 1984; Koike et al., 1984). The role of the integrated sequences in tumour development also remains to be defined.

Less is known about the structure of integrated DNAs found in normal liver tissue. The only integrated viral sequence studied in detail lacked rearrangements except for a small deletion (Rogler and Summers, 1984). However, Rogler and Summers (1982) and Marion et al. (1982) have obtained evidence consistent with the existence of nonintegrated, high molecular weight forms of viral DNA in chronically infected ground squirrel and woodchuck livers. These DNAs contained inversions, deletions and/or duplications characteristic of integrated DNAs. Rogler and Summers (1982) have pointed out that such nonintegrated DNAs might exist as precursors to the highly rearranged integrated forms and have suggested that these forms might arise from the unit-sized supercoiled DNAs found in the nuclei of productively infected hepatocytes. Mechanisms involved in generation of rearranged DNAs and/or integrated DNAs may serve to deplete the pool of supercoiled DNA. If these species are necessary as templates for synthesis of pre-genomic RNA and other RNAs essential for virus production, virus synthesis would eventually stop, as is observed in man.

EXTRAHEPATIC INFECTIONS

Shusterman and London (1984) have reviewed the evidence linking HBV infection with rare cases of immune-complex diseases in man. It is not known if any of these diseases are associated with an extrahepatic replication of HBV. For present purposes, we will review very recent evidence for infections, by hepadna viruses, of tissues outside the liver. The pathologic effects of such infections, if any, remain a subject for future investigation.

Evidence for extrahepatic infections in man and ducks have been obtained by several procedures, including immunofluorescent microscopy, DNA and RNA blotting, and hybridization *in situ* . In considering the subsequent discussion, it should be borne in mind that results obtained for a particular virus/host system might not extrapolate to the remaining systems.

The clearest examples of extrahepatic infections with manifestations similar to infection of hepatocytes have been obtained in Pekin ducks. Studies with congenitally and experimentally DHBV-infected ducks have revealed, in both the pancreas and kidney, all the forms of viral DNA and RNA characteristic of active synthesis of virus (Halpern *et al.*, 1983). Immunofluorescent microscopy has revealed a correlation between the presence of these nucleic acids in tissue extracts and accumulation of DHBV surface and core antigens in scattered acinar cells of the pancreas and convoluted proximal tubular epithelium of the kidney (Halpern *et al.*, 1983; Mason *et al.*, 1984). Moreover, core particles identical to liver cores have been isolated from the pancreas (J.-M. Lien and W. S. Mason, unpublished).

Accumulation of DHBV surface and core antigens has also been found in α and β cells of the endocrine pancreas (Halpern *et al.*, 1983, 1984; Mason *et al.*, 1984). This accumulation has not yet been correlated with synthesis of replication-specific forms of viral nucleic acids. However, the presence of viral core antigen is compatible with the idea that virion DNA synthesis occurs in these cells. In contrast, only DHBV surface antigen has been detected in the adrenal cortex (Halpern *et al.*, 1984), and we have been unable to detect replication-specific forms of viral DNA in adrenal gland extracts, raising the possibility that antigen accumulation in this tissue reflects either uptake from the blood or an inability of these cells to support the complete replication cycle of DHBV.

Several extrahepatic sites of infection have been identified in man. The most likely site for extrahepatic replication in man appears to be the pancreas. Shimoda *et al.* (1981) have detected both HBV surface and core antigen accumulation in scattered acinar cells, in a pattern similar to that observed in Pekin ducks. It has not yet been reported, however, whether this pattern of accumulation correlates with synthesis of replication-specific forms of viral nucleic acids.

Separate studies, also generally based on a single technique, have confirmed and extended this initial observation of extrahepatic infection. Using hybridization *in situ*, Blum *et al.* (1984a) detected HBV DNA in germinal centers of the spleen. Pontisso *et al.* (1984) used the Southern (1975) blotting technique to detect either free monomeric or integrated HBV DNA in peripheral blood lymphocytes of some chronically infected individuals. Likewise, Siddiqui (1983) detected an episomal HBV DNA, with an electrophoretic mobility somewhat lower than that of monomers, in Kaposi's sarcoma and skin of one patient without other signs of HBV infection and in Kaposi's sarcoma of another patient with both surface antigen and virus in the blood.

Finally, Dejean *et al.* (1984a) reported integrated HBV DNA in the pancreas, kidney, and skin of two HBV carriers. All of these reports, based upon detection of HBV DNA, are consistent with the idea that HBV infects cells outside the liver; however, none of these investigators reported evidence for forms of DNA characteristic of active virus production. Evidence has not yet been reported indicating viral gene expression in association with these examples of integrated and episomal DNA in extrahepatic locations in man. The detection of integrated DNA by the Southern technique suggests extensive cell proliferation subsequent to integration, raising the possibility that infection occurred at an early stage of cell differentiation. A recent report by Romet-Lemonne *et al.* (1983) revealed HBV surface antigen in 1–2 % of bone marrow cell aspirates in five patients with surface antigenemia. Bone marrow aspirate from one patient was developed into a lymphoblastoid cell culture that expressed surface and core antigens and contained relaxed circular and supercoiled monomers of HBV DNA (Elfassi *et al.*, 1984); however, the forms of intracellular HBV DNA present in original bone marrow aspirates have not yet been described.

A variety of reports, therefore, support the concept of extrahepatic infections. The possibility remains that, in some or all cases, virus replication may be abortive.

OUTLOOK

The hepadna viruses remain a potentially fertile area of investigation for pathologists, immunologists and virologists. The basis for virus-induced liver disease is uncertain and the mechanisms by which a single virus can cause both acute and chronic infections is unknown. Moreover, recent observations suggest that hepadna virus infections and associated disease states should be evaluated with reference to the possibility of direct effects of virus infection on tissues outside the liver. Although it is often assumed that development of tissue culture systems permissive for hepadna virus replication will be essential to future progress, the basic problems and fascinations in studying hepadna viruses remain those of understanding infections as they actually occur in man and animals.

We are grateful to Lori Bartasavich for help in preparing this manuscript. This work was supported by grants from the National Institutes of Health, and the American Cancer Society, and an appropriation from the Commonwealth of Pennsylvania.

REFERENCES

BEASLEY, R. P. (1982). Hepatitis B virus as the etiologic agent in hepatocellular carcinoma. Epidemiologic considerations. *Hepatology*, **2**, 21S–26S.

BLUM, H. E., GEBALLE, A. P., STOWRING, L., FIGUS, A., HAASE, A. T. and VYAS, G. N. (1984a). Hepatitis B virus in nonhepatocytes: Demonstration of viral DNA in spleen, bile duct epithelium, and vascular elements by *in situ* hybridization. In *Viral Hepatitis and Liver Disease*, ed. G. N. Vyas, J. L. Dienstag and J. H. Hoofnagle, p. 634. New York, Grune and Stratton.

BLUM, H. E., HAASE, A. T., HARRIS, J. D., WALKER, D. and VYAS, G. N. (1984b). Asymmetric replication of HBV DNA in human liver demonstrated by blot analysis and *in situ* hybridization. In *Viral Hepatitis and Liver Disease*, ed. G. N. Vyas, J. L. Dienstag and J. H. Hoofnagle, p. 634, New York, Grune and Stratton.

BRECHOT, C., SCOTTO, J., CHARNAY, P., HADCHOUEL, M., DEGOS, F., TREPO, C. and TIOLLAIS, P. (1981). Detection of hepatitis B virus DNA in liver and serum: A direct appraisal of the chronic carrier state. *Lancet*, **ii**, 765–8.

CHAKRABORTY, P. R., RUIZ-OPAZO, N. and SHAFRITZ, D. A. (1981). Transcription of human hepatitis B core antigen gene sequences in an *in vitro* HeLa cellular extract. *Virology*, **111**, 647–52.

CHAMPOUX, J., GILBOA, E. and BALTIMORE, D. (1984). Mechanism of RNA primer removal by the ribonuclease H activity of avian myeloblastosis virus reverse transcriptase. *Journal of Virology*, **49**, 686–91.

CHEN, D.-S., HOYER, B. H., NELSON, J., PURCELL, R. H. and GERIN, J. L. (1982). Detection and properties of hepatitis B viral DNA in liver tissues from patients with hepatocellular carcinoma. *Hepatology*, **2**, 42S–46S.

DANE, D. S., CAMERON, C. H. and BRIGGS, M. (1970). Virus-like particles in serum of patients with Australia-antigen associated hepatitis. *Lancet*, i, 695–8.

DEJEAN, A., LUGASSY, C., ZAFRANI, S., TIOLLAIS, P. and BRECHOT, C. (1984a). Detection of hepatitis B virus DNA in pancreas, kidney, and skin of two human carriers of the virus. *Journal of General Virology*, **65**, 651–5.

DEJEAN, A., SONIGO, P., WAIN-HOBSON, S. and TIOLLAIS, P. (1984b). Specific hepatitis B virus integration in hepatocellular carcinoma DNA through a viral 11 base pair direct repeat. In *Viral Hepatitis and Liver Disease*, ed. G. N. Vyas, J. L. Dienstag and J. H. Hoofnagle, p. 632. New York, Grune and Stratton.

DEJEAN, A., VITVITSKI, L., BRECHOT, C., TREPO, C., TIOLLAIS, P. and CHARNAY, P. (1982). Presence and state of woodchuck hepatitis virus DNA in liver and serum of woodchucks: Further analogies with human hepatitis B virus. *Virology*, **121**, 195–9.

ELFASSI, D., ROMET-LEMONNE, J. L., SUREAU, C., McLANE, M. F., HASELTINE, W. and ESSEX, M. (1984). Hepatitis B viral DNA in cultured human lymphoblastoid cells. In *Viral Hepatitis and Liver Disease*, ed. G. N. Vyas, J. L. Dienstag and J. H. Hoofnagle, p. 627. New York, Grune and Stratton.

GALIBERT, F., CHEN, T. N. and MANDART, E. (1982). Nucleotide sequence of a cloned woodchuck hepatitis B virus genome: Comparison with the hepatitis B virus sequence. *Journal of Virology*, **41**, 51–65.

GALIBERT, F., MANDART, E., FITOUSSI, F., TIOLLAIS, P. and CHARNAY, P. (1979). Nucleotide sequence of the hepatitis B genome (subtype ayw) cloned in *E. coli*. *Nature*, **281**, 646–50.

GANEM, D., GREENBAUM, L. and VARMUS, H. E. (1982a). Viral DNA of ground squirrel hepatitis virus: Structural analysis and molecular cloning. *Journal of Virology*, **44**, 374–83.

GANEM, D., WEISER, B., BARCHUK, A., BROWN, J. R. and VARMUS, H. E. (1982b). Biological characterization of acute infection with ground squirrel hepatitis virus. *Journal of Virology*, **44**, 366–73.

GERLICH, W. H. and ROBINSON, W. S. (1980). Hepatitis B virus contains protein attached to the 5' terminus of its complete strand. *Cell*, **21**, 801–9.

GILBOA, E., MITRA, S. W., GOFF, S. and BALTIMORE, D. (1979). A detailed model of reverse transcription and tests of crucial aspects. *Cell*, **18**, 93–100.

HALPERN, M. S., EGAN, J., MASON, W. S. and ENGLAND, J. M. (1984). Viral antigen in endocrine cells of the pancreatic islets and adrenal cortex of Pekin ducks infected with duck hepatitis B virus. *Virus Research*, **1**, 213–23.

HALPERN, M. S., ENGLAND, J. M., DEERY, D. T., PETCU, D. J., MASON, W. S. and MOLNAR-KIMBER, K. L. (1983). Viral nucleic acid synthesis and antigen accumulation in pancreas and kidney of Pekin ducks infected with duck hepatitis B virus. *Proceedings of the National Academy of Sciences, USA*, **80**, 4865–9.

HINO, O., KITAGAWA, T., KOIKE, K., KOBAYASHI, M., HARA, M., MORI, W., NAK-ASHIMA, T., HATTORI, N. and SUGANO, H. (1984). Detection of hepatitis B virus DNA in hepatocellular carcinomas in Japan. *Hepatology*, **4**, 90–5.

KAM, W., RALL, L. B., SMUCKLER, E. A., SCHMID, R. and RUTTER, W. J. (1982). Hepatitis B viral DNA in liver and serum of asymptomatic carriers. *Proceedings of the National Academy of Sciences, USA*, **79**, 7522–6.

KOCH, S., KOSHY, R., VON LORINGHOVEN, R. F. and HOFSCHNEIDER, P. H. (1984). Analysis of cloned integrated HBV sequences from the human hepatoma cell line PLC/PRF/5. In *Viral Hepatitis and Liver Disease*, ed. G. N. Vyas, J. L. Dienstag and J. H. Hoofnagle, p. 628. New York, Grune and Stratton.

KOIKE, K., MIZUSAWA, H., KOBAYASHI, M., TAIRA, M., YOSHIDA, E. and YAGINUMA, K. (1984). Integrated hepatitis B virus DNA and cellular flanking sequences are inversely repeated in the human hepatoma-derived cell lines. In *Viral Hepatitis and Liver Disease*, ed. G. N. Vyas, J. L. Dienstag and J. H. Hoofnagle, p. 627. New York, Grune and Stratton.

KOSHY, R., KOCH, S., VON LORINGHOEVEN, A. F., KAHMANN, R., MURRAY, K. and HOFSCHNEIDER, P. H. (1983). Integration of hepatitis B virus DNA: evidence for integration in the single-stranded gap. *Cell*, **34**, 215–23.

MANDART, E., KAY, A. and GALIBERT, F. (1984). Nucleotide sequence of a cloned duck hepatitis virus genome: Comparison with the human and woodchuck hepatitis B virus sequences. *Journal of Virology*, **49**, 782–92.

MARION, P. L., KNIGHT, S. S., HO, B.-K., GUO, Y.-Y., ROBINSON, W. S. and POPPER, H. (1984). Liver disease associated with duck hepatitis B virus infection of domestic ducks. *Proceedings of the National Academy of Sciences, USA*, **81**, 898–902.

MARION, P. L., KNIGHT, S. S., SALAZAR, F. H., POPPER, H. and ROBINSON, W. S. (1983). Ground squirrel hepatitis virus infection. *Hepatology*, **3**, 519–27.

MARION, P. L., OSHIRO, L. S., REGNERY, D. C., SCULLARD, G. H. and ROBINSON, W. S. (1980). A virus in Beechey Ground Squirrels which is related to hepatitis B virus of man. *Proceedings of the National Academy of Sciences, USA*, **77**, 2941–5.

MARION, P. L. and ROBINSON, W. S. (1983). Hepadna viruses: Herpatitis B and related viruses. *Current Topics in Microbiology and Immunology*, **105**, 99–121.

MARION, P. L., ROBINSON, W. S., ROGLER, C. E. and SUMMERS, J. (1982). High molecular weight GSHV-specific DNA in chronically-infected ground squirrel liver. *Journal of Cellular Biochemistry, Supplement*, **6**, 203.

MASON, W. S., ALDRICH, C., SUMMERS, J. and TAYLOR, J. M. (1982). Asymmetric replication of duck hepatitis B virus DNA in liver cells: Free minus-strand DNA. *Proceedings of the National Academy of Sciences, USA*, **79**, 3997–4001.

MASON, W. S., HALPERN, M. S., ENGLAND, J. M., SEAL, G., EGAN, J., COATES, L., ALDRICH, C. and SUMMERS, J. (1983). Experimental transmission of duck hepatitis B virus. *Virology*, **131**, 375–84.

MASON, W. S., NEWBOLD, J., SEAL, G., ALDRICH, C. E., COATES, L., ENGLAND, J. M., SUMMERS, J. and HALPERN, M. S. (1984). Expression of duck hepatitis B virus in congenitally and experimentally infected Pekin ducks. In *Viral Hepatitis and Liver Disease*, ed. G. N. Vyas, J. L. Dienstag and J. H. Hoofnagle, pp. 443–50. New York, Grune and Stratton.

MASON, W. S., SEAL, G. and SUMMERS, J. (1980). Virus of Pekin ducks with structural and biological relatedness to human hepatitis B virus. *Journal of Virology*, **36**, 829–36.

MASON, W. S., TAYLOR, J. M., SEAL, G. and SUMMERS, J. (1981). An HBV-like virus of domestic ducks. In *Viral Hepatitis 1981 International Symposium*, ed. W. Szmuness, H. J. Alter and J. E. Maynard, pp. 107–16. Philadelphia, Franklin Institute Press.

MILLER, R. H., TRAN, C.-T., MARION, P. L. and ROBINSON, W. S. (1984). Replication of hepatitis B virus DNA. In *Viral Hepatitis and Liver Disease*, ed. G. N. Vyas, J. L. Dienstag and J. H. Hoofnagle, p. 629. New York, Grune and Stratton.

MITAMURA, K., HOYER, B. H., PONZETTO, A., NELSON, J., PURCELL, R. H. and GERIN, J. L. (1982). Woodchuck hepatitis virus DNA in woodchuck liver tissues. *Hepatology*, **2**, 47S–50S.

MOLNAR-KIMBER, K. L., SUMMERS, J. and MASON, W. S. (1984). Mapping of the cohesive overlap of duck hepatitis B virus DNA and of the site of initiation of reverse-transcription. *Journal of Virology*, **51**, 181–91.

MOLNAR-KIMBER, K. L., SUMMERS, J., TAYLOR, J. M. and MASON, W. S. (1983). Protein covalently bound to minus-strand DNA intermediates on duck hepatitis B virus. *Journal of Virology*, **45**, 165–72.

MONJARDINO, J., FOWLER, M. J. F., MONTANO, L., WELLER, I., TSIQUAYE, K. N., ZUCKERMAN, A. J., JONES, D. M. and THOMAS, H. C. (1982). Analysis of hepatitis virus DNA in liver and serum of HBe antigen-positive chimpanzee carriers. *Journal of Medical Virology*, **9**, 189–99.

NEWBOLD, J. E., MASON, W. S. and SUMMERS, J. (1984). Purification and characterization of DHBV liver cores. In *Viral Hepatitis and Liver Disease*, ed. G. N. Vyas, J. L. Dienstag and J. H. Hoofnagle, p. 653. New York, Grune and Stratton.

O'CONNELL, A. P., URBAN, M. K. and LONDON, W. T. (1983). Naturally occurring infection of Pekin duck embryos by duck hepatitis B virus. *Proceedings of the National Academy of Sciences, USA*, **80**, 1703–6.

OGSTON, C. W., JONAK, G. J., ROGLER, C. E., ASTRIN, S. M. and SUMMERS, J. (1982). Cloning and structural analysis of integrated woodchuck virus sequences from hepatocellular carcinomas of woodchucks. *Cell*, **29**, 385–94.

OMATA, M., UCHIUMI, K., ITO, Y., YOKOSUKA, O., MORI, J., TERAO, K., WEI-FA, Y., O'CONNELL, A. P., LONDON, W. T. and OKUDA, K. (1983). Duck hepatitis B virus and liver diseases. *Gastroenterology*, **85**, 260–7.

PANGANIBAN, A. T. and TEMIN, H. M. (1983). The terminal nucleotides of retrovirus DNA are required for integration but not virus production. *Nature*, **306**, 155–60.

PASEK, M., GOTO, T., GILBERT, W., ZINK, B., SCHALLER, H., MACKAY, P., LEADBETTER, G. and MURRAY, K. (1979). Hepatitis B virus genes and their expression in *E. coli. Nature*, **282**, 575–9.

PONTISSO, P., DEJEAN, A., POON, M. C., TIOLLAIS, P. and BRECHOT, C. (1984). Detection of hepatitis B virus DNA in human blood mononuclear cells. In *Viral Hepatitis and Liver Disease*, ed. G. N. Vyas, J. L. Dienstag and J. H. Hoofnagle, p. 630. New York, Grune and Stratton.

POPPER, H., SHIH, J. W.-K., GERIN, J. L., WONG, D. C., HOYER, B. H., LONDON, W. T., SLY, D. L. and PURCELL, R. H. (1981). Woodchuck hepatitis and hepatocellular carcinoma: Correlation of histologic with virologic observations. *Hepatology*, **1**, 91–8.

POURCEL, C. and SUMMERS, J. (1984). Expression of duck hepatitis B virus in culture hepatocytes. In *Viral Hepatitis and Liver Disease*, ed. G. N. Vyas, J. L. Dienstag and J. H. Hoofnagle, p. 653. New York, Grune and Stratton.

RALL, L. B., STANDRING, D. N., LAUB, O. and RUTTER, W. J. (1983). Transcription of hepatitis B virus by RNA polymerase II. *Molecular and Cellular Biology*, **3**, 1766–73.

ROGLER, C. E. and SUMMERS, J. (1982). Novel forms of woodchuck hepatitis virus DNA isolated from chronically infected woodchuck liver nuclei. *Journal of Virology*, **44**, 852–63.

ROGLER, C. E. and SUMMERS, J. (1984). Cloning and structural analysis of integrated woodchuck hepatitis virus sequences from a chronically infected liver. *Journal of Virology*, **50**, 832–7.

ROMET-LEMONNE, J.-L., McLANE, M. F., ELFASSI, E., HASELTINE, W. A., AZOCAN, J. and ESSEX, M. (1983). Hepatitis B virus infection in cultured human lymphoblastoid cells. *Science*, **221**, 667–9.

RUIZ-OPAZO, N., CHAKRABORTY, P. R. and SHAFRITZ, D. A. (1982). Evidence for supercoiled hepatitis B virus DNA in chimpanzee liver and serum Dane particles: Possible implications in persistent HBV infections. *Cell*, **29**, 129–38.

SEEGER, C., GANEM, D. and VARMUS, H. E. (1984a). The nucleotide sequence of an infectious, molecularly cloned genome of the ground squirrel hepatitis virus. *Journal of Virology*, **51**, 367–75.

SEEGER, C., GANEM, D. and VARMUS, H. E. (1984b). Toward a genetic approach to hepatitis B viruses: The cloned genome of ground squirrel hepatitis B virus (GSHV) is infectious in animals. In *Viral Hepatitis and Liver Disease*, ed. G. N. Vyas, J. L. Dienstag and J. N. Hoofnagle, p. 654. New York, Grune and Stratton.

SHAFRITZ, D. A., SHOUVAL, D., SHERMAN, H. I., HADZIYANNIS, S. J. and KEW, M. C. (1981). Integration of hepatitis B virus DNA into the genome of liver cells in chronic liver disease and hepatocellular carcinoma. *New England Journal of Medicine*, **305**, 1067–73.

SHIMODA, T., SHIKATA, T., KARASAWA, T., TSUKAGOSHI, S., YOSHIMURA, M. and SAKURAI, I. (1981). Light microscopic localization of hepatitis B virus antigens in the human pancreas. *Gastroenterology*, **81**, 998–1005.

SHUSTERMAN, N. and LONDON, W. T. (1984). Hepatitis B and immune complex disease. *New England Journal of Medicine*, **310**, 43–6.

SIDDIQUI, A. (1983). Hepatitis B virus DNA in Kaposi sarcoma. *Proceedings of the National Academy of Sciences, USA*, **80**, 4861–4.

SMITH, J. K., CYWINSKI, A and TAYLOR, J. M. (1984). Initiation of plus-strand DNA synthesis during reverse transcription of an avian retrovirus genome. *Journal of Virology*, **49**, 200–4.

SNYDER, R. L. and SUMMERS, J. (1980). Woodchuck hepatitis virus and hepatocellular carcinoma. In *Viruses in Naturally Occurring Cancer*, Cold Spring Harbour Conferences on Cell Proliferation, vol. 7, ed. M. Essex, G. Todaro and H. zur Hausen, pp. 447–57. New York, Cold Spring Harbor Laboratories.

SNYDER, R. L., TYLER, G. and SUMMERS, J. (1982). Chronic hepatitis and hepatocellular carcinoma associated with woodchuck hepatitis virus. *American Journal of Pathology*, **107**, 422–5.

SOUTHERN, E. M. (1975). Detection of specific sequences among DNA fragments separated by gel electrophoresis. *Journal of Molecular Biology*, **98**, 503–17.

STRINGER, J. R. (1981). Integrated simian virus 40 DNA: Nucleotide sequences at cell–virus recombinant junctions. *Journal of Virology*, **38**, 671–9.

SUMMERS, J. (1981). Three recently described animal models for human hepatitis B virus. *Hepatology*, **1**, 179–83.

SUMMERS, J. and MASON, W. S. (1982a). Replication of the genome of a hepatitis B-like virus by reverse transcription of an RNA intermediate. *Cell*, **29**, 403–15.

SUMMERS, J. and MASON, W. S. (1982b). Properties of the hepatitis B-like viruses related to their taxonomic classification. *Hepatology*, **2**, 61S–66S.

SUMMERS, J., O'CONNELL, A. and MILLMAN, I. (1975). Genome of hepatitis B virus: Restriction enzyme cleavage and structure of DNA extracted from Dane particles. *Proceedings of the National Academy of Sciences, USA*, **72**, 4597–601.

SUMMERS, J., SMOLEC, J. and SNYDER, R. (1978). A virus similar to human hepatitis B virus associated with hepatitis and hepatoma in woodchucks. *Proceedings of the National Academy of Sciences, USA*, **75**, 4533–7.

SZMUNESS, W. (1975). Recent advances in the epidemiology of hepatitis B. *American Journal of Pathology*, **81**, 629–50.

TIOLLAIS, P., CHARNAY, P. and VYAS, G. N. (1981). Biology of hepatitis B virus. *Science*, **213**, 406–11.

TOH, H., HAYASHIDA, H. and MIYATA, T. (1983). Sequence homology between retroviral reverse transcriptase and putative polymerases of hepatitis B virus and cauliflower mosaic virus. *Nature*, **305**, 829–31.

WEISER, B., GANEM, D., SEEGER, C. and VARMUS, H. E. (1983). Closed circular viral DNA and asymmetric heterogenous forms in livers from animals infected with ground squirrel hepatitis virus. *Journal of Virology*, **48**, 1–9.

VALENZUELA, P., GRAY, P., QUIROGA, M., ZALDIVAR, J., GOODMAN, H. M. and RUTTER, W. J. (1979). Nucleotide sequence of the gene coding for the major protein of hepatitis B virus surface antigen. *Nature*, **280**, 815–19.

ZHOU, Y. Z. (1980). A virus possibly associated with hepatitis and hepatoma in ducks. *Shanghai Medical Journal*, **3**, 641–4.

BIOLOGY AND EPIDEMIOLOGY OF HEPATITIS B VIRUS

ARIE J. ZUCKERMAN

Department of Medical Microbiology and WHO Collaborating Centre for Reference and Research on Viral Hepatitis, London School of Hygiene and Tropical Medicine (University of London), London WC1E 7HT, UK

Viral hepatitis is a major public health problem occurring endemically in all parts of the world. The term human viral hepatitis refers to infections caused by four or more different viruses or groups of viruses. These infections are: hepatitis A; hepatitis B; the more recently identified non-A, non-B hepatitis (caused by more than two viruses); and epidemic non-A hepatitis (previously referred to as epidemic non-A, non-B hepatitis). Also identified is the delta agent, a defective virus which replicates in individuals infected with hepatitis B virus. Hepatitis A and hepatitis B can be differentiated by sensitive laboratory tests for specific antigens and antibodies, and the viruses have been characterised. Specific laboratory tests are also available for the delta agent. Laboratory tests are under development for epidemic non-A hepatitis, which is not caused by the recognised serotype of hepatitis A. This epidemic and endemic strain(s) of virus is commonly transmitted by contaminated water, causing a hepatitis A-like illness particularly in the subcontinent of India, Burma, the eastern USSR, parts of the Middle East and North Africa. At present, however, there are no precise virological criteria or specific laboratory tests for non-A, non-B hepatitis.

It is clearly impossible to review the vast literature on this topic and the following account is, therefore, only a summary of the biology and epidemiology of human hepatitis B virus. Reference is also made to hepatitis B-like viruses that infect some species of animals other than the higher primates. The discovery of Australia antigen by Blumberg, Alter and Visnich (1965) and the subsequent demonstration of the association of Australia antigen with hepatitis B resulted in rapid progress in the understanding of this infection. Australia antigen, now referred to as hepatitis B surface antigen, was the first of a series of complex structural antigens of the virus to be identified. The availability of specific serological tests for markers of the infection unravelled the complexity of hepatitis B, established the global distribution of the infection and led to remarkable advances in our knowledge of the pathogenesis of hepatitis B and its associated chronic liver disorders.

THE BIOLOGY OF HEPATITIS B VIRUS

Structural and antigenic analysis of the virus

Examination by electron microscopy of plasma containing hepatitis B surface antigen (reviewed by Zuckerman, 1975) reveals the presence of small spherical particles measuring on average 22 nm in diameter, tubular forms of varying length but with a diameter close to 22 nm (Bayer, Blumberg, and Werner, 1968; Almeida *et al.*, 1969), and large double-shelled or solid particles approximately 42 nm in diameter described by Dane, Cameron and Briggs (1970). The 42 nm particle (the Dane particle) is the complete virion and contains a core, or nucleocapsid, about 27 nm in diameter surrounded by an envelope approximately 14 nm in thickness. The core particles may be released from the virions by detergent treatment (Almeida, Rubenstein and Stott, 1971). The core contains a partially double-stranded circular DNA molecule (Robinson, Clayton and Greenman, 1974). The molecular weight of the DNA is about 2.3×10^6 and the DNA is approximately 3.2 kb in length, with a single-stranded gap varying from 0.6 to 2.1 kb. The core particle also contains a DNA-dependent DNA polymerase which is closely associated with the DNA template (Kaplan *et al.*, 1973) and a protein kinase which phosphorylates the major virus-specified core polypeptide (Albin and Robinson, 1980). Another antigen, hepatitis B *e* antigen (Magnius and Espmark, 1972), is closely associated with the core and its antigenic reactivity. The core antigen can be converted into *e* antigen by proteolytic degradation under dissociating conditions (Mackay, Lees and Murray, 1981); this confirms that, at the molecular level, *e* antigen is a component of the core antigen. The small spherical 22 nm particles and the tubular forms found in the plasma are non-infectious surplus virus coat protein and contain a variable amount of lipid and carbohydrate (reviewed in detail by Zuckerman and Howard, 1979; Deinhardt and Deinhardt, 1983; Marion and Robinson, 1983).

Mondelli and Eddleston (1984) and Thung and Gerber (1984) reviewed the experimental evidence that suggests the presence of specific structural receptors for polyalbumin on the complete virion and on purified hepatitis B surface antigen particles and polypeptides, particularly if these are derived from *e* antigen plasma. Polyalbumin receptors are also present on the surface of hepatocytes with polymerised serum albumin acting as a linker molecule between the surface antigen and the cell. The virion polyalbumin receptors are species and ligand specific and they react only with polymerised serum albumin, whereas the hepatocyte-associated polyalbumin receptors are not ligand specific and they react with polymeric and monomeric albumins from different species. The evidence reviewed by Thung and Gerber (1984) suggests that the polyalbumin receptors are encoded by the genome of hepatitis B virus.

After the virus attaches to the surface of the hepatocyte, penetration of the virus into the cells may occur by two mechanisms; endocytosis of intact virions with subsequent release from endosomes or fusion between the viral envelope and the liver cell plasma membrane with penetration of the nucleocapsid into the cytoplasm. Replication of the virus in liver cells results in the production of viral proteins and the assembly of the complete virion. Hepatitis B surface antigen and core antigen are expressed on the plasma membrane of infected cells and subsequently large amounts of the surface antigen and virus are released into the circulation.

Antigenic heterogeneity of hepatitis B surface antigen has been demonstrated by serological analysis. The surface antigen particles share a common group-specific antigen *a* (which has several subspecificities) and generally carry at least two mutually exclusive subdeterminants, *d* or *y* and *w* or *r*. The subtypes are the phenotypic expression of distinct genotype variants of hepatitis B virus. Four principal phenotypes are recognised, *adw*, *adr*, *ayw* and *ayr*, but other complex permutations of these subdeterminants and new variants have been described, all apparently on the surface of the same physical particles. The major subtypes have differing geographical distributions. For example, in northern Europe, the Americas and Australia subtype *adw* predominates. Subtype *ayw* occurs in a broad zone which includes northern and western Africa, the eastern Mediterranean, eastern Europe, northern and central Asia and the Indian subcontinent. Both *adw* and *adr* are found in Malaysia, Thailand, Indonesia and Papua New Guinea, while subtype *adr* predominates in other parts of South-East Asia including China, Japan and the Pacific islands. The subtypes provide useful epidemiological markers of hepatitis B virus.

Ben-Porath and Wands (1984) summarised their studies on the use of monoclonal antibodies against hepatitis B surface antigen for subtyping the surface antigen by radioimmunoassay. Their monoclonal antibodies recognised different determinants, as demonstrated by the absence of competitive inhibition in surface antigen binding studies. Eight monoclonal antibodies reacted with all subtypes of the surface antigen while four other monoclonal antibodies bound to most but not all of them. Another six antibodies bound selectively to certain subtypes of the virus and it was possible to identify subsets of particles within a subtype of the surface antigen which were not recognised previously by polyclonal antibodies. There were also quantitative differences in binding by all monoclonal antibodies to known subtypes which could not be explained on the basis of antigen titres or binding affinity for antigen epitopes. It was therefore possible to define differences in epitope concentrations on surface antigen particles by monoclonal antibodies. The results suggest that there are distinct determinants which reside on a domain common to all subtypes but that there are also quantitative and qualitative differences in epitope density among the various subsets. This implies that the surface antigen particles are much more heterogeneous than described hitherto. It is, therefore now possible to 'finger-print' hepatitis B virus and this will permit fine analysis of the evolution of the virus and its epidemiology in different parts of the world.

Immune response to acute infection with hepatitis B virus

Infection with the virus leads to the appearance in the plasma during the incubation period of hepatitis B surface antigen about 2–8 weeks before biochemical evidence of liver dysfunction or the onset of jaundice. This antigen persists during the acute illness and is usually cleared from the circulation during convalescence. Next to appear in the circulation is the viral DNA polymerase associated with the core of the virus and at about the same time another antigen, the e antigen, becomes detectable, again preceding serum aminotransferase elevations. The e antigen is a distinct soluble antigen which is located within the core and correlates closely with the number of virus particles and relative infectivity. Antibody to the hepatitis core antigen is found in the serum 2–4 weeks after the appearance of the surface antigen and it is always detectable during the early acute phase of the illness. Core antibody of the IgM class becomes undetectable within some months of the onset of uncomplicated acute infection, but IgG antibody persists for many years after recovery and possibly for life. The next antibody to appear in the circulation is directed against the e antigen and there is evidence that, in general terms, anti-e indicates relatively low infectivity of serum, although a better measure of infectivity is the presence of hepatitis B virus DNA in serum. Antibody to the surface antigen component, hepatitis B surface antibody, is the last marker to appear late during convalescence. Precipitating antibodies reacting with antigenic determinants on the complete virus particle have also been described and these antibodies may be relevant to the clearance of circulating hepatitis B virions. Cell-mediated immunity is also important in hepatitis B infection and the interrelationship between the two types of immune response is summarised below.

Evidence accumulated more recently indicates, as discussed by Mondelli and Eddleston (1984), that at the time of replication of hepatitis B virus in the liver cells the surface antigen and core antigen are expressed on the plasma membrane and both cellular and immune responses are initiated. The release of large amounts of surface antigen into the circulation which follows may induce high tolerance and rapid disappearance of the immune response to this antigen. Virions carrying polyalbumin receptors also stimulate the formation of antibodies against polymerised human serum albumin which prevent attachment and penetration of the virus into uninfected liver cells by reacting with the receptor on the surface of liver cells. Elimination of the virus thus depends on a combined cellular and humoral response, with both receptor-neutralising polymerised human serum albumin antibodies and effective cytotoxic T-cells. Failure of either of these mechanisms would lead to chronic liver damage and viral persistence. The extent of liver damage then depends on a number of factors which include autoimmune reactions directed at hepatocyte membrane antigens and modulation of lysis by T-cells of infected hepatocytes expressing core antigen on the surface of the cells. This would result eventually in termination of active viral replication with seroconversion to e antibody, with clinical and histological remission. On the other hand, cells with integrated viral genomes do not express core antigen on their surface and are protected from T-cell lysis. The destruction of hepatocytes containing integrated hepatitis B virus DNA may be dependent, as reviewed by

Thomas and Lok (1984), on an immune response to hepatitis B surface antigen and failure of this process of elimination results in the persistence of clones of cells with the potential for transformation.

EPIDEMIOLOGY OF HEPATITIS B VIRUS

Much has been written during the last decade on all aspects of hepatitis B infection and an enormous bibliography has accumulated on the complex epidemiological patterns of hepatitis B. The entire subject has been extensively reviewed elsewhere (Zuckerman, 1975; World Health Organisation, 1977; Zuckerman and Howard, 1979; Szmuness, Alter and Maynard, 1982; Deinhardt and Deinhardt, 1983). A summary of the epidemiology of hepatitis B therefore follows.

In the past, hepatitis B was defined principally on the basis of infection occurring several months after transfusion of human blood or plasma fractions or the use of inadequately sterilised syringes and needles. The availability of specific laboratory tests for markers of hepatitis B has altered significantly the epidemiological concepts of this infection and demonstrated the global importance of hepatitis B.

The incubation period varies from 14 days to 180 days and the clinical picture of the infection varies widely in its presentation from asymptomatic or subclinical infection, through mild gastrointestinal symptoms and the anicteric form of the disease, to acute illness with jaundice, severe prolonged jaundice, and acute fulminant hepatitis. Hepatitis B may lead to persistent infection and the carrier state and may progress to chronic liver disease including hepatocellular carcinoma (World Health Organisation, 1983). Reliable information on the incidence and long-term trend of hepatitis B is only available in a few countries; many infections are not recognised or are not reported and the extent of chronic infection with hepatitis B virus became evident recently as tests for hepatitis B surface antigen were introduced widely. It is estimated that there are between 175 and 200 million carriers in the world providing a huge reservoir of infection (reviewed by Zuckerman, 1982a). Survival and persistence of hepatitis B virus in the population is ensured by the reservoir of carriers, prolonged shedding of the virus by a proportion of carriers, varied mechanisms and routes of transmission including perinatal infection and the relative stability of the virus in the environment.

Mode of spread

Specific laboratory tests for hepatitis B confirmed the importance of the parenteral routes of transmission and infectivity appears to be especially related to blood. However, the infection is not spread exclusively by blood and blood products. There are observations that under certain circumstances the virus is infective by mouth and the infection may be endemic in closed and semi-closed institutions and in institutions for the mentally-handicapped. It is more prevalent in adults in urban communities and among those living in poor socio-economic conditions. Considerable differences in prevalence of the infection and of the carrier state exist in different geographical regions and between different ethnic and socio-economic groups.

There is much evidence for the transmission of hepatitis B by intimate contact and by the sexual route. The sexually promiscuous, particularly male homosexuals who change partners frequently, are at very high risk of infection with hepatitis B virus. Hepatitis B surface antigen has been found in blood and in various body fluids such as saliva, menstrual and vaginal discharges, seminal fluid, colostrum and breast milk and serous exudates, and these have been implicated as vehicles of transmission of infection. Contact-associated hepatitis is thus of major importance. Transmission of the infection may result from accidental inoculation of minute amounts of blood or body fluids contaminated with blood, such as may occur during medical, surgical and dental procedures, immunization with inadequately sterilised syringes and needles, intravenous and percutaneous drug abuse, tattooing, ear-piercing and nose-piercing, acupuncture, laboratory accidents and accidental inoculation with razors, shared tooth brushes, bath brushes, towels and similar objects which have been contaminated with blood. Additional factors may be important for the transmission of hepatitis B infection in the tropics and in warm-climate countries. These include traditional tattooing and scarification, blood letting, ritual circumcision and repeated biting by blood-sucking arthropod vectors. Results of investigations into the role which biting insects may play in the spread of hepatitis B are conflicting. Hepatitis B surface antigen has been detected in several species of mosquito and in bed-bugs which have either been trapped in the wild or fed experimentally on infected blood, but no convincing evidence for replication of the virus in insects has been obtained. Mechanical transmission of the infection, however, is a possibility, particularly as a result of interrupted feeding in high prevalence areas.

The presence of hepatitis B surface antigen has been reported in faeces, bile and urine, usually as a result of contamination with blood. Hepatitis B does not appear to be transmitted by the faecal–oral route and urine is probably not infectious unless contaminated with blood. There is no convincing evidence that airborne infections occur.

Clustering of hepatitis B also occurs within family groups, but on the whole it is not related to genetic factors and does not reflect maternal or venereal transmission. The mechanisms of intrafamilial spread of hepatitis B infection are not known.

Mother-to-infant transmission

Transmission of hepatitis B from carrier mothers to their babies can occur during the perinatal period and appears to be the single most important factor in determining the prevalence of the infection in some regions, particularly in China and South-East Asia. The risk of infection in the infant may reach 50–60% although it varies from country to country and appears to be related to ethnic groups. Infection of infants is especially important because a large proportion of these infants will become carriers. Infectivity is directly related to the presence of high titres of hepatitis B surface antigen and/or hepatitis B e antigen in the mother's circulation. When e antigen is present, as many as 95% of their newborn children are infected, usually in the perinatal period. The prevalence of e antigen among surface antigen maternal carriers, and thus the infectivity of mothers for their infants, varies markedly in different geographical areas and in different ethnic groups. In some parts of Asia, particularly in eastern Asia, 30–50% of surface antigen carrier women of childbearing age also carry e antigen in their blood, and perinatal infections may account for about half the carriers in the population (World Health Organisation, 1983). Perinatal transmission is of intermediate frequency in mothers of West-Asian or Afro-Caribbean origin. In contrast, the carrier state and perinatal transmission are uncommon in Caucasian mothers. The pattern of mother-to-infant transmission and establishment of the carrier state is different in Africa, where e antigen is less frequent in carrier mothers and infection of infants occurs most commonly during early childhood. Another mode of transmission of hepatitis B is infection of children of non-carrier mothers by contact with children who had been infected by their carrier mothers.

It should also be noted that there is a substantial risk of perinatal infection if the mother has acute hepatitis B, particularly during the third trimester of pregnancy or within two months after delivery. Intrauterine infections are uncommon, since the virus does not cross the placenta and the few infections which occur *in utero* are probably the result of a leakage of maternal blood into the foetal circulation associated with a tear in the placenta. Finally, the precise mechanism of perinatal infection is uncertain but it probably occurs during or shortly after birth as a result of a leak of maternal blood into the baby's circulation or its ingestion or inadvertent inoculation. Most of the children infected during the perinatal period become persistent carriers.

The carrier state

The carrier state is defined on the basis of longitudinal studies as persistence of the hepatitis B surface antigen in the circulation for more than six months. The carrier state may be life-long and may be associated with liver damage varying from minor changes in the nuclei of hepatocytes to persistent hepatitis, chronic active hepatitis, cirrhosis and hepatocellular carcinoma (Arias and Shafritz, 1982). In carriers of hepatitis B virus with or without histological evidence of liver disease, integration of the viral DNA may be at many sites or at a unique site in the host genome. Most of these carriers have the surface antigen and *e* antibody and it has been suggested that the continued expression of the surface antigen in patients may result from integrated viral DNA. Some patients, however, may have hepatitis B viral DNA in their liver without expression of the surface antigen (latent viral infection). All the available evidence is consistent with the interpretation that integration of the viral DNA into the genome of the liver cell precedes the development of hepatocellular carcinoma by months or years. However, the exact time at which integration occurs, the relationship between integration and serological markers of hepatitis B, and the frequency with which integration leads to the development of liver cancer require further investigation. From the epidemiological evidence, it appears that the carrier state precedes by many years the development of cancer (World Health Organisation, 1983; Zuckerman *et al.*, 1983).

Several risk factors have been identified in relation to the development of the carrier state. It is commoner in males, more likely to follow infections acquired in childhood, as described above, than those acquired in adult life and more likely to occur in patients with natural or acquired immune deficiencies. A carrier state becomes established in approximately 5–10% of infected adults. The prevalence of carriers among apparently healthy adults, particularly among blood donors, in northern Europe, North America and Australia is 0.1% or less. In central and eastern Europe it is up to 5% with a higher frequency in southern Europe, in countries bordering the Mediterranean, and in parts of Central and South America, while in some parts of Africa, Asia and the Pacific region as many as 15% or more of the population may be carriers (reviewed by McCollum and Zuckerman, 1981; Zuckerman, 1982a; Deinhardt and Gust, 1983).

Age distribution and prevalence of infection

Two different patterns of age distribution of infection are recognised. In populations with a high prevalence of hepatitis B virus, infection is usually acquired early in life, and the highest infection and carrier rates are seen amongst children and young adults, with lower prevalence among older age groups. The e antigen has been found more commonly in young than in adult carriers, while the prevalence of e antibody is higher in older age groups. These findings suggest that young carriers could be the most infective. In countries in which infection with hepatitis B virus is relatively uncommon, the highest prevalence of hepatitis B surface antigen is found in the 20 to 40 age group. The highest rates of infection are found among groups who have an increased risk of contact with blood or blood products, as outlined above, including health care personnel, certain categories of patients, intravenous drug abusers and male homosexuals who change partners frequently.

In summary, the overall pattern of the global epidemiology of hepatitis B infection is as follows. The prevalence of the carrier state may be conveniently divided, as outlined above, into three categories. Low endemic areas such as northern Europe, North America and Australia; areas with intermediate prevalence which include eastern Europe, the Mediterranean, Central America and parts of South America and South West Asia; and high endemic areas such as China, South-East Asia and tropical Africa. Evidence of infection by hepatitis B virus, as measured by the prevalence of hepatitis B surface antibody in plasma, shows a similar geographical distribution, 4–6% in areas of low endemic disease, 20–55% in intermediate areas and 70–95% in areas of high endemic disease. Infection with hepatitis B virus is thus indeed a universal problem and prevention of infection by immunisation with effective vaccines would have a considerable impact on health, on the prevention of the carrier state and associated chronic liver disease in a proportion of carriers, and on the long-term risk of progression to primary liver cancer.

HEPATITIS-B-LIKE VIRUSES IN ANIMALS: THE HEPADNA VIRUSES

The identification of a number of human hepatitis-B-like viruses in lower animals (other than the great apes) resulted in important advances in the knowledge of the mode of replication, comparative pathology, viral persistence and molecular biology of this unique group of DNA viruses.

Snyder (1968) reported the presence of liver cancer in 22 out of 76 eastern woodchucks (*Marmota monax*) that lived longer than four years in an established colony. In addition, the lesions of chronic active hepatitis and sometimes cirrhosis were usually found in the non-tumour tissue. Examination by electron microscopy of sera collected from the captive woodchucks revealed virus particles that closely resemble human hepatitis B virus (Summers, Smolec and Snyder, 1978). Human hepatitis B virus and the woodchuck hepatitis virus share the following characteristics. Infection with either virus results in the accumulation in the blood of large amounts of excess virus coat protein in the form of spherical and tubular particles measuring 20–25 nm in diameter. Both viruses are 40–45 nm double-shelled or solid particles, with a nucleocapsid containing double-stranded circular DNA with a gap, and both contain a viral DNA polymerase. Each virus is associated with chronic hepatitis and hepatocellular carcinoma. Werner *et al.* (1979) reported antigenic cross-reactivity between the cores of the two viruses, but only minor common antigenic determinants were identified on the virus surface protein. A small region of 100–150 base pairs of nucleic acid homology, measured by liquid hybridisation, was found in the genomes of the two viruses. It seems likely that this 3–5% nucleic acid homology represents one or two regions of nearly identical nucleotide sequence. Cummings *et al.* (1980) cloned the DNA of human hepatitis

B virus and the DNA of woodchuck hepatitis virus in the vector λgt.WES. This was then subcloned into the kanamycin resistance plasmid pAO1. Comparison of the recombinant DNAs with authentic virus DNAs by specific hybridisation, size and restriction enzyme analysis showed that the recombinants contained the complete genome of each virus. The nucleic acid homology between the two viral DNAs was confirmed with the cloned DNAs. So, the woodchuck hepatitis virus and the human hepatitis virus are phylogenetically related. The analogy between the two viruses is even stronger when judged by their adaptation in their respective hosts; both cause persistent infection and have a close association with chronic liver disease and hepatocellular carcinoma (Snyder, Tyler and Summers, 1982).

Another virus which is related to human hepatitis B has been described in Beechey ground squirrels (*Spermophilus beecheyi*) in northern California (Marion *et al.*, 1980*a*). Features held in common with the human hepatitis B virus include virus morphology, size and structure of the viral DNA, a virion DNA polymerase which repairs a single-stranded region in the partially double-stranded circular genome, cross-reacting surface viral antigens, antigen–antibody systems similar to hepatitis B *e* antigen and the core antigen and persistent infection with viral antigen present continuously in the blood. The antigenic and structural relationships between the surface antigens of the human hepatitis B virus, the ground squirrel hepatitis B virus and the woodchuck hepatitis virus have been described in more detail by Feitelson, Marion and Robinson (1981, 1982) and reviewed by Marion and Robinson (1983). The differences in pathogenicity of these three viruses are notable. While persistent infection in infected ground squirrels is common in areas where the disease is endemic and the titre of the virus, as measured by viral DNA polymerase activity, is high, there is little or no evidence of hepatitis in infected ground squirrels. In persistently infected squirrels followed-up in the laboratory for over two years, only the mildest form of inflammation of the liver was found in some animals and none of 25 ground squirrels developed cirrhosis or liver cancer (Marion and Robinson, 1983). The complete DNA sequence of the ground squirrel hepatitis virus has not been published at the time of writing. However, cross-hybridisation studies show a significant degree of homology between the DNA of human hepatitis B virus and the ground squirrel hepatitis virus (Siddiqui, Marion and Robinson, 1981). The results to date indicate that the coding sequences for the major surface antigen polypeptide and the major core polypeptide of the three mammalian hepatitis viruses have homologous regions located similarly in relation to the unique physical features of the virion DNA.

The fourth member of this group of viruses was discovered in some domesticated ducks (*Anas domesticus*) in the People's Republic of China following the observation of frequent liver cancer in Pekin ducks. Mason, Seal and Summers (1980) reported that approximately 10% of Pekin ducks in some commercial flocks in the United States carry a hepatitis-B-like virus which was named duck hepatitis B virus. This species of duck was originally imported from China in the nineteenth century. The duck hepatitis B virus is similar morphologically to the three mammalian viruses, although the spherical particles are larger and the more pleomorphic and tubular forms have not been found.

The viral genome is circular and partially single-stranded and an endogenous DNA polymerase can convert the DNA genome to a complete double-stranded circular form with a size of approximately 3 kb. Examination of viral DNA in the organs of infected birds revealed preferential localisation in the liver. The virus is transmitted vertically and infected ducklings may develop persistent viraemia (Mason *et al.*, 1980). Studies on antigenic analysis and nucleotide sequences are in progress (reviewed by Summers, 1981; and Marion and Robinson, 1983).

Robinson (1980) coined the term hepadnaviruses (hepatic DNA viruses) for human hepatitis B virus and the three similar viruses of lower animals, which may well be members of a larger group of related viruses.

HEPATITIS B VIRUS AND HEPATOCELLULAR CARCINOMA

Hepatocellular carcinoma is one of the ten most common cancers in the world and one of the most prevalent cancers in developing countries, with over 250 000 new cases of liver cancer each year. The actual age-adjusted incidence of hepatocellular carcinoma is over 30 new cases per 100 000 of the population each year in some parts of Asia and Africa, whereas it is less than 5 new cases per 100 000 per year in most countries in Europe, in North America and in Australia. Nevertheless, there appears to be an upward time trend in the majority of the low incidence countries. This form of cancer is more common in males than females and it is well established that the incidence of liver cancer increases with age, but in high risk populations the disease occurs in younger age groups. There is a marked increase in incidence in certain ethnic groups (World Health Organisation, 1983).

Epidemiological correlation between hepatitis B infection and hepatocellular carcinoma

Many studies in different parts of the world, particularly in Africa, Asia and the Mediterranean region (especially in Greece), show a highly significant excess of surface antigen, core antibody and surface antibody in patients with hepatocellular carcinoma (reviewed by Szmuness, 1978; Maupas and Melnick, 1981). An important factor in the aetiological association between hepatitis B and liver cell carcinoma may lie in an early age of infection (Zuckerman, 1974). In areas of the world where the prevalence of macronodular cirrhosis and hepatocellular carcinoma is high, infection with hepatitis B virus and development of the persistent carrier state occur most frequently in infants and children.

Observations that suggest that infection with hepatitis B virus, probably as a result of infection acquired perinatally or early in life, precedes the development of liver cell cancer were obtained in studies in Senegal and Taiwan. Larouze et al. (1976) carried out a case-control study of patients with hepatocellular carcinoma and their families in Dakar, Senegal. Nearly all the 28 patients with hepatocellular carcinoma (93%) had serological evidence of infection with hepatitis B virus, but the patients with carcinoma were less likely to have surface antibody. Hepatitis B surface antigen was detected in 71% of the mothers of the patients, whereas only 14.5% of the mothers of the control subjects were antigen-positive. Similar findings were obtained in Taiwan (Beasley, 1982).

The results of a large prospective survey were reported by Beasley et al. (1981). Between 1975 and 1978, 22 707 male Chinese civil servants in Taiwan were enrolled in a prospective study to establish the relative risk of hepatocellular carcinoma among carriers of hepatitis B surface antigen and to determine whether the carrier state precedes the development of this cancer. There were 2454 (15.2%) carriers in the group, of whom 40 subsequently died of hepatocellular carcinoma. Only one of the 19 253 antigen-negative men died of liver cell cancer, giving a relative risk of 223 for the development of this form of cancer in surface antigen carriers. There was also a large excess mortality from cirrhosis among antigen-positive men; 17 out of 19 men who died from cirrhosis were antigen-positive.

Liver cell cancer and cirrhosis together accounted for 57 out of 105 deaths among the antigen-positive men (54.3%), compared with three out of 202 deaths amongst those without the antigen (1.5%). Together, hepatocellular carcinoma and cirrhosis accounted for 20% of all deaths in this study, but mortality from other causes was similar in the two groups. Follow-up of the two groups has been extended to 97 000 man years. The number of patients who died from hepatocellular carcinoma increased to 70 in the antigen-positive group but no further cases occurred in the control group, which gives a relative risk of 390 (Beasley, 1982). A study among railway workers in Japan showed a similar risk of hepatocellular carcinoma in carriers (Sakuma et al., 1982). Supporting evidence was obtained from other studies in Japan (Okuda et al., 1982) and from longitudinal serological surveys in Japan (Obata et al., 1980) and Alaska (Heyward et al., 1982) and prospective studies are in progress in China, Senegal, Singapore, the United States of America and elsewhere. The results published to date clearly establish that the carrier state of hepatitis B surface antigen commonly precedes hepatocellular carcinoma and suggest that the virus is closely associated with the process leading to this type of cancer and that it is not simply a risk factor.

Production of hepatitis B surface antigen by continuous cell lines derived from human hepatocellular carcinoma

The establishment of continuous cell lines from hepatocellular carcinoma which produce hepatitis B surface antigen provided a laboratory model for the study of various aspects of the biology of hepatitis B. The first of these cell lines, the PLC/PRF/5 cell line described by Alexander et al. (1976) produces hepatitis B surface antigen similar in size, morphology and polypeptide composition to the form which occurs in the serum of naturally infected individuals (Stannard and Alexander, 1977; Skelly et al., 1979). Aden et al. (1979) described the second hepatitis B surface antigen-producing cell line, Hep 3B, which differs in some ways from PLC/PRF/5. A third hepatocellular carcinoma cell line, DELSH-5, was established by Das et al. (1980). This cell line released hepatitis B surface antigen into the medium from the thirteenth passage onwards and, like the

Hep 3B cells, the cells synthesised albumin and alpha-foetoprotein. Other similar cell lines have been derived in other laboratories.

Integration of hepatitis B viral DNA in the PLC/PRF/5 cell line

Marion et al. (1980b) reported that the PLC/PRF/5 cells contained approximately four copies of viral DNA per haploid mammalian cell DNA equivalent. Evidence was obtained that DNA from all regions of the viral genome is present in these cells, suggesting that they contain most, and possibly all, of the viral genome. Furthermore, the results indicated that the viral DNA is integrated in high molecular weight DNA at three different sites and that there is no viral DNA in an episomal form. Cellular RNA radiolabelled with ^{32}P was found to hybridise with all restriction fragments of hepatitis B virus DNA; this suggests that most, and possibly all, of the viral DNA in these cells is transcribed.

Chakraborty et al. (1980) also presented evidence for integration of the DNA of hepatitis B, virus into the host genome of the PLC/PRF/5 cells and for expression of three RNA molecules containing specific sequences of hepatitis B virus. In order to study the expression of viral RNA in this cell line cytoplasmic and nuclear RNAs were isolated and separated into polyA$^+$ and polyA$^-$ fractions; hepatitis B virus sequences were identified by hybridisation with a ^{32}P-labelled hepatitis B virus probe. The cytoplasmic polyA$^+$ RNA contained two virus-specific transcripts of 21.5S and 19.5S. There were three transcripts in the nuclear RNA of 27S, 21.5S and 19.5S. The 21.5S and 19.5S transcripts probably represent the same components as found in the cytoplasm. The biological activity and function of these viral RNAs are not known. The presence of a polyA tail in the 21.5S and 19.5S RNA species suggests that they represent specific viral mRNAs. Either the 21.5S or 19.5S polyA$^+$ RNA most probably codes for hepatitis B surface antigen. The finished polypeptide of molecular weight 25 000 may be derived by post-translational processing of a larger precursor.

Using a modification of the Southern blot transfer hybridisation technique and cloned hepatitis B virus DNA as a probe, Brechot et al. (1980) demonstrated the integration of viral DNA in the cellular genome of human hepatocellular carcinoma tissue and in the PLC/PRF/5 cell line. The results suggest the existence of a limited number of integration sites in the cellular DNA, a proposition consistent either with the development of the liver tumour from one clone with several integration sites or with its development from a few clones each having particular integration sites. Other observations suggest the presence of two or more viral genomes inserted in a tandem head-to-tail array.

Edman *et al.* (1980) also showed that sequences of hepatitis B virus are integrated into a minimum of six different sites in the DNA of the PLC/PRF/5 cells. The results of the experiments suggest that the nicked cohesive end region of hepatitis B virus coincides with the region of integration of at least several viral fragments.

Heterotransplantability of the PLC/PRF/5 cells

Desmyter *et al.* (1980) induced tumours in 80% or more of nude mice (Pfd: NMRI/nu-nu) injected subcutaneously with $5–10 \times 10^6$ cells. The tumours usually became detectable after two weeks and grew up to 15–30% of the body weight of the mice. Metastases were not found. When the cells were inoculated intraperitoneally, multiple tumours developed in various abdominal organs. The histology of the tumours was that of a well-differentiated human hepatocellular carcinoma. Hepatitis B surface antigen was demonstrated by immunofluorescence in 0.1–10% of the tumour cells. Hepatitis B surface antigen, but no other serological marker of the virus, was found in the serum of the tumour-bearing mice. Newborn and weanling nude mice were equally susceptible.

The tumours were serially transplantable three to five times but the take rates decreased with each passage, although there were no obvious differences between first-passage and later-passage tumours. The take rates were better when pieces of tumour were transplanted rather than trypsinised tumour cells. It was also possible to clone cells from the tumours and their progeny induced tumours. Similar tumours were obtained in nude, athymic rats (Rowett Pfd: WIST-rnu-rnu), although the rats were less susceptible than nude mice. Similar results have been reported by Bassendine *et al.* (1980) and by other investigators.

Integration of hepatitis B virus DNA into the genome of liver cells

Lutwick and Robinson (1977) and Summers *et al.* (1978) reported the finding of hepatitis B virus DNA in hepatocellular carcinoma tissue. Shafritz and Kew (1981) examined DNA from hepatocellular carcinomas from 13 patients for hepatitis B virus DNA sequences by molecular hybridization using ^{32}P-labelled cloned HBV-DNA. Hepatitis B surface antigen was present in the serum of eight of these patients and viral DNA sequences were found to be integrated into the host genome in carcinomatous tissues in each patient. The integration pattern was unique for each tumour. Hepatitis B virus DNA was not found in DNA from the tumours of five patients who were not carriers of the surface antigen. Subsequently Shafritz et al. (1981) reported the results from 20 patients with hepatocellular carcinoma, all of whom had at least one marker of current or past infection with hepatitis B virus. In all 12 patients with surface antigen, integrated viral DNA was present in the tumour. Integrated hepatitis B virus DNA was also found in 3 out of 8 patients with liver cell cancer in whom surface antibody was present but who were serologically negative for the surface antigen. This finding indicates that there is no simple correlation between serum markers of hepatitis B and viral DNA in the tumour. Furthermore, in some specimens of liver tissue adjacent to tumours, extra-chromosomal and integrated viral DNA were present. The presence of integrated DNA in non-tumour tissue from patients with hepatocellular carcinoma suggests that integration precedes development of gross neoplasia.

In carriers of hepatitis B virus with or without histological evidence of chronic liver disease, integrated DNA was not found in most of the carriers with relatively recent history of liver disease. However, in several patients who were long-term carriers, there was diffuse hybridisation in the high molecular weight regions of the gel. Free viral DNA was not identified. In these cases it is possible that viral DNA is integrated at different sites in different cells. Such integration might precede a stage in persistent infection with hepatitis B virus during which a specific subpopulation of hepatocytes undergoes cellular division into a clonal focus containing integrated viral DNA in one or a few specific sites. Additional factors may then be involved, perhaps years later, in the development of neoplasia from such a clonal focus (Shafritz et al. 1981; Shafritz, 1982).

Brechot et al. (1981) used the Southern blot transfer hybridisation technique to examine DNA extracted from tissues. Viral DNA was found to be integrated into cellular DNA in both liver tumour and non-tumour tissue in patients with hepatocellular carcinoma, as demonstrated by hybridisation of high molecular weight DNA after digestion with HindIII and EcoR1 endonucleases. Integrated viral DNA was also found in patients with cirrhosis with or without chronic active hepatitis. Free hepatitis B virus DNA was found in the liver in two patients with chronic persistent hepatitis and one patient with chronic active hepatitis. Restriction endonuclease patterns in two patients with acute hepatitis B strongly suggested viral DNA integration. If these findings are confirmed by the examination of a large number of patients with acute hepatitis B, then viral integration seems to occur early in the course of infection.

Although many questions remain to be answered, molecular biology is providing definitive evidence for covalent integration of the genome of hepatitis B virus into cellular DNA in hepatocellular carcinomatous tissue, suggesting a possible sequence of events after infection of hepatocytes culminating in malignant transformation.

In summary, therefore, various studies, including comparative pathology and comparative virology of infected eastern woodchucks and Pekin ducks and the mode of replication of the hepadnaviruses, have established that there is a strong and specific association between persistent hepatitis B infection and hepatocellular carcinoma, and it is likely that this association is causal in up to 80% of such cancers (World Health Organisation, 1983). However, factors other than hepatitis B virus may be implicated. It is possible that hepatocellular carcinoma is the cumulative result of several co-factors or hepatocarcinogens including genetic, immunological, nutritional and hormonal factors, mycotoxins, particularly aflatoxin, chemical carcinogens and other environmental influences including alcohol, and that hepatitis B virus acts either as a carcinogen or as a co-carcinogen in persistently infected hepatocytes (reviewed by Zuckerman, 1982b; Arthur, Hall and Wright, 1984).

The hepatitis research programme at the London School of Hygiene and Tropical Medicine is supported by generous grants from the Medical Research Council, the Department of Health and Social Security, the Wellcome Trust, the World Health Organisation and Organon B.V.

The hepatitis B vaccine development at the London School of Hygiene and Tropical Medicine is generously supported by the British Technology Group (formerly the National Research Development Corporation), the Department of Health and Social Security and the Wellcome Trust.

REFERENCES

Aden, D. P., Fogel, A., Plotkin, S., Damjanov, I. and Knowles, B. B. (1979). Controlled synthesis of HBsAg in a differentiated human liver carcinoma-derived cell line. *Nature*, **282**, 615–16.

Albin, C. and Robinson, W. S. (1980). Protein kinase activity in hepatitis B virus. *Journal of Virology*, **34**, 297–302.

ALEXANDER, J. J., BEY, E. M., GEDDES, E. W. and LECATSAS, G. (1976). Establishment of a continuously growing cell line from primary carcinoma of the liver. *South African Medical Journal*, **50**, 2124–8.

ALMEIDA, J. D., RUBENSTEIN, D. and STOTT, E. J. (1971). New antigen–antibody system in Australia antigen-positive hepatitis. *Lancet*, **2**, 1225–7.

ALMEIDA, J. D., ZUCKERMAN, A. J., TAYLOR, P. E. and WATERSON, A. P. (1969). Immune electron microscopy of the Australia-SH (serum hepatitis) antigen. *Microbios*, **2**, 117–23.

ARIAS, I. M. and SHAFRITZ, D. A. (eds.) (1982). Hepatitis B virus, chronic liver disease, and primary carcinoma of the liver. *Hepatology*, **2** (Suppl.), 1–133.

ARTHUR, M. J. P., HALL, A. J. and WRIGHT, R. (1984). Hepatitis B, hepatocellular carcinoma and strategies for prevention. *Lancet*, **1**, 607–10.

BASSENDINE, M. F., ARVORGH, B. A. M., SHIPTON, N., MONJARDINO, J., ARANGUIBEL, F., THOMAS, H. C. and SHERLOCK, S. (1980). Hepatitis B surface antigen and alpha-fetoprotein secreting human primary liver cancer in athymic mice. *Gastroenterology*, **79**, 528–32.

BAYER, M. E., BLUMBERG, B. S. and WERNER, B. (1968). Particles associated with Australia antigen in sera of patients with leukaemia, Down's syndrome and hepatitis. *Nature*, **218**, 1057–9.

BEASLEY, R. P. (1982). Hepatitis B virus as the etiologic agent in hepatocellular carcinoma – epidemiologic considerations. *Hepatology*, **2** (Suppl.), 21–6.

BEASLEY, R. P., HWANG, L-Y., LIN, C-C. and CHIEN, C-S. (1981). Hepatocellular carcinoma and hepatitis B virus: A prospective study of 22 707 men in Taiwan. *Lancet*, **2**, 1129–33.

BEN-PORATH, E. and WANDS, J. R. (1984). Monoclonal antibodies as diagnostic probes in the aetiology of hepatitis. In *Seminars in Liver Disease*, vol. 4, ed. M. A. Gerber, pp. 76–88. New York, Thieme-Stratton.

BLUMBERG, B. S., ALTER, H. J. and VISNICH, S. (1965). A 'new' antigen in leukaemia sera. *Journal of the American Medical Association*, **191**, 541–6.

BRECHOT, C., HADCHOUEL, M., SCOTTO, J., FONCK, M., POTET, F., VYAS, G. N. and TIOLLAIS, P. (1981). State of hepatitis B virus DNA in hepatocytes of patients with hepatitis B surface antigen-positive and -negative liver diseases. *Proceedings of the National Academy of Sciences, USA*, **78**, 3906–10.

BRECHOT, C., PURCEL, C., LOUISA, A., RAIN, B. and TIOLLAIS, P. (1980). Presence of integrated hepatitis B virus DNA sequences in cellular DNA of human hepatocellular carcinoma. *Nature*, **286**, 533–5.

CHAKRABORTY, P. R., RUIZ-OPAZO, N., SHOUVAL, D. and SHAFRITZ, D. (1980). Identification of integrated hepatitis B virus DNA and expression of viral RNA in an HBsAg-producing hepatocellular carcinoma cell line. *Nature*, **286**, 531–3.

CUMMINGS, I. W., BROWNE, J. K., SALSER, W. A., TYLER G. V., SNYDER, R. L., SMOLEC, J. M. and SUMMERS, J. (1980). Isolation, characterisation and comparison of recombinant DNAs derived from genomes of human hepatitis B virus and woodchuck hepatitis virus. *Proceedings of the National Academy of Sciences, USA*, **77**, 1842–6.

DANE, D. S., CAMERON, C. H. and BRIGGS, M. (1970). Virus-like particles in serum of patients with Australia antigen-associated hepatitis. *Lancet*, **2**, 695–8.

DAS, P. K., NAYAK, N. C., TSIQUAYE, K. N. and ZUCKERMAN, A. J. (1980). Establishment of a human hepatoma carcinoma cell line releasing hepatitis B surface antigen. *British Journal of Experimental Pathology*, **61**, 648–54.

DEINHARDT, F. and DEINHARDT, J. (eds.) (1983). *Viral Hepatitis: Laboratory and Clinical Science*. New York, Marcel Dekker.

DEINHARDT, F. and GUST, I. D. (1983). Viral hepatitis. *Bulletin of the World Health Organization*, **60**, 661–91.

DESMYTER, J., DE GROOTE, J., RAY, M. B., BRADBURNE, A. F., DESMET, V., DE SOMER, P. and ALEXANDER, J. (1980). HBsAg-producing human hepatoma cell line: tumours in nude mice and interferon properties. In *Virus and the Liver*, ed. L. Bianchi, W. Gerok, K. Sinkinger and G. A. Stalder, pp. 217–21. Lancaster, MTP Press.

EDMAN, J. C., GRAY, P., VALENZUELA, P., RALL, L. B. and RUTTER, W. J. (1980). Integration of hepatitis B virus sequences and their expression in a human hepatoma cell. *Nature*, **286**, 535–8.

FEITELSON, M. A., MARION, P. L. and ROBINSON, W. S. (1981). Antigenic and structural relationship of the surface antigens of hepatitis B virus, ground squirrel hepatitis virus, and woodchuck hepatitis virus. *Journal of Virology*, **39**, 447–54.

FEITELSON, M. A., MARION, P. L. and ROBINSON, W. S. (1982). The core particles of HBV and GSHV. 1. Relationship between HBcAg-and GSHcAg-associated poly-peptides by SDS–PAGE and tryptic peptide mapping. *Journal of Virology*, **43**, 687–96

HEYWARD, W., BENDER, T. R., LANIER, A. P., FRANCIS, D. P., MCMAHON, B. J. and MAYNARD, J. E. (1982). Serological markers of hepatitis B virus and alpha-feto-protein levels preceding primary hepatocellular carcinoma in Alaskan Eskimos. *Lancet*, **2**, 889–91.

KAPLAN, P. M., GREENMAN, R. L., GERIN, J. L., PURCELL, R. H. and ROBINSON, W. S. (1973). DNA polymerase associated with human hepatitis B antigen. *Journal of Virology*, **12**, 995–1005.

LAROUZE, B., LONDON, W. T., SAIMOT, G., WERNER, B. G., LUSTBADER, E. D., PAYET, M. and BLUMBERG, B. S. (1976). Host responses to hepatitis B infection in patients with primary hepatic carcinoma and their families: A case-control study in Senegal, West Africa. *Lancet*, **2**, 534–8.

LUTWICK, L. I. and ROBINSON, W. S. (1977). DNA synthesized in the hepatitis B Dane particle DNA polymerase reaction. *Journal of Virology*, **21**, 96–104.

MCCOLLUM, R. W. and ZUCKERMAN, A. J. (1981). Viral hepatitis: Report on a WHO informal consultation. *Journal of Medical Virology*, **8**, 1–29.

MACKAY, P., LEES, J. and MURRAY, K. (1981). The conversion of hepatitis B core antigen synthesized in *E. coli* into e antigen. *Journal of Medical Virology*, **8**, 237–43.

MAGNIUS, L. O. and ESPMARK, J. A. (1972). New specificities in Australia antigen-positive sera distinct from Le Bouvier determinants. *Journal of Immunology*, **109**, 1017–21.

MARION, P. L., OSHIRO, L., REGNERY, D. C., SCULLARD, G. H. and ROBINSON, W. S. (1980a). A virus in Beechey ground squirrels which is related to hepatitis B virus of man. *Proceedings of the National Academy of Sciences, USA*, **77**, 2941–5.

MARION, P. L. and ROBINSON, W. S. (1983). Hepadna viruses: hepatitis B and related viruses. In *Current Topics in Microbiology and Immunology*, **105**, ed. M. Cooper, P. H. Hofschneider, H. Koprowski, F. Melchers, R. Rott, H. G. Schweiger, P. K. Vogt and R. Zinkernagel, pp. 99–121. Berlin, Springer-Verlag.

MARION, P. L., SALAZAR, F. H., ALEXANDER, J. J. and ROBINSON, W. S. (1980b). State of hepatitis B viral DNA in a human hepatoma cell line. *Journal of Virology*, **33**, 795–806.

MASON, W. S., SEAL, G. and SUMMERS, J. (1980). Virus of Pekin ducks with struc-tural and biological relatedness to human hepatitis B virus. *Journal of Virology*, **36**, 829–36.

MAUPAS, P. and MELNICK, J. L. (eds.) (1981). Hepatitis B virus and primary hepato-cellular carcinoma. *Progress in Medical Virology*, **27**, 1–210.

MONDELLI, M. and EDDLESTON, A. L. W. F. (1984). Mechanisms of liver cell injury in acute and chronic hepatitis B. In *Seminars in Liver Disease*, vol. 4, ed. M. A. Gerber, pp. 47–58. New York, Thieme-Stratton.

OBATA, H., HAYASHI, N., MOTOIKE, Y., HISUMITSU, T., OKUDA, H., KOBAYASHI, S. and NISHIOKA, K. (1980). A prospective study on the development of hepato-cellular carcinoma from liver cirrhosis and persistent hepatitis B virus infection. *International Journal of Cancer*, **25**, 741–7.

OKUDA, K., NAKASHIMA, T., SAKAMOTO, K., HIDAKA, H., KUBO, Y., SAKUMA, K., MOTOIKE, Y., OKUDA, H. and OBATA, H. (1982). Hepatocellular carcinoma arising in non-cirrhotic and highly cirrhotic livers. *Cancer*, **49**, 450–5.

ROBINSON, W. S. (1980). Genetic variation among hepatitis B and related viruses. *Annals of the New York Academy of Science*, **354**, 371–8.

ROBINSON, W. S., CLAYTON, D. N. and GREENMAM, R. L. (1974). DNA of a human hepatitis B virus candidate. *Journal of Virology*, **14**, 384–391.

SAKUMA, K., TAKAHARA, T., OKUDA, K., TSUDA, F. and MAYUMI, M. (1982). Prognosis of hepatitis B virus surface antigen carriers in relation to routine liver function tests: A prospective study. *Gastroenterology*, **83**, 114–17.

SHAFRITZ, D. A. (1982). Integration of HBV-DNA into liver and hepatocellular carcinoma cells during persistent HBV infection. *Journal of Cell Biochemistry*, **20**, 303–16.

SHAFRITZ, D. A. and KEW, M. C. (1981). Identification of integrated hepatitis B virus DNA sequences in human hepatocellular carcinomas. *Hepatology*, **1**, 1–8.

SHAFRITZ, D. A., SHOUVAL, D., SHERMAN, H. I., HADZIYANNIS, S. J. and KEW, M. C. (1981). Integration of hepatitis B virus DNA into the genome of liver cells in chronic liver disease and hepatocellular carcinoma: Studies in percutaneous liver biopsies and post-mortem tissue specimens. *New England Journal of Medicine*, **305**, 1067–73.

SIDDIQUI, A., MARION, P. L. and ROBINSON, W. S. (1981). Ground squirrel hepatitis virus DNA: Molecular cloning and comparison with hepatitis B virus DNA. *Journal of Virology*, **38**, 393–7.

SKELLY, J., COPELAND, J. A., HOWARD, C. R. and ZUCKERMAN, A. J. (1979). Hepatitis B surface antigen produced by a human hepatoma cell line. *Nature*, **282**, 617–18.

SNYDER, R. L. (1968). Hepatomas of captive woodchucks. *American Journal of Pathology*, **52**, 32.

SNYDER, R. L., TYLER G. and SUMMERS, J. (1982). Chronic hepatitis and hepato-cellular carcinoma associated with woodchuck hepatitis virus. *American Journal of Pathology*, **107**, 422–5.

STANNARD, L. M. and ALEXANDER, J. (1977). Electron microscopy of HBsAg from human hepatoma cell line. *Lancet*, **2**, 713.

SUMMERS, J. (1981). The recently described animal virus models for human hepatitis B virus. *Hepatology*, **1**, 179–83.

SUMMERS, J., O'CONNELL, A., MAUPAS, P., GOUDEAU, A., COURSAGET, P. and DRUCKER, J. (1978). Hepatitis B virus DNA in primary hepatocellular carcinoma. *Journal of Medical Virology*, **2**, 207–14.

SUMMERS J., SMOLEC, M. J. and SNYDER, R. L. (1978). A virus similar to human hepatitis B virus associated with hepatitis and hepatoma in woodchucks. *Proceedings of the National Academy of Sciences, USA*, **75**, 4533–7.

SZMUNESS, W. (1978). Hepatocellular carcinoma and the hepatitis B virus: Evidence for a causal association. *Progress in Medical Virology*, **24**, 40–69.

SZMUNESS, W., ALTER, H. J. and MAYNARD, J. E. (eds.) (1982). *Viral Hepatitis. 1981 International Symposium*. Philadelphia, The Franklin Institute Press.

THOMAS, H. C. and LOK, A. S. F. (1984). The immunopathology of autoimmune and hepatitis B virus-induced chronic hepatitis. In *Seminars in Liver Disease*, vol. 4, ed. M. A. Gerber, pp. 36–46. New York, Thieme-Stratton.

THUNG, S. N. and GERBER, M. A. (1984). Polyalbumin receptors: Their role in the attachment of hepatitis B virus to hepatocytes. In *Seminars in Liver Disease*, vol. 4, ed. M. A. Gerber, pp. 69–75. New York, Thieme-Stratton.

WERNER, B., SMOLLEC, J. M., SNYDER, R. L. and SUMMERS, J. (1979). Serological relationship of woodchuck hepatitis virus and human hepatitis B virus. *Journal of Virology*, **32**, 314–22.

WORLD HEALTH ORGANISATION (1977). Advances in viral hepatitis. *Report of the WHO Expert Committee on Viral Hepatitis. World Health Organisation Technical Report Series, No. 602*, Geneva.

WORLD HEALTH ORGANISATION (1983). Prevention of primary liver cancer. *Report of a WHO Meeting. World Health Organisation Technical Report Series, No. 691*, Geneva.

ZUCKERMAN, A. J. (1974). Viral hepatitis, the B antigen and liver cancer. *Cell*, **1**, 65–7.

ZUCKERMAN, A. J. (1975). *Human Viral Hepatitis*, 2nd edn. Amsterdam, North-Holland/American Elsevier.

ZUCKERMAN, A. J. (1982a). Persistence of hepatitis B virus in the population. In *Virus Persistence*, Symposium 33 of the Society for General Microbiology, ed. B. W. J. Mahy, A. C. Minson and G. K. Darby, pp. 39–56. Cambridge, Cambridge University Press.

ZUCKERMAN, A. J. (1982b). Primary hepatocellular carcinoma and hepatitis B virus. *Transactions of the Royal Society of Tropical Medicine and Hygiene*, **76**, 711–18.

ZUCKERMAN, A. J. and HOWARD, C. R. (1979). *Hepatitis Viruses of Man*. London, Academic Press.

ZUCKERMAN, A. J., SUN, T-t., LINSELL, A. and STJERNSWARD, J. (1983). Prevention of primary liver cancer. *Lancet*, **1**, 463–5.

THE MOLECULAR BIOLOGY OF BOVINE PAPILLOMAVIRUSES

PETER M. HOWLEY, YU-CHUNG YANG and
MICHAEL S. RABSON

*Laboratory of Pathology, National Cancer Institute, National Institutes
of Health, Bethesda, Maryland 20205, USA*

The papillomaviruses are small DNA viruses which induce benign squamous epithelial tumors (warts and papillomas) of cutaneous or mucosal epithelium. The first papillomavirus described was the cottontail rabbit papillomavirus (Shope, 1933). Subsequently, papillomaviruses have been isolated and characterized from many higher vertebrate species including man. The viral etiology of warts was first suggested by Ciuffo (1907), who demonstrated their transmission by cell-free filtrates. The papillomaviruses, and in particular the human papillomaviruses, have remained refractory to the standard techniques of virology because they have not yet been successfully propagated *in vitro*. This failure may in part be due to the fact that the productive functions of the papillomaviruses are expressed only in fully differentiated squamous epithelial cells which cannot be grown in culture.

Vegetative viral DNA synthesis can be detected by *in situ* hybridization techniques in the squamous epithelial cells of the stratum spinosum and of the granular layer but not in the basal layer of the epidermis nor in the underlying dermal fibroblasts. Viral capsid protein production and virion assembly occur in the upper stratum spinosum and in the granular layer. Although it has not yet been unequivocally demonstrated, it is believed that the viral genome is present in the cells of the basal layer. It is also generally thought that the expression of viral DNA in the basal layer is responsible for the epithelial cell proliferation characteristic of a wart or papilloma. The epithelial cells of the basal layer are non-permissive for the papillomavirus vegetative functions. These cells undergo a program of differentiation as they migrate upward through the stratum spinosum and into the granular layer. The control of papillomavirus late gene expression, therefore, appears to be linked to the state of differentiation of the squamous epithelial cell. The molecular basis of this control is not known. To date, attempts at using the epidermal cell culture systems that permit expression of some of the differentiated functions of squamous epithelial cells have not been successful at propagating the papillomaviruses (LaPorta and Taichman, 1982).

The papillomaviruses contain a double-stranded, covalently closed, circular DNA genome of approximately 8 kb. They were originally grouped together with the polyomaviruses to form a family of viruses called the papovaviruses (Melnick, 1962). The papillomaviruses, however, are larger than the polyomaviruses (55 nm compared with 40 nm) and contain larger genomes (8 kb compared with 5 kb). Members of the polyomavirus genus such as SV40 and the murine polyoma virus have been studied in great detail, due in large part to the fact that they can easily be propagated in the laboratory. But it was only with the advent of recombinant DNA technology which permitted the cloning of the various papillomavirus genomes that detailed studies of the molecular biology and genomic organization of the papillomaviruses have been feasible. Molecular cloning has provided sufficient quantities of viral DNA to begin a systematic study of the papillomaviruses and has permitted the standardization of viral genomic reagents.

TRANSFORMING PAPILLOMAVIRUSES

Certain papillomaviruses are capable of inducing fibroblastic tumors in hamsters and of transforming rodent tissue culture cell lines *in vitro*. The subgroup of papillomaviruses which are oncogenic in hamsters includes the bovine papillomaviruses associated with cutaneous fibropapillomas in cattle (BPV-1 and BPV-2) (Friedman *et al.*, 1963), the deer papillomavirus (Koller and Olson, 1972), the sheep (ovine) papillomavirus (Gibbs, Smale and Lauman, 1975), and the European elk papillomavirus (Stenlund *et al.*, 1983). With the exception of the ovine papillomavirus, each of these viruses has been demonstrated to be capable also of transforming susceptible rodent cells in culture.

BPV-1 is the most extensively studied of the group and has served as the prototype for unraveling the molecular biology of the papillomaviruses. In addition to its ability to induce fibroblastic tumors in hamsters it is capable of inducing such tumors in rabbits and pika (Puget, Favre and Orth, 1975; Breitburd *et al.*, 1981). Studies demonstrating the ability of BPV to transform cells in tissue culture date back to 1963, when the transformation of bovine tissue culture cells was reported (Black *et al.*, 1963). Subsequent studies have reported the transformation of mouse, hamster and rat cells in culture by BPV-1. Although the transformed cells are not productively infected by this virus, cellular transformation provides a biological system for studying the papillomavirus/host cell interaction. Quantitative assays for transformation by BPV have been developed using NIH3T3 cells and C127 cells (Dvoretzky, Shober and Lowy, 1980). The transformation focus assay developed for BPV-1 infectious virions can also be used for BPV-1 viral DNA or molecularly cloned BPV-1 DNA (Howley *et al.*, 1980). Stable transformation of rodent and bovine cells has been reported using both calcium phosphate co-precipitation of DNA and spheroplast fusion using *E. coli* containing cloned BPV-1 DNA (Lowy *et al.*, 1980; Binétruy *et al.*, 1982). The entire BPV-1 genome is not required for transformation; a specific 5.5 kb *Hind*III to *Bam*HI restriction fragment of BPV-1 DNA which comprises 69% of the viral genome (BPV_{69T}) is sufficient (Lowy *et al.*, 1980). The cloned DNAs of bovine papillomavirus types 2 and 4 (Moar *et al.*, 1981; Campo and Spandidos, 1983), the cottontail rabbit papillomavirus (CRPV; Watts *et al.*, 1983), and the human papillomavirus (HPV) type 5 (Watts *et al.*, 1983) have also been reported to transform rodent cells in an analogous manner.

The properties of mouse C127 cells or mouse NIH3T3 cells transformed by BPV-1 virus, by the full-length cloned BPV-1 DNA, or by the specific 69% subgenomic transforming segment are the same. The cells are anchorage independent, tumorigenic in nude mice, grow to high saturation density and do not synthesize virus capsid proteins, virus particles, or 'late' RNA. A novel feature of BPV-1 transformed cells that sets them apart from other virus transformation systems such as SV40, polyomavirus, adenoviruses, or retroviruses, is that BPV transformed cells contain multiple copies of non-integrated, supercoiled circular viral DNA molecules in their nuclei (Law *et al.*, 1981; Lancaster, 1981). The DNA exists as a stable multicopy plasmid in cells transformed by virions, the full-length cloned BPV-1 DNA, or BPV_{69T}. This feature of extrachromosomal plasmid maintenance seems to be characteristic of papillomavirus transformation in that it is also a feature of BPV-2 transformed cells and CRPV transformed cells (Moar *et al.*, 1981; Watts *et al.*, 1983). This transformation system therefore provides two functions of the papillomaviruses which can be identified and studied: cellular transformation and plasmid maintenance. Since BPV_{69T} is capable of transforming susceptible mouse cells and because the BPV DNA exists as a plasmid within these cells, it necessarily follows that all viral *cis*- and *trans*-acting functions required for cellular transformation and for autonomous plasmid replication and maintenance must be located within that segment of the genome.

The mechanism by which BPV-1 transforms cells is currently unknown. The evidence is convincing, however, that it encodes gene products that are required for this transformation. Indirect evidence comes from the rapid appearance of transformed foci following infection of mouse cells by BPV-1 and from experiments which have established that the continued presence of the BPV-1 genome within cells is required for the maintenance of the transformed state. The copy number of BPV-1 plasmids in transformed cells varies between 20 and 400 per diploid genome in different cell lines examined, and remains relatively constant in independent clonal transformed lines. Spontaneous revertants are rare and were not detected in early passage BPV-1 transformed C127 cells (Turek *et al.*, 1982). By treating clonal lines of BPV-1 transformed cells with mouse L-cell interferon it was possible to demonstrate a decrease in the average number of plasmid viral genomes present within the transformed cells. In cell lines carried for 60 generations in the continued presence of 200 units of interferon per millilitre, between 1 and 10% of the cells reverted to a flat, non-transformed morphology. These flat revertants had the biological characteristics of untransformed C127 cells and could be retransformed by BPV-1. DNA hybridization revealed that the revertant lines had been cured of the viral DNA; the sensitivity of the assay was sufficient to detect 0.2 copies per cell, establishing that the continued presence of the viral genome, and presumably therefore its expression, is required for papillomavirus cellular transformation (Turek *et al.*, 1982).

GENOMIC ORGANIZATION OF PAPILLOMAVIRUSES

BPV-1 has served as the prototype for the study of the molecular biology and genomic organization of papillomaviruses. Its complete 7945 bp genome has been sequenced (Chen *et al.*, 1982) and analysis of these sequence data, in conjunction with the known transcriptional data for the virus in transformed cells (Heilman *et al.*, 1982) and in productively infected bovine fibropapillomas (Amtmann and Sauer, 1982; Engel, Heilman and Howley, 1983), has provided a basis for determining the organization of the papillomavirus genome. All the major open reading frames (ORFs) greater than 380 bases in size are located on the same DNA strand. The region transcribed in BPV-1 transformed cells contains ORFs in all three translation frames. An additional region is transcribed only in productively infected bovine fibropapillomas and is characterized by two large ORFs (L1 and L2) separated by a single translation stop codon (Chen *et al.*, 1982; Engel *et al.*, 1983). A schematic representation of the genomic organization of BPV-1 derived from these transcriptional data and from the sequence analysis is shown in Fig. 1. The ORFs indicated as E1 through E8 are located in the region expressed in transformed cells, whereas the ORFs L1 and L2 are in the region expressed only in productively infected warts. From the data of Engel *et al.* (1983) it is clear that the mRNAs detected in transformed cells are a subset of those RNAs detected in productively infected cells. A series of at least five transcripts have been mapped in transformed cells (Heilman *et al.*, 1982) with a common 3' end at 0.53 map units. There is a polyadenylation site (AATAAA) located at base 4180 in the genome. Sequences characteristic of promoter elements are present in a 1 kb segment located 5' to the E ORFs. Potential TATAAA promoter elements are located at bases 7108 and 58 in this non-coding segment. The promoter element at base 58 appears to be the major promoter active in transformed cells in that the 5' ends of most of the spliced messages found in transformed cells map to base 89 (Ahola *et al.*, 1983; Yang, Okayama and Howley, 1985).

Multiple virus-specific polyadenylated RNA species have also been identified in bovine fibropapillomas infected by BPV-1 (Engel *et al.*, 1983). All of the RNA species are transcribed from the same DNA strand and, as mentioned above, each of the RNA species found in BPV-1 transformed cells appears also to be present in productively infected bovine fibropapillomas. Additional RNA species with a common 3' terminus at 0.90 map units as indicated in Fig. 1 are also found in productively infected warts. Preliminary evidence indicates that, as with the RNA species found in transformed cells, the RNA species found in productively infected bovine fibropapillomas are also generated by differential splicing. At the present time there are no data mapping the promoter(s) functioning in the biosynthesis of the wart-specific RNAs. It is not known yet whether the same or different promoters function in the biosynthesis of the set of RNAs with 3' ends at 0.53 map units and the set of RNAs with their 3' ends at 0.90 map units. Two non-abundant species of RNA of 6.7 kb and 8 kb, each with a 3' end at 0.90 map units, were found in productively infected fibropapillomas. The 5' ends of these species

72

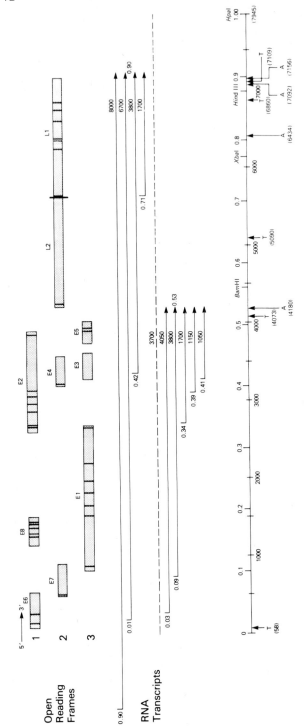

Fig. 1. Map of the genomic organization of BPV-1 DNA. The full-length linear molecule (7945 bp), opened at the unique *Hpa*I site (base 1), is marked off in bases and map units at the bottom of the figure (Chen *et al.*, 1982; Engel *et al.*, 1983; Danos *et al.*, 1983). Potential polyadenylation recognition sites (AATAAA) are indicated by the letter A and potential TATAAA and TATATA promoter elements by the letter T. The bodies of the BPV-1 specific polyadenylated RNA transcripts and their sizes as found in transformed cells (Heilman *et al.*, 1982) and productive bovine fibropapillomas (Engel *et al.*, 1983) are drawn immediately above. The stippled horizontal bars represent the ORFs and therefore the potential coding regions for virus-specific proteins in each of the three reading frames. Superimposed on these ORFs are vertical lines which indicate the positions of the in-frame methionine initiation codon ATG. ORFs within the 69% transforming region BPV_{69T} have been designated E1 to E8. The two ORFs within the 31% region not essential for transformation are designated L1 and L2, and are separated by a single stop codon. This figure is a slight modification of figures previously published (Engel *et al.*, 1983; Howley, 1983). The sequence information used to generate the ORFs is from the original sequence (Chen *et al.*, 1982), with base changes noted by Danos *et al.* (1983) and with an additional G added at base 3445 (U. Pettersson and A. Stenlund, unpublished).

are located in the non-coding region, indicating that their promoter elements may be located in the same region as the promoter elements which function in the biosynthesis of the set of RNAs with 3' ends at 0.53 map units. The differential biosynthesis of these two sets of RNAs, each characterized by polyadenylation recognition sequences at either base 4180 or base 7146, may result from the utilization of different promoter elements. Thus the biosynthesis of RNA species required for the full productive cycle, i.e. capsid protein synthesis, would correlate with the activation of a wart-specific promoter element which through the presence of specific 5' leader sequences would direct the use of the 3' processing signals at 0.90 map units. Alternatively, the biosynthesis of the wart-specific RNAs which terminate at 0.90 map units may not result from the use of a different set of promoter elements from those used in generating the species ending at 0.53 map units, but rather from alternative processing. The wart-specific RNAs with 3' ends at 0.90 map units could be derived from the same promoter as that used for the RNAs with 3' ends at 0.53 map units if mechanisms exist which result in repression of the transcriptional termination sequences and polyadenylation recognition sequences located at 0.53 map units. Mechanisms directing the choice of the 3' termini could be under viral control or alternatively be provided by the differentiated squamous epithelial cell.

The genomic organization of BPV-1 is quite similar to that of other papillomaviruses which have been sequenced. The sequences of HPV-1a (Danos, Katinka and Yaniv, 1982; Danos et al., 1983), HPV-6 (Schwarz et al., 1983) and CRPV (Danos et al., 1984) DNAs have been determined and compared with that of BPV-1. Each of the sequenced papillomaviruses contains ORFs on only one strand. There are characteristic homologies in the nucleotide and deduced amino acid sequences, particularly within the four largest ORFs (E1, E2, L1 and L2). Using these conserved sequences it is possible to align co-linearly each of the papillomavirus genomes. In each, 5' to the putative transforming (E) ORFs is a non-coding region varying in size up to 1 kb. The strongest regions of homology among the papillomaviruses lie within the L1 ORF and in the carboxy-terminus of the E1 ORF. The E6 ORF for each of these papillomaviruses contains a four-fold repetition and homologous spacing for the tetrapeptide sequence Cys-X-X-Cys. Thus it seems reasonable to extrapolate from the functional data that are emerging for the BPV genome to those of other papillomaviruses including human papillomaviruses, for which no functional data are available.

FUNCTIONAL ANALYSIS OF PAPILLOMAVIRUS GENOMES

In the absence of a tissue culture system *in vitro* for productively propagating BPV-1 virus, the only biological system available to be studied functionally is the transformation of tissue culture cells. The transforming region of the BPV-1 genome has previously been localized to a specific fragment (BPV$_{69T}$) comprising 69% of the genome, which also contains the elements sufficient for extrachromosomal plasmid maintenance (Howley *et al.*, 1980; Lowy *et al.*, 1980). To define more precisely the viral DNA sequences which are involved in cellular transformation we have tested the ability of defined deletion mutants of BPV-1 DNA to morphologically transform C127 cells. Cells containing the mutated DNA have been examined for anchorage independence and also for tumorigenicity in nude mice. Several distinct regions of the BPV-1 genome have been found to influence the expression of the viral transformation functions. A transcriptional control element which maps in the 5′ non-coding region between the *Hinc*II site at base 7143 and the *Hpa*I site at base 1 is essential and can be functionally reconstituted by a segment of SV40 DNA which has sequences for the promotion and enhancement of RNA transcription (Sarver *et al.*, 1984). Similarly, this region can be replaced by a retrovirus long terminal repeat (LTR) which also has promotional and enhancement activity (Nakabayashi, Chattopadhyay and Lowy, 1983). A transcriptional enhancer element, located 3′ to the polyadenylation site for the viral RNA expressed in transformed cells, has previously been shown to be essential for transformation (Campo *et al.*, 1983; Lusky *et al.*, 1983).

The BPV-1 genome contains at least two coding segments which affect transformation. One of these has been previously localized to a 2.3 kb *Eco*RI to *Bam*HI segment (Nakabayashi *et al.*, 1983). By defined deletion mutagenesis we have localized this region to the E2 ORF which could by itself encode a 48 kD protein. The E2 ORF is apparently transcribed intact in transformed cells in that a polyadenylated RNA species, the 5′ end of which maps near the amino-terminus of the E2 ORF, has been described (Heilman *et al.*, 1982). The amino-terminal half of the protein encoded by the E2 ORF is well conserved among different papillomaviruses including HPV-1a and HPV-6 (Danos *et al.*, 1983; Schwarz *et al.*, 1983). In addition there is structural and amino acid homology within the 86 amino acids at the carboxy-terminus of the E2 ORFs of the BPV-1, HPV-1a, HPV-6b and CRPV genomes, and limited amino acid homology between this region and the human c-*mos* gene has been described (Danos and Yaniv, 1984). The E2 gene product has not yet been described as a protein and its function in benign warts may involve the induction of the dermal fibroblastic proliferation seen in bovine fibropapillomas and possibly in the induction of the epidermal cell proliferation characteristic of all papillomavirus-induced lesions.

Expression of the E2 ORF by itself is not sufficient for the fully transformed phenotype. Recombinants expressing the E2 ORF under the control of the Harvey murine sarcoma virus LTR (Nakabayashi *et al.*, 1983) or of the SV40 early promoter segment can induce transformed foci; however, the resultant transformed mouse cells are less tumorigenic in nude mice in that the tumors are smaller and appear later. In addition, the cells are quantitatively as well as qualitatively less anchorage independent when assayed for soft agar growth. In the presence of an intact E2 ORF, deletion mutation analysis maps a function essential for the fully transformed phenotype between the *Hpa*I (base 1) and *Sma*I (base 945) sites. Deletion mutants which leave the E2 ORF intact as well as this *Hpa*I to *Sma*I region have a fully transformed phenotype as assayed in nude mice and soft agar. Deletion mutants lacking the E6 and E7 ORFs are still able to induce transformation but at a lower efficiency (6–8% compared to that of wild-type) and, as mentioned above, these transformants have an altered phenotype. Deletion mutants lacking the E2 ORF, particularly the amino-terminal half, are significantly impaired in their ability to transform (efficiency down to 1% that of wild-type); however, they are not totally defective in their ability to transform. Thus two coding regions, the E2 ORF and the E6–E7 ORF region, are required for BPV-1 transformation.

Deletion mutagenesis has also localized a region important for autonomous plasmid maintenance but which is not essential for transformation. Mutations localized within the E1 ORF that do not affect any other ORFs are not significantly altered in their transformation efficiencies; however, the transformants contain integrated copies of the BPV genome. Thus the E1 gene product is implicated in stable plasmid replication and maintenance (Lusky and Botchan, 1984; Sarver *et al.*, 1984).

cDNA CLONING OF TRANSFORMING RNAs

None of the non-virion proteins encoded by any papillomavirus has yet been described. Early investigators were unable to define any tumor antigens for any of the transforming papillomaviruses or with the Shope papillomavirus using antisera from tumor-bearing animals and standard immunological techniques including indirect immuno-florescence and complement fixation. The inability to define any papillomavirus tumor antigens or early gene products has continued to perplex researchers in this field. The genetic evidence from the BPV-1 studies clearly indicates that virus-encoded functions are required for the transformed phenotype and for stable plasmid main-tenance (Nakabayashi *et al.*, 1983; Lusky and Botchan, 1984; Sarver *et al.*, 1984). Another difficulty is the characterization of the viral RNAs in virus transformed cells, as they are of exceedingly low abundance (Heilman *et al.*, 1982). In a subsequent study, in which attempts were made to clone full-length cDNAs from BPV-1 trans-formed cells, 200 virus-specific clones were found in a total pool of 100 000 cDNA clones, indicating an abundance of 0.2% (Yang, Okayama and Howley, 1985). As mentioned above, there are mul-tiple viral RNA species that are found within transformed cells and for BPV-1 these viral RNAs all map to the transforming segment BPV$_{69T}$ and have a common 3' end mapping to the poly(A) recogni-tion site at 0.53 map units (Heilman *et al.*, 1982).

From the full-length cDNA clones from BPV-1 transformed cells, a number of classes of RNAs were identified. The cDNAs were cloned using the schema outlined in Fig. 2 (Okayama and Berg, 1983). This approach has the advantage of insuring enrichment of full-length copies of the RNA species that are present in the RNA population being cloned. Sequence analysis of these cDNA clones has provided structural information which defines the 5' ends, the splice donor and acceptor sites, and the coding exons for mRNA species present in the BPV-1 transformed cells. This is important because deletion mutagenesis only defines regions important in transformation and extrachromosomal replication, and does not identify how these ORFs or coding exons may be placed together to encode specific viral gene products. It is, however, possible to use the cDNA data in conjunction with the sequence data defining ORFs for BPV-1 to predict the structure of the virus encoding proteins which could potentially be translated in BPV-1 transformed cells.

A schematic representation of the various cDNA clones initially recognized and characterized in BPV-1 transformed cells is shown in Fig. 3. It was found that a major 5' end for the RNAs in transformed cells maps to base 89, in agreement with the data of Ahola *et al.* (1983). From the sequence of the cDNAs obtained, it was found

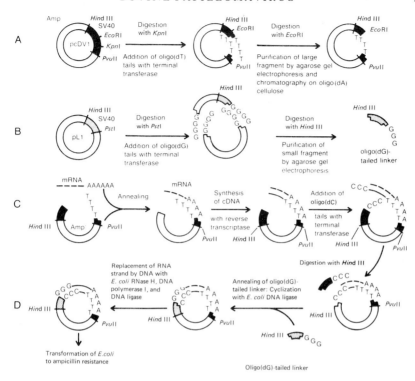

Fig. 2. Schema used in the generation of full-length cDNA clones from BPV-1 transformed cells (Yang, Okayama and Howley, 1985). Preparation of plasmid primer (A) and oligo(dG)-tailed linker fragment (B) is shown. The steps (C) in the construction of the plasmid–cDNA recombinants (D) are also indicated. Detailed maps of pcDV1 and pL1 have been described (Okayama and Berg, 1983). Stippled areas represent SV40 early promoter and a modified SV40 late region intron, black areas SV40 late region polyadenylation signal.

that RNAs existed which would direct the synthesis of an E6 protein, an E6/E7 fusion protein (spliced together from bases 304 to 527), an E6/E4 fusion protein (spliced together from bases 304 to 3224), an E7 protein and an E2 protein. Since each of these cDNAs was cloned into an expression vector being driven by the SV40 early promoter, as described in Fig. 2, it was possible to test each of them independently for their ability to transform mouse cells. Of the five classes tested and represented in Fig. 3, two were found to be able to transform mouse C127 cells, albeit at a low frequency. These were the cDNAs which expressed E2 intact and E6 intact. Cell lines transformed with each of these DNAs contained the BPV cDNA clone integrated into the host virus chromosome – as would be expected since the sequences encoding E1 protein are lacking in these clones. A more detailed analysis of these transformants and an analysis of the ability of the

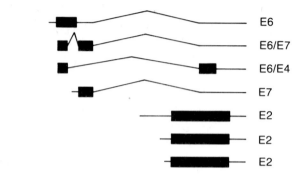

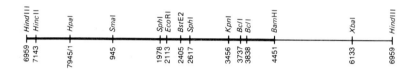

Fig. 3. Representation of structural features of classes of BPV-1 cDNA clones isolated from BPV-1 transformed mouse C127 cells. The structures of various cDNA clones are indicated with their coding regions (■■■), 5′ and 3′ non-coding regions (——), and spliced-out intervening segments (⌒). Splice donor sites have been localized to bases 304, 864 and 2505. Splice acceptor sites have been localized to bases 527 and 3225.

transforming proteins to complement each other as well as other defined transforming proteins is under way.

The expression vector utilized in generating these full-length cDNA clones contains the SV40 origin of DNA replication. This permits amplification of these plasmids in monkey cells expressing the SV40 large T-antigen (Cos-1 cells). By placing these cDNAs under the control of the SV40 late promoter, we hope to generate sufficient gene products in these cells to permit the definition and biochemical characterization of the BPV-1 E proteins involved in cellular transformation and in stable plasmid maintenance.

This work was supported in part by Grant No. 1597 from the Council for Tobacco Research, USA, Inc., to M.S.R. We are grateful to Sue Hostler for editorial assistance in preparing this manuscript.

REFERENCES

AHOLA, H., STENLUND, A., MORENO-LOPEZ, J. and PETTERSSON, U. (1983). Sequences of bovine papillomavirus type 1 DNA: functional and evolutionary implications. *Nucleic Acids Research*, **11**, 2639–50.

AMTMANN, E. and SAUER, G. (1982). Bovine papillomavirus transcription: poly-adenylated RNA species and assessment of the direction of transcription. *Journal of Virology*, **43**, 59–66.

BINÉTRUY, B. MENEGUZZI, G., BREATHNACH, R. and CUZIN F. (1982). Recombinant DNA molecules comprising bovine papilloma virus type 1 DNA linked to plasmid DNA are maintained in a plasmidial state both in rodent fibroblasts and in bacterial cells. *EMBO Journal*, **1**, 621–8.

BLACK, P. H., HARTLEY, J. W., ROWE, W. P. and HUEBNER, R. J. (1963). Transformation of bovine tissue culture cells by bovine papilloma virus. *Nature, London*, **199**, 1016–18.

BREITBURD, F., FAVRE, M., ZOOROB, R., FORTIN, D. and ORTH, G. (1981). Detection and characterization of viral genomes and search for tumoral antigens in two hamster cell lines derived from tumors induced by bovine papillomavirus type 1. *International Journal of Cancer*, **27**, 693–702.

CAMPO, M. S. and SPANDIDOS, D. A. (1983). Molecularly cloned bovine papillomavirus DNA transforms mouse fibroblasts *in vitro*. *Journal of General Virology*, **64**, 549–57.

CAMPO, M. S. SPANDIDOS, D. A., LANG, J. and WILKIE, N. M. (1983). Transcription control signals in the genome of bovine papillomavirus type 1. *Nature*, **303**, 77–80.

CHEN, E. Y., HOWLEY, P. M., LEVINSON, A. D. and SEEBURG, P. H. (1982). The primary structure and genetic organization of the bovine papillomavirus (BPV) type 1 genome. *Nature, London*, **299**, 529–34.

CIUFFO, G. (1907). Imnesto positivo con filtrato di verruca volgare. *Giornale Italiano della Malattie Venereologia*, **48**, 12–17.

DANOS, O., ENGEL, L. W., CHEN, E. Y., YANIV, M. and HOWLEY, P. M. (1983). Comparative analysis of the human type 1a and bovine type 1 papillomavirus genomes. *Journal of Virology*, **46**, 557–566.

DANOS, O., GIRI, I., THIERRY, F. and YANIV, M. (1984). Papillomavirus genomes: sequences and consequences. *Journal of Investigative Dermatology*, in press.

DANOS, O., KATINKA, M. and YANIV, M. (1982). Human papillomavirus 1a complete DNA sequence: a novel type of genome organization among Papovaviridae. *EMBO Journal*, **1**, 231–6.

DANOS, O. and YANIV, M. (1984). An homologous domain between the c-*mos* gene product and a papillomavirus polypeptide with putative role in cellular transformation. In *Cancer Cells*, part 1: *The Transformed Phenotype*, ed. A. J. Levine *et al.*, pp. 291–4. New York, Cold Spring Harbor Laboratory.

DVORETZKY, I., SHOBER, R. and LOWY, D. R. (1980). Focus assay in mouse cells for bovine papilloma virus. *Virology*, **103**, 369–75.

ENGEL, L. W., HEILMAN, C. A. and HOWLEY, P. M. (1983). Transcriptional organization of the bovine papillomavirus type 1. *Journal of Virology*, **47**, 516–28.

FRIEDMAN, J. C., LEVY, J. P., LASNERET, J., THOMAS, M., BOIRON, M. and BERNARD, J. (1963). Induction de fibromes sous-cutanés chez le hamster doré par inoculation d'extraits acellulaires de papillomes bovins. *Comptes Rendus hébdomadaires des Séances de l'Académie des Sciences, Paris*, **257**, 2328–31.

GIBBS, E. P. J., SMALE, C. J. and LAUMAN, M. J. P. (1975). Warts in sheep. *Journal of Comparative Pathology*, **85**, 327–34.

HEILMAN, C. A., ENGEL, L., LOWY, D. R. and HOWLEY, P. M. (1982). Virus specific transcription in bovine papillomavirus transformed mouse cells. *Virology*, **119**, 22–34.

HOWLEY, P. M. (1983). The molecular biology of papillomavirus transformation. *American Journal of Pathology*, **113**, 414–21.

HOWLEY, P. M., LAW, M.-F., HEILMAN, C. A., ENGEL, L. W., ALONSO, M. C., LANCASTER, W. D., ISRAEL, M. A. and LOWY, D. R. (1980). Molecular characterization of papillomavirus genomes. In *Viruses in Naturally Occurring Cancers*, ed. M. Essex, G. Todaro and H. zur Hausen, pp. 233–47. New York, Cold Spring Harbor Laboratory.

KOLLER, L. D. and OLSON, C. (1972). Attempted transmission of warts from man, cattle, and horses, and of deer fibroma, to selected hosts. *Journal of Investigative Dermatology*, **58**, 366–8.

LANCASTER, W. D. (1981). Apparent lack of integration of bovine papillomavirus DNA in virus-induced equine and bovine tumor cells and virus-transformed mouse cells. *Virology*, **108**, 251–5.

LAPORTA, R. F. and TAICHMAN, L. B. (1982). Human papilloma viral DNA replicates as a stable episome in cultured epidermal keratinocytes. *Proceedings of the National Academy of Sciences, USA*, **79**, 3393–7.

LAW, M.-F., LOWY, D. R., DVORETZKY, I. and HOWLEY, P. M. (1981). Mouse cells transformed by bovine papillomavirus contain only extrachromosomal viral DNA sequences. *Proceedings of the National Academy of Sciences, USA*, **78**, 2727–31.

LOWY, D. R., DVORETZKY, I., SHOBER, R., LAW, M.-F., ENGEL, L. and HOWLEY, P. M. (1980). *In vitro* tumorigenic transformation by a defined subgenomic fragment of bovine papilloma virus DNA. *Nature, London*, **287**, 72–4.

LUSKY, M., BERG, L., WEIHER, H. and BOTCHAN, M. (1983). The bovine papilloma virus contains an activator of gene expression at the distal end of the transcriptional unit. *Molecular and Cellular Biology*, **3**, 1108–22.

LUSKY, M. and BOTCHAN, M. R. (1984). Characterization of the bovine papilloma virus plasmid maintenance sequences. *Cell*, **36**, 391–401.

MELNICK, J. L. (1962). Papova virus group. *Science*, **135**, 1128–30.

MOAR, M. H., CAMPO, M. S., LAIRD, H. and JARRETT, W. F. H. (1981). Persistence of non-integrated viral DNA in bovine cells transformed *in vitro* by bovine papillomavirus type 2. *Nature, London*, **293**, 749–51.

NAKABAYASHI, Y., CHATTOPADHYAY, S. K. and LOWY, D. R. (1983). The transforming function of bovine papillomavirus DNA. *Proceedings of the National Academy of Sciences, USA*, **80**, 5832–6.

OKAYAMA, H. and BERG, P. (1983). A cDNA cloning vector that permits expression of cDNA inserts in mammalian cells. *Molecular and Cellular Biology*, **3**, 280–9.

PUGET, A., FAVRE, M. and ORTH, G. (1975). Induction de tumeurs fibroblastiques cutanées ou sous-cutanées chez l'Ochotone afghan (*Ochotono rufescens*) par l'inoculation du virus du papilloma bovin. *Comptes Rendus hébdomadaires des Séances de l'Académie des Sciences, Paris*, **280**, 2813–16.

SARVER, N., RABSON, M. S., YANG, Y.-C., BYRNE, J. C. and HOWLEY, P. M. (1984). Localization and analysis of bovine papillomavirus type 1 transforming functions. *Journal of Virology*, **52**, 377–88.

SCHWARZ, E., DÜRST, M., DEMANKOWSKI, C., LATTERMAN, O., ZECH, R., WOLFSPER-GER, E., SUHAI, S. and ZUR HAUSEN, H. (1983). DNA sequence and genome organization of genital human papillomavirus type 6B. *EMBO Journal*, **2**, 2341–8.

SHOPE, R. E. (1933). Infectious papillomatosis of rabbits. *Journal of Experimental Medicine*, **58**, 607–24.

STENLUND, A., MORENO-LOPEZ, J., AHOLA, H. and PETTERSSON, U. (1983). European elk papillomavirus: characterization of the genome, induction of tumors in animals, and transformation *in vitro*. *Journal of Virology*, **48**, 370–6.

TUREK, L. P., BYRNE, J. C., LOWY, D. R., DVORETZKY, I., FRIEDMAN, R. M. and HOWLEY, P. M. (1982). Interferon induces morphologic reversion with elimination of extrachromosomal viral genomes in bovine papillomavirus transformed mouse cells. *Proceedings of the National Academy of Sciences, USA*, **79**, 7914–18.

WATTS, S. L., OSTROW, R. S., PHELPS, W. C., PRINCE, J. T. and FARAS, A. J. (1983). Free cottontail rabbit papillomavirus DNA persists in warts and carcinomas of infected rabbits and in cells in culture transformed with virus or viral DNA. *Virology*, **125**, 127–38.

YANG, Y. C., OKAYAMA, H. and Howley, P. M. (1985). Bovine papillomavirus contains multiple transforming genes. *Proceedings of the National Academy of Sciences*, *USA*, **82**, in press.

GENITAL PAPILLOMAVIRUS INFECTIONS

HARALD ZUR HAUSEN

*Deutsches Krebsforschungszentrum, Im Neuenheimer Feld 280, D-6900
Heidelberg 1, Federal Republic of Germany*

The first observations pointing to an infectious component in the etiology of human genital cancer date back to 1843 when Rigoni-Stern in Italy noted significant differences in celibate women when compared to prostitutes. A number of epidemiological studies conducted during the past decades confirmed this observation, showing that promiscuity represents the prime risk factor for the development of cervical cancer (reviewed in zur Hausen, 1983). Other studies demonstrated the existence of marital clusters and a geographical correlation between the incidence rates of cervical and penile cancer (although the latter is much less frequent than cervical malignancies); another important parameter is an early onset of sexual activity.

The role of viruses in human genital cancer has been under investigation since 1968, when Rawls and his associates described higher antibody titers against *Herpes simplex* (HSV) type 2 infections in patients with cervical cancer in comparison to appropriately matched controls (Rawls *et al.*, 1968). In addition, the percentage of seroreactive women was consistently higher in the patient group than the control group.

These findings were subsequently confirmed in many laboratories, although remarkable variations in the percentage of seroreactive patients were noted in some of these studies, depending on the socioeconomic background of the patient population analyzed (reviewed in zur Hausen, Gissmann and Schlehofer, 1984). Because in some areas (e.g. the city of Toronto) only 35% of the patients revealed evidence of past HSV exposure, the interpretation that HSV would represent the major risk factor in the development of genital cancer seemed less likely.

Since 1970, our own studies have concentrated on attempts to demonstrate HSV nucleic acid by nucleic acid hybridizations, within tumor biopsy tissues, as a footprint for viral genome persistence (zur Hausen *et al.*, 1974a,b). In spite of the analysis of more than 100 individual biopsies, we were entirely unsuccessful, although a substantial number of tests performed in recent years were rather sensitive, allowing the detection of gene-size fragments of the viral genome. Since all other tumor viruses known today persist, with their genetic material, within the tumorous tissue, these findings were difficult to reconcile with a more 'conventional' role of HSV in the induction of human genital cancer.

Subsequent studies, performed in our laboratory mainly by J. Schlehofer and B. Matz (Schlehofer and zur Hausen, 1982; Matz, Schlehofer and zur Hausen, 1984) showed that infections by *Herpes simplex* viruses may give rise to effects also exhibited by chemical or physical carcinogens. HSV infection may lead to an accumulation of mutations within specific sites of the host cell genome. Similarly, gene amplification effects, as observed in some cell lines treated with chemical or physical carcinogens, were also noted after HSV infection. Thus, this virus may interact with at least some host cells, in a manner similar to that of other carcinogens, without maintenance of its DNA within the modified host cell genome.

It is interesting to note that recent epidemiological studies provide strong evidence that heavy and prolonged smoking represents a significant risk factor for cervical cancer development (Winkelstein, 1977; Vonka *et al.*, 1984). These data suggest that events related to mutagenesis and to initiation do play a role in the genesis of human genital cancer, although cervical cancer is the only genital type thoroughly analyzed thus far.

The possible involvement of papillomavirus infections (HPV) in the etiology of human genital cancer has been under investigation in my laboratory since 1972. The research was stimulated by a number of anecdotal reports on malignant conversion of genital warts (*Condylomata acuminata*) and by our failure to detect HSV sequences in cervical cancer cells. Initial hybridization experiments (zur Hausen *et al.*, 1974a) performed with nucleic acid derived from human plantar wart virus and cervical cancer biopsies and *Condylomata acuminata* provided entirely negative results. In 1976 we established the plurality of papillomavirus types (Gissmann and zur Hausen, 1976; Gissmann, Pfister and zur Hausen, 1977; Orth, Favre and Croissant, 1977; Orth *et al.*, 1977), and up to 30 individual types of human papillomaviruses are now recognized. Since all these identifications are based on nucleic acid analysis, the various types should be denoted as genotypes. It seems likely that the number of genotypes will increase, pointing to a remarkable heterogeneity of the papillomavirus group.

Individual HPV types produce lesions in skin or mucosa which show type-specific macroscopic and histological features. Distinct viruses cause common warts, flat warts, genital warts or leukoplakias of the mucosa. Among each individual group of lesions, specific histological patterns may be assigned to specific types of HPV infection.

Thus far, none of the papillomaviruses has been successfully propagated in tissue culture, and this group of agents has only been thoroughly investigated within the last few years when genetic engineering techniques offered a new approach. Papillomavirus particle production has never been observed in rapidly growing tissue, e.g. in basal cells of the epidermis or mucosa. Infection of these cells leads to the uptake of viral DNA which then persists as a circular episome without being integrated into the host cell chromosomal DNA. The persisting viral genome stimulates host cell DNA replication and induces changes characteristic of the particular type of virus infection. The multiplying cell, on the other hand, prevents independent virus replication. In carcinomas resulting from specific types of papillomavirus infection, we would therefore not expect to be able to demonstrate HPV particles by electron microscopy. However, the cells should contain viral DNA, which can be revealed by biochemical procedures.

In the differentiating and keratinizing cell, the situation is different. Independent viral DNA replication and particle maturation take place and, depending on the type of infection, large quantities of viral particles may be formed in the stratum spinosum and can be visualized in the stratum corneum. This biological behaviour explains the data obtained with peroxidase-labelled antisera against papillomavirus group-specific antigens which reveal viral antigens only in the superficial keratinized parts of the papillomas (Jenson et al., 1980). This is frequently, but mistakenly, interpreted as indicating a superficial virus infection, rather than a consequence of non-productive infection of underlying basal layer cells, which contain viral DNA but no structural proteins. Underlining this statement, HPV particle production in non-dividing differentiating tissue is always associated with infection of the underlying virus-particle-free basal cell layer.

Specific types of papillomaviruses in animals and man have been shown to be involved in the development of specific malignant tumors. This was initially demonstrated for the Shope papillomavirus in cotton-tail rabbits and subsequently demonstrated in a number of additional animal systems (Orth et al., 1977). It is an interesting feature of almost all of these systems that a specific papillomavirus infection by itself is insufficient for malignant conversion. An additional interaction of the infected cell with chemical or physical carcinogens seems to be required before the initial papilloma turns into a carcinoma.

In man, at least two conditions have been noted which reveal a similar pattern. A rare skin disease, epidermodysplasia verruciformis, predisposes the patient for HPV infections. Although a large number of distinct HPV types may be revealed within one individual patient, only infections by specific types, preferentially HPV-5 and HPV-8, lead to the development of squamous cell carcinomas (Orth *et al.*, 1978). Interestingly, carcinomas appear almost exclusively in areas of skin exposed to the sun.

Laryngeal papillomas of childhood, now known to be primarily caused by genital papillomavirus types 11 and 6, have been exposed to X-irradiation in past decades for therapeutic reasons. Many of the unfortunate children developed squamous cell carcinomas with latency periods of 5 to 40 years following X-irradiation (reviewed in zur Hausen, 1977). Spontaneous malignant conversion of these tumors is an extremely rare event, although it has been noted in a small number of such patients.

This brings us back to genital papillomavirus infections. For years we encountered great difficulties in isolating and characterizing viral DNA from genital warts, as a result of the low copy number of viral DNA in the genital papillomas. In 1979 and 1980 my co-workers Lutz Gissmann and Ethel-Michele de Villiers succeeded in isolating and charactertizing HPV-6, the most common virus type in *Condylomata acuminata* (Gissmann and zur Hausen, 1980; de Villiers, Gissmann and zur Hausen, 1981). Shortly thereafter we were able to isolate HPV-11 from a laryngeal papilloma and further demonstrated the presence of this virus in a number of genital warts (Gissmann *et al.*, 1982). HPV-6 and HPV-11 are closely related agents. The former is found as several subtypes in about 60% of genital warts, the latter being identified in close to 30% of this type of tumor. In addition, some genital warts appear to be caused by other types of HPV infection. Occasionally HPV-1 or HPV-2 viruses are identified in genital warts, and it is likely that some additional types may contribute to the total spectrum of HPV associated with such lesions.

Although the majority of cervical cancers were negative when tested for DNA of HPV types 6 and 11, a small number of biopsies reacted in hybridization experiments under conditions of low stringency, permitting the detection of related but not identical sequences. By applying molecular cloning techniques, two new types of HPV DNA were isolated directly from carcinoma tissue. They are now designated as HPV-16 (Durst *et al.*, 1983) and HPV-18 (Boshart *et al.*, 1984). They are very distantly related to HPV-6 and HPV-11 and differ substantially from each other.

On the basis of cytological observations of koilocytotic cells, Meisels and Fortin (1976) reported that mild dysplastic lesions of the cervix may represent condylomatous changes. Subsequently electron-microscopic as well as serological studies supported this observation, indicating that about 50% of cervical dysplasias, mainly those showing koilocytotic changes, are associated with HPV infections (Meisels *et al.*, 1981).

In collaboration with many clinical colleagues, we have recently analyzed a large number of dysplastic lesions for the presence of HPV types. The data are summarized as follows: close to 40% of cervical dysplasias reveal evidence for HPV-11 or HPV-6 infections or are infected by closely related agents. A few additional dysplasias contain HPV-10 or HPV-3 related viruses, which are otherwise found in flat warts of the skin. In general, the virus-positive proliferations containing koilocytotic cells are regarded as mild or moderate dysplasias. In about 20%, infections with HPV-16 or HPV-18 were noted. The vast majority of these lesions fail to reveal evidence for a high degree of koilocytosis. They commonly reveal atypia corresponding to cervical intraepithelial neoplasia II or III and, as revealed by a cooperative study with Richart and Crum from New York, they usually show an aneuploid DNA pattern (Crum, Ikenberg and Richart, 1984). Thus, HPV-16 and HPV-18 infections appear to be associated with a clearly premalignant histological stage. This conforms with data obtained with carcinomata-in-situ, penile or vulval Bowen's disease and Bowenoid papulosis. The latter proliferations obtained from more than 50 individual patients contained in 80% of the cases HPV-16 related sequences (Ikenberg *et al.*, 1984). This suggests that the primary manifestations of HPV-16 and HPV-18 infections are advanced atypias of the cervix and Bowenoid lesions at external genital sites.

The analysis of a large number of cervical carcinoma biopsies and a small number of penile and vulval cancer biopsies led to the following conclusions: close to 50% of all cervical cancer biopsies contain HPV-16 DNA. Close to 20% of tumors are positive for HPV-18, occasional tumors contain HPV-11 or HPV-10 and an additional 15% contain sequences crossreacting with HPV-16 or HPV-18 DNA only under conditions of low stringency, suggesting the presence of related but different HPVs in these biopsies. Taken together, up to 90% of all cervical cancer biopsies appear to be positive for HPV DNA. In penile and vulval cancer the situation seems to be similar although the smaller number of tumors tested thus far does not permit firm conclusions. The viral DNA is evenly distributed within the tumor cells as revealed by hybridization *in situ* (E. Grussendorf, personal communication). Three cell lines obtained from cervical cancer, among them the well-known HeLa line, contain HPV-18 DNA (Boshart *et al.*, 1984). It is interesting to note that, in two of these lines, HPV-18 DNA contains a similar deletion. HPV-16 and HPV-18 DNA is actively transcribed in all tumor biopsies and cell lines investigated so far. This points to a specific role for these DNAs within the tumor cells (Schwarz *et al.*, 1984).

All the available data point to the existence of two risk groups of genital papillomavirus infections, those representing a low risk, including HPV-6, HPV-11 and HPV-10 and those associated with a high risk, particularly HPV-16 and HPV-18 and possibly some others. As noted previously, interaction with mutagenic 'initiating' events may represent an important co-factor in genital carcinogenesis and smoking and *Herpes simplex* virus infections may also play a role in this respect (zur Hausen, 1982).

The mode of interaction is presently poorly understood. Some experimental data hint, however, to a facilitation of integration of episomal papillomavirus DNA by chemical and physical carcinogens. The type of HPV DNA as well as the site of integration into the host cell genome may then be of pivotal importance. This is, however, at present a working hypothesis which requires experimental documentation.

REFERENCES

BOSHART, M., GISSMANN, L., IKENBERG, H., KLEINHEINZ, A., SCHEURLEN, W. and ZUR HAUSEN, H. (1984). A new type of papillomavirus DNA, its presence in genital cancer biopsies and in cell lines derived from cervical cancer. *EMBO Journal*, 3, 1151–7.

CRUM, L. P., IKENBERG, H. and RICHART, R. M. (1984). Human papillomavirus type 16 and early cervical neoplasia. *New England Journal of Medicine*, 310, 880–3.

DE VILLIERS, E.-M., GISSMAN, L. and ZUR HAUSEN, H. (1981). Molecular cloning of viral DNA from human genital warts. *Journal of Virology*, 40, 932–5.

DÜRST, M., GISSMANN, L., IKENBERG, H and ZUR HAUSEN, H. (1983). A new type of papillomavirus DNA from a cervical carcinoma and its prevalence in genital cancer biopsies from different geographic regions. *Proceedings of the National Academy of Sciences, USA*, **80**, 3812–15.

GISSMAN, L., BICKEL, V., SCHULZ-COULON, A. and ZUR HAUSEN, H. (1982). Molecular cloning and characterization of human papilloma-virus DNA from a laryngeal papilloma. *Journal of Virology*, **44**, 393–400.

GISSMANN, L., PFISTER, H. and ZUR HAUSEN, H. (1977). Human papillomaviruses (HPV): Characterization of four different isolates. *Virology*, **76**, 569–80.

GISSMANN, L. and ZUR HAUSEN, H. (1976). Human papilloma viruses: Physical mapping and genetic heterogeneity. *Proceedings of the National Academy of Sciences, USA*, **73**, 1310–13.

GISSMANN, L. and ZUR HAUSEN, H. (1980). Partial characterization of viral DNA from human genital warts (*Condylomata acuminata*). *International Journal of Cancer*, **25**, 605–9.

IKENBERG, H., GISSMANN, L., GROSS, G., GRUSSENDORF-CONEN, E.-I. and ZUR HAUSEN, H. (1984). Human papillomavirus type 16-related DNA in genital Bowen's disease and in Bowenoid papulosis. *International Journal of Cancer*, **32**, 563–5.

JENSON, A. B., ROSENTHAL, J. R., OLSON, C., PASS, F., LANCASTER, W. D. and SHAH, K. V. (1980). Immunological relatedness of papillomaviruses from different species. *Journal of the National Cancer Institute*, **64**, 495–500.

MATZ, B., SCHLEHOFER, J. R. and ZUR HAUSEN, H. (1984). Identification of a gene function of *Herpes simplex* virus type 1 essential for amplification of simian virus 40 sequences in transformed hamster cells. *Virology*, **134**, 328–37.

MEISELS, A. and FORTIN, R. (1976). Condylomatous lesions of the cervix and vagina. I. Cytological patterns. *Acta Cytologica*, **20**, 505–9.

MEISELS, A., ROY, M., FORTIER, M., MORIN, C., CASAS-CORDERO, C., SHAH, K. V. and TURGEON, H. (1981). Human papillomavirus infection of the cervix: the atypical papilloma. *Acta cytologica*, **25**, 7–16.

ORTH, G., BREITBURD, F., FAVRE, M. and CROISSANT, O. (1977). Papillomaviruses: Possible role in human cancer. In *Origins of Human Cancer*, ed. H. H. Hiatt, J. D. Watson and J. A. Winsten, pp. 1043–68. Cold Spring Harbour Laboratories, New York.

ORTH, G., FAVRE, M. and CROISSANT, O. (1977). Characterization of a new type of human papillomavirus that causes skin warts. *Journal of Virology*, **24**, 108–20.

ORTH, G., JABLONSKA, S., FAVRE, M., CROISSANT, O., OBALLEK, S., JARZABEK-CHORZELSKA, M. and RZESA, G. (1978). Characterization of two types of human papillomaviruses in lesions of epidermodysplasia verruciformis. *Proceedings of the National Academy of Sciences, USA*, **75**, 1537–41.

RAWLS, W. E., TOMPKINS, W. A. F., FIGUEROA, M. E. and MELNICK, J. L. (1968). *Herpes simplex* virus type 2: Association with carcinoma of the cervix. *Science*, **161**, 1255–6.

RIGONI-STERN, D. (1842). Fatti statistici relativi alle mallattie cancrosi che servirono de base alle poche cose dette dal dott. *Giornale service progr. pathol. terap. ser.* 2, **2**, 507–17.

SCHLEHOFER, J. R. and ZUR HAUSEN, H. (1982). Induction of mutations within the host genome by partially inactivated *Herpes simplex* virus type 1. *Virology*, **122**, 471–5.

SCHWARZ, E., FREESE, U. K., GISSMANN, L., MAYER, W., ROGGENBUCK, B. and ZUR HAUSEN, H. (1984). Structure and transcription of human papillomavirus 18 and 16 sequences in cervical carcinoma cells. *Nature*, in press.

VONKA, V., KANKA, J., JELLINEK, J., SUBRT, J., SUCHANEK, A., HAVRANKOVA, A., VACHAL, M., HIRSCH, I., DOMORAZKOVA, E., ZAVADOVA, A., RICHTEROVA, V., NAPRSTKOVA, J., DRORAKOVA, V. and SVOBODA, B. (1984). Prospective study on the relationship between cervical neoplasia and *Herpes simplex* type 2 virus. I. Epidemiological characteristics. *International Journal of Cancer*, **33**, 49–60.

WINKELSTEIN, W., JR (1977). Smoking and cancer of the uterine cervix. *American Journal of Epidemiology*, **106**, 257–9.

ZUR HAUSEN, H. (1977). Human papillomaviruses and their possible role in squamous cell carcinomas. *Current Topics in Microbiology and Immunology*, **78**, 1–30.

ZUR HAUSEN, H. (1982). Human genital cancer: Synergism between two virus infections or synergism between a virus infection and initiating events. *Lancet*, **ii**, 1370–2.

ZUR HAUSEN, H. (1983). *Herpes simplex* virus in human genital cancer. *International Reviews of Experimental Pathology*, **25**, 307–26.

ZUR HAUSEN, H., GISSMANN, L. and SCHLEHOFER, J. R. (1984). Viruses in the etiology of human genital cancer. *Progress in Medicine and Virology*, **30**, 170–86.

ZUR HAUSEN, H., MEINHOF, W., SCHEIBER, W. and BORNKAMM, G. W. (1974a). Attempts to detect virus-specific sequences in human tumors. I. Nucleic acid hybridizations with complementary RNA of human wart virus. *International Journal of Cancer*, **13**, 650–6.

ZUR HAUSEN, H., SCHULTE-HOLTHAUSEN, H., WOLF, A., DÖRRIES, K. and EGGER, H. (1974b). Attempts to detect virus-specific DNA in human tumors. II. Nucleic acid hybridizations with complementary RNA of human herpes group viruses. *International Journal of Cancer*, **13**, 657–64.

IMMORTALISING GENE(S) ENCODED BY EPSTEIN–BARR VIRUS

BEVERLY E. GRIFFIN†, LORAINE KARRAN†,
DAVID KING† and SIDNEY E. CHANG*

†*Imperial Cancer Research Fund, PO Box 123, Lincoln's Inn Fields,
London WC2A 3PX, UK and *Marie Curie Memorial Foundation
Research Institute,
The Chart, Oxted, Surrey RH8 0TL, UK*

The causal role of Epstein–Barr virus (EBV) in infectious mononucleosis and its association with two human malignancies, Burkitt's lymphoma and nasopharyngeal carcinoma, are well-documented (Henle, Henle and Diehl, 1968; Klein *et al.*, 1974; de Thé *et al.*, 1978; Epstein and Achong, 1979; Klein, 1983). Less well-defined and of more recent interest is the recognition of diffuse, polyclonal B-cell lymphomas also associated with EBV. This type of disease, originally noted in studies with the cotton-top marmoset, has now been observed in a number of cases in man, most notably in patients with genetic disorders or receiving immunosuppressive treatment (Miller, 1984). To date, expression of a viral gene product has not been definitively linked to any of the diseases, although an EBV nuclear antigen(s), EBNA, is ubiquitous in EBV-infected cells and is tacitly assumed to be intimately involved in cellular transformation (Reedman and Klein, 1973; Klein and Klein, 1984). It is not clear in every case that viral gene expression need be a prerequisite for the transformation event, however, in that the chromosomal translocations detected in Burkitt's lymphoma cell lines at least allow the possibility of some sort of 'hit and run' mechanism mediated by the virus (Galloway and McDougall, 1983; Klein, 1983). Moreover, for a virus as large as EBV, more than one pathway to cellular immortalisation and transformation could exist. For example, alternative modes of transformation may be involved in the case of B-lymphocytes and epithelial cells.

This paper describes an attempt to resolve some of the questions about EBV transformation and in particular to answer the question of whether cellular alterations can be correlated with any particular region of the EBV genome. Studies focussing on the interaction of EBV with epithelial cells suggest that they can, and reinforce the notion that B-lymphocytes and epithelial cells may be immortalised by different routes.

B-LYMPHOCYTE IMMORTALISATION

Before proceeding to our experimental results on epithelial cells, it seems pertinent to review briefly the literature on EBV-mediated B-cell proliferation. This is a highly efficient process. B-lymphocytes contain receptors for the virus, and infection of cells derived from human peripheral or umbilical cord blood, or from several species of non-human primates, leads to immortalisation (Falk *et al.*, 1974; Frank, Andiman and Miller, 1976). Recently, Katamine *et al.* (1984) have shown that cells can be immortalised by EBV at different stages of the B-cell differentiation pathway, even before the immuno-globulin gene rearrangements occur. Cells immortalised *in vitro* contain multiple copies of an episomal form of the complete viral genome, the precise copy number being characteristic of an individual cell line. The processes that control viral genome amplification are not known. Nor has the question of integration into the host chromosome been experimentally resolved, although it is assumed that in some cell lines, at least, integration must occur (Anderson and Lindahl, 1976). EBNA can be detected as early as 6–8 h post-infection, reaching a peak at 36 h if high multiplicities of infection are used (Robinson and Smith, 1981).

In attempting to understand the role of EBV in B-cell immortalis-ation much attention has been focussed on the non-transforming P3HR1 strain of the virus. This variant induces EBNA but is not a mitogen for normal B-lymphocytes and does not immortalise them (Miller *et al.*, 1974; Menezes, Leibold and Klein, 1975). DNA isolated from P3HR1 has been found to have a large deletion (in the *Bam*HI Y and H fragments, see Fig. 1) relative to transforming strains of EBV (Bornkamm *et al.*, 1980; Heller, Dambaugh and Kieff, 1981). Naturally, sequences removed by this deletion have been assumed to play a key role in the process(es) leading to immor-talisation and there are data which support this (Stoerker and Glaser, 1983). Unfortunately, studies on P3HR1 have not turned out to be straightforward, since the virus produced by this cell line is extremely heterogeneous and numerous alterations, in addition to that cited above, can be detected by restriction enzyme analysis alone (Rabson *et al.*, 1982). Use of cloned cell lines (Heston *et al.*, 1982) should in

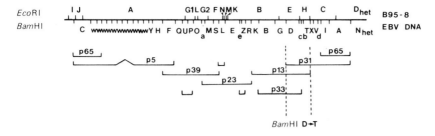

Fig. 1. A cosmid library of EBV DNA. EB virion DNA from B95-8 cells was partially digested with *Bam*HI, fragments about 40 kb in size ligated to *Bam*HI cleaved cosmid pHC79 DNA (from which the 5′-phosphates had been removed), and the resulting hybrid DNAs cloned according to the procedure of Hohn and Collins (1980). Individual recombinant DNAs were isolated and analysed as previously described (Griffin and Karran, 1984). Those that yielded results consistent with the known physical maps of EBV DNA (as shown) were selected to make up a library of largely overlapping fragments which consisted of six clones designated p65, p5, p39, p23, p13 and p31, as indicated. (p33 was selected originally as part of the library because 'miniprep' analysis showed it to contain the *Bam*HI R fragment, necessary to provide an overlap of p23 and p13. Further characterisation failed to confirm this. The heterogeneous terminal fragments, N_{het}, were presumably not represented within this library.)

time produce meaningful results with this strain of virus. In support of some role in immortalisation for the region of the genome deleted in P3HR1, the well-characterised Burkitt's lymphoma line, Daudi, also has a deletion that overlaps that of P3HR1; Daudi cells can be induced to produce virus which also appears to be non-transforming (Jones *et al.*, 1984; D. H. Crawford, personal communication). As Daudi is a primary tumour isolate, it is therefore important to study it further in this context.

In spite of the very considerable amount of data that has accumulated from studies on the immortalisation of B-lymphocytes with EBV, only a small part of which has been described above, there is still very little of significance that can be said regarding the role of the viral genome, except that (*a*) it seems to be highly relevant to the process, (*b*) it is retained, and (*c*) viral gene products are expressed to various degrees in immortalised cells. In a recent review, Miller (1984) has focussed on five interesting and potentially important regions of the genome including that covered by the P3HR1 deletion and that coding for EBNA. The others encompass the large internal repeat (*Bam*HI W), the termini, and a region (*Eco*RI J) that encodes two small, abundant RNA molecules. Whether any, or all, of these regions are relevant to the induction of continuous cell growth remains to be seen.

It is important to stress the difference between immortalisation *in vitro* and tumourigenesis *in vivo*, since the first process need not, and frequently does not, lead to the other. This point has been recently emphasised elsewhere (Klein, 1981, 1983). Thus, EBV-immortalised lymphoblastoid cell lines are normally diploid, are not clonable in agarose and *do not* produce tumours in nude mice when injected subcutaneously (Nilsson *et al.*, 1977). On the other hand, cells derived from Burkitt's lymphomas are chromosomally abnormal, grow in agarose and *do* produce tumours *in vivo* in nude mice. The principle role of EBV appears to be to convert the normal mortal cells to continuously proliferating (immortalised) cells; it has been argued by Klein and others that the tumourigenic phenotype of the Burkitt's lymphoma cell is a consequence of the specific chromosomal translocations in these cells (Klein, 1979) and may be related to the altered location and expression of cellular oncogenes, especially c-*myc*, normally found on chromosome 8 (for instance, see Dalla-Favera *et al.*, 1982; Taub *et al.*, 1982). If it is independent of viral gene expression, the nature of the co-factor that results in chromosomal rearrangement has not been elucidated. It may be a consequence of a more or less specific recombinational event (a genetic accident) that accompanies the continuously dividing immortalised B-cell, or possibly a specific external stimulus. It is perhaps noteworthy that *in vitro* such chromosomal translocations rarely occur spontaneously, so that the concept of an active co-factor in the B-cell malignancy must continue to receive serious consideration.

IMMORTALISATION OF EPITHELIAL CELLS

The undifferentiated, or poorly differentiated, carcinomas of the nasopharynx (so-called nasopharyngeal carcinoma, NPC) associated with EBV are a major health problem in certain areas of Southern China and to a lesser extent in other parts of the world (de Thé *et al.*, 1975; Zeng *et al.*, 1980). One of the uncertainties regarding this disease is the mode of entry by the virus, since no epithelial cells with receptors for EBV have yet been found. Whilst it cannot at this stage be concluded that such cells do not exist, an alternative mode of entry might involve a fusion event between an EBV-bearing B-cell and an epithelial cell of the nasopharynx (Sixbey *et al.*, 1983; Li *et al.*, 1984). The studies we have carried out on epithelial cells (Griffin and Karran, 1984, see below) have been designed to investigate the existence of a possible immortalising function(s) within the viral genome and to set up a model system *in vitro* for NPC.

To this end, virion EBV DNA isolated from a lymphoblastoid marmoset line, B95-8, was subjected to partial digestion with the endonuclease *Bam*HI; fragments about 40 kb in size were selected and cloned in the cosmid vector pHC79. In all, 780 clones (out of a total of about 40 000 EBV-positive clones produced) were analysed, from which the EBV library shown in Fig. 1 was derived. Individual recombinant EBV DNAs from this library were used to transfect a mixed population of epithelial and fibroblast cells derived from African green monkey (*Cercopithecus aethiops*) kidneys (AGMK) using the calcium phosphate co-precipitation method, essentially as described by van der Eb and Graham (1980), in the presence of carrier calf thymus DNA. Untreated (control) AGMK cells could be passaged for five or six generations in culture before reaching senescence. After several months in culture, following treatment with specific fragments of EBV DNA, small 'foci' were observed in a number of experiments, notably those in which cells had been transfected with DNA from the clone p65, with a mixture of all the cloned DNAs, or with control calf thymus DNA; large distinct 'foci' were observed in the AGMK cells transfected with DNA from clones p13 plus p33, and p31 (Fig. 2). Although none of the other foci could be propagated in culture, a number of cell lines were established from foci selected from the p13/p33 and p31 dishes. The following questions were asked with regard to cells that were propagated from foci: (1) Were they immortalised? That is, would they grow continuously in culture? (2) What type of cells were they? (3) Would they grow in the presence of low serum? (4) Would they grow in soft agar or produce tumours in susceptible animals, such as nude mice? (5) Did they contain integrated EBV DNA? (6) Did they express EBNA? On the basis of what is already known about the interaction of EBV with its natural hosts, the answers obtained from studies on these cells were perhaps not wholly surprising. That is, lines established from AGMK cells transfected with either of the overlapping fragments, p13/p33 and p31, could be subcultured weekly at a 1:10 dilution and grown continuously in culture for over $1\frac{1}{2}$ years. (At this stage, it was assumed that the cells were immortalised and they were not further propagated.) Morphologically, they appeared to be more like epithelial than fibroblast cells (Fig. 3*a*), and this was confirmed by immunofluorescent staining using a monoclonal antibody, LE61 (Lane, 1982), specific for epithelial cells (Fig. 3*b*). Further studies suggested that although the cells had properties, such as the ability to grow in low serum, characteristic of transformed or potentially

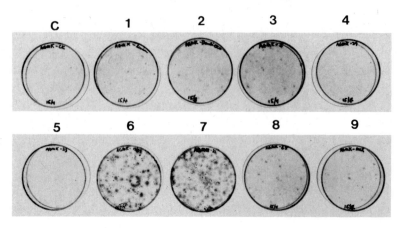

Fig. 2. Transfection studies. Dense focus assays of AGMK cells transfected with different fragments of EBV DNA. Sub-confluent dishes (50 mm) of early passage AGMK cells were transfected with EBV DNA in the presence of calf thymus carrier DNA and calcium phosphate for 6 h at 37°C (van der Eb and Graham, 1980), then media (3% DMEM/3% glutamine, 5% foetal calf serum) removed and fresh media added. After three days in culture, cells were trypsinised and replated at 1:5 dilution. Thereafter, they were grown at 37°C with weekly changes of media. Individual dishes contain AGMK cells transfected with calf thymus DNA only (C), or with linear virion DNA (1); episomal DNA from Daudi cells (2); recombinant DNA from clones p5 (3); p39 (4); p23 (5); p13 + p33 (6); p31 (7); p65 (8); and a mixture of recombinant DNAs (9), as noted. Foci were visualised using Leishman's stain and the data obtained with dishes that had been in culture for about two months are shown. (From Griffin and Karran, 1984.)

tumourigenic cells, they lacked others. Notably lacking was the ability to form colonies in semi-solid media. That is, in soft agar the immortalised cells were found to proliferate for several weeks, after which time no further growth was observed. Neither did the cells die since, even a month after growth had ceased, cells could be re-established in tissue culture on plastic dishes and proceeded to grow normally. Cells injected subcutaneously into nude mice failed to produce tumours (Griffin and Karran, 1984). The conclusions from studies at this juncture were that a minor population of kidney cells that resembled simple cuboidal epithelium had been induced to grow continuously in culture by sub-genomic fragments of EBV, but that the cells *per se* were not capable of inducing tumours *in vivo*. These findings were reminiscent of those obtained following infection of B-lymphocytes with EB virions (see Nilsson *et al.*, 1977). In marked contrast to studies *in vitro* on cellular transformation with *Herpes simplex* virus (Galloway and McDougall, 1983), EBV DNA 'footprints' were found in the chromosomal DNA of the immortalised AGMK cells, as exemplified in Fig. 4. At early passage (Fig. 4A) the pattern of exogenous DNA in the cells was complex and not only

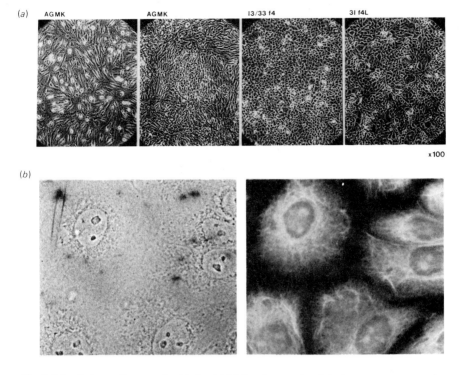

Fig. 3. Morphology of immortalised cells. (*a*) Light micrographs of the heterogeneous population of cells observed in (*left to right*) a sparse and a dense early culture of AGMK cells, and
homogeneous cuboidal epithelial cells present in dense cultures of the cell lines 13/33f4 and
31f4L after 20 weeks in culture. 13/33f4 or 31f4L cells examined at earlier (4 weeks) or later
(34–76 weeks) periods showed no obvious morphological differences from the 20-week culture.
Living cells shown by phase-contrast optics.

(*b*) Sub-confluent 31f4L cells fixed and stained with the monoclonal antibody LE61 that
specifically recognises kidney epithelium (Lane, 1982). Cells observed (*left*) by phase-contrast
optics and (*right*) by immunofluorescence to show the tonofilament organisation.

were sequences characteristic of EBV DNA observed but those
derived from the cosmid vector used in DNA cloning were also
present. At later passage (Fig. 4B, C) the latter were lost, and the
DNA carried by the cells was largely to be found in a single band that
hybridised to EBV DNA. Although exhaustive analyses have not
been carried out, the size of this fragment differed among cell lines
(Fig. 4D) and the pattern on further passaging of any particular line
appeared to remain constant. The single aspect of our data that was
surprising is that in none of the cell lines was EBNA expression
observed.

At this stage, it seemed important for a number of reasons to
ascertain whether experiments such as those described above could
be repeated in other systems. First and foremost was the need to

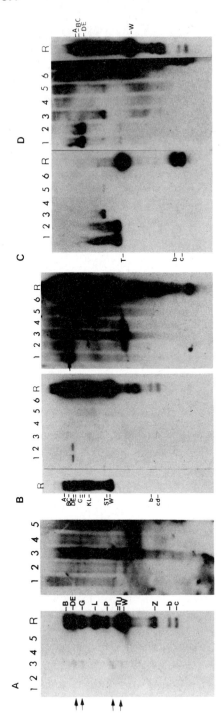

Fig. 4. Autoradiograms of *Bam*HI restriction enzyme-cleaved chromosomal DNA from immortalised AGMK cells hybridised against ^{32}P-labelled nick-translated EBV DNA.

Panel A, two exposures of results obtained with DNA isolated from cells passaged for only four weeks in culture following selection and establishment from individual foci: Panel B, two exposures of a similar analysis of DNA from cells passaged continuously for six months in culture. Whole cell DNA was cleaved with *Bam*HI, fragments separated by electrophoresis on 0.8% agarose slab gels, transferred to nitrocellulose filters and probed with ^{32}P-labelled DNA from a mixture

of clones p13 and p33 under standard conditions. Tracks 1–5 contain DNA fragments from cell lines 13/33f2, f3, f4, f4L and f5, respectively, and track R contains EB virion DNA. Prominently hybridising bands are denoted by arrows.

Panel C, illustrates the patterns obtained when the probe was a nick-translated recombinant DNA containing the EBV fragment *EcoRI* H (which hybridises to EBV DNA *Bam*HI fragments T, b and c, as shown) or, panel D, EBV DNA from virions. (Data adapted from Griffin and Karran, 1984.)

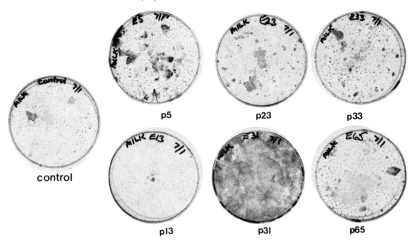

Fig. 5. Transfection studies on human cells. Cells from human breast milk epithelium (Chang *et al.*, 1982) were transfected with recombinant DNA from the EBV cosmid library (see Fig. 1), as illustrated. These dishes have been maintained for about four months with occasional sub-culturing and are visualised with Leishman's stain.

determine to what extent the immortalisation was a property of the AGMK cells themselves and/or how important was the contribution of EBV DNA to the cell phenotype. Secondly, there was the need to monitor chromosomal changes and alterations in the expression of genetic information so as to assess the role of co-factors in generating a fully malignant phenotype. If chromosomal alterations such as translocations can be used as a marker for the latter (Klein, 1981), the fact that AGMK cells have a diploid number of 60 could complicate any analysis of marker chromosomes in these cells.

   Two different studies were set up to answer these questions. One involved transfection of cells from human milk with individual fragments from the recombinant EBV DNA library and the other involved similar types of experiments using primary kidney cells from common marmosets (*Callithrix jacchus*). (In contrast to AGMK cells, common marmoset cells, like human cells, have a diploid chromosome number of 46; moreover, the ready availability of marmoset colonies at the Royal College of Surgeons (Wedderburn *et al.*, 1984) will in the long run allow studies to be carried out *in vivo*). In both experiments, the fragment of EBV DNA found in the recombinant p31 was found to have a marked effect on cell growth. For a finite time (that is, over a six-month period) the human cells, as illustrated (four months) in Fig. 5, were stimulated to grow. Unlike experiments reported earlier with SV40 (Chang *et al.*, 1982), we have

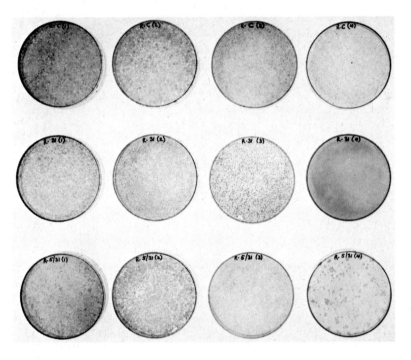

Fig. 6. Transfection studies on common marmoset kidney cells. Cells from primary cultures of common marmoset (*Callithrix jacchus*) kidneys were transfected with recombinant DNA from the EBV cosmid library (see Fig. 1) and sub-cultured at three-weekly intervals. Illustrated here (cells visualised with Leishman's stain) are: row A, control cells at passages 1 to 4 (*left to right*); row B, cells transfected with p31 recombinant DNA, passages 1 to 4, and similarly, row C, cells co-transfected with a mixture of p31 and p5 recombinant DNAs. By the fifth passage, the latter were completely confluent, resembling the fourth passage of cells seen in row B. The prefix 'R' on each dish means that cells were transfected again, about a week after the initial transfection, and before the sub-culture protocol was adopted. Cells transfected only a single time (not shown) were immortalised much more slowly, supporting the notion that gene dosage may be important in this process.

not, however, yet been successful in establishing a continuously growing line. More success has accompanied the studies with marmoset cells (D. King, L. Karran and B. E. Griffin, unpublished). Again, p31 DNA proved to be a growth stimulant. Although no densely growing foci were ever observed (Fig. 6), after about three months (or five passages) in culture, the epithelial population of cells had outgrown the fibroblasts and, by the tenth passage, a homogeneous population of epithelial cells (see Fig. 7A) was evident. From this time on, cells were sub-cultured weekly and have now been continuously in culture for nearly a year. As observed in the case of AGMK cells, these marmoset cells have only a finite capacity to grow in soft agar, which is nonetheless superior to that observed with

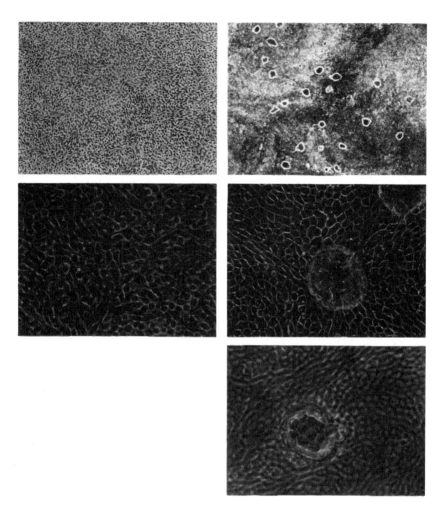

Fig. 7. Common marmoset kidney cells growth stimulated by sub-genomic fragments of EBV DNA. Cells at the tenth passage subsequent to transfection with recombinant DNA from the cosmid clone p31, or p31 and p5, examined by light microscopy. Panel A: Dense culture of cells designated R-31 (see Fig. 6) at two magnifications. Panel B: Dense cultures of cells designated R-5/31 (see Fig. 6). The presence of 'domes' or hemicysts is clearly evident in R-5/31 cultures. In bottom figure of B only cells comprising the dome are in focus.

Immunofluorescent staining of these cells with the monoclonal antibody LE61 (see Fig. 3b) showed the epithelial nature of all cells in both Panel A and Panel B (data not given).

control cells. Moreover, whereas the latter had completely senesced after two months in the semi-solid support, the p31 DNA–treated cells could be removed from soft agar and re-established on plastic. Finally, 'footprints' of the complete EBV DNA used in the transfection procedure were still evident in the marmoset cells after ten passages, as had previously been observed in the AGMK cell experiments. Although these became simpler in late passage cells, no alterations in chromosomes were observed.

A very interesting phenotypic alteration of marmoset kidney cells was observed when they were transfected simultaneously with p31 recombinant DNA and a region of EBV DNA encompassed in another cloned fragment, p5 (see Fig. 1). Instead of a flat homogeneous population of epithelial cells becoming established, a much more mobile, apparently more highly differentiated population was observed. These have now also been in culture nearly a year but they grow slowly and, unlike cells transfected with p31 DNA alone, cannot be passaged weekly. Moreover, when individual dishes of cells are kept in culture past confluence, they develop 'domes' or hemicysts characteristic of actively transporting renal epithelium (Fig. 7B), similar to those observed, for example, in Madin–Darby canine kidney (MDCK) cells, a continuously growing line established from an apparently normal dog kidney (Lever, 1979; Jefferson *et al.*, 1980; Cereijido, 1984). Cells with the potential to form domes can be actively stimulated by certain solvents, such as DMSO, that are potent inducers of cell differentiation; conversely, dome formation can be abolished by ouabain, which inhibits transepithelial ion transport (Birek *et al.*, 1982). Similar effects in the presence of these agents were observed with the marmoset cells that had been co-transfected with p5 and p31 DNAs, suggesting that information obtained from the second fragment of EBV (contained in p5) might have relevance to active ion transport and/or cell differentiation. It may be noteworthy that the p5 recombinant DNA encodes three of the five regions of the EBV genome that Miller (1984) has suggested to be of especial importance to the biology of EBV.

These data taken together strongly support the notion of a function(s) within the EBV genome that, directly or indirectly, stimulates cell growth and at least in non-human primates can result in the immortalisation of sub-classes of epithelial cells. Based on the studies with the mixture of fragments (p31 and p5), it would appear that, in the presence of the immortalising function(s), other consequences of virus–cell interactions can be observed. Very preliminary data on some of the immortalised cells suggest that mRNAs with homology to EBV are being expressed. If this can be confirmed, it will stimulate us in our search for EBV-coded proteins that play important roles in the overall process of cellular immortalisation. Further experiments aimed at defining the minimum size of fragment which, when transfected into a suitable cell, will produce continuously growing lines appear promising. Cleavage of the recombinant DNA from p31 with the restriction enzymes *Eco*RI, *Bgl*II and *Hind*III appear to abolish this activity, but cleavage with certain other enzymes apparently does not. These types of studies thus suggest that a minimum immortalising region may in time be identified and, with the aid of the total DNA sequence (Baer *et al.*, 1984), a viral gene localised.

CONCLUSIONS AND UNSOLVED PROBLEMS

Our data on immortalisation of primate epithelial cells in culture by specific sub-genomic fragments of EBV DNA are consistent with the large volume of information that has accumulated over a number of years from studies on the diseases associated with this virus. The latter support the notion that EBV can induce continuous proliferation of cells but that other stimuli (such as tumour promoters, chemical carcinogens, or cellular 'oncogenes') may be required for full expression of malignancy. By introducing viral DNA directly into the cell we have avoided the question of whether epithelial cells with receptors for EBV exist. Thus, at this juncture, we would suggest that, using genetic information initially supplied by the virus, we have obtained model epithelial cell lines that can be used for the investigation of agents needed in subsequent stages along the pathway to a malignant phenotype; that is, we are part of the way towards setting up a model system *in vitro* for nasopharyngeal carcinoma in non-human primates. Whether chromosomal alterations are involved in this process, as is the case with Burkitt's lymphoma, is not clear, although Mittelman *et al.* (1983) have observed a haploid change in chromosome 3 in two cases of NPC tumours. So far, our attempts to produce phenotypic alterations in immortalised monkey kidney cells by continuous passage in culture have not been successful. This essentially passive approach, which presupposes a 'genetic accident', may be an important step *in vivo* but is probably too slow to yield meaningful results *in vitro*. For example, passaging immortalised AGMK epithelial cells for nearly two years in culture failed to produce any marked alteration in the cells. Therefore, to accelerate the process, we have begun studies designed to determine if gene products from other viruses or cellular oncogenes can convert immortalised cells to potentially malignant ones; alternatively, we are investigating whether addition of chemicals observed to be active in tumour promotion or cell differentiation can accomplish this effect. In this regard, two recently published papers are of considerable interest in that they study environmental factors that, together with EBV, may be involved in the development of NPC. Using the induction of viral early antigens in Raji cells as an assay, Ito *et al.* (1983) found that diterpene esters from plants of the Euphorbiacae family were highly effective in their ability to initiate expression of some EBV functions. Similarly, a group of scientists in Beijing (Zeng *et al.*, 1983) have investigated a large number of commonly used medicinal herbs to determine whether factors that can alter EBV gene expression, and presumably host gene expression, are present in Chinese medicines. Five species of plants from the Thymelacacae family were strongly positive in this respect, especially if used together with sodium butyrate, a chemical that stimulates cells to differentiate (Leder and Leder, 1975).

Our immortalisation studies focus attention on a portion of the viral genome that contains none of the 'five interesting and potentially important regions' of EBV discussed in a recent review (Miller, 1984). Thus, we are forced to conclude that although we find 'footprints' of EBV DNA in the continually growing monkey-epithelial cells, viral gene expression may not be an essential feature of the cellular immortalisation process. Alternatively, in the absence of good tissue culture systems for the virus, and considering its large size, it seems more likely that a very important viral function has up to now escaped detection.

Further, it is difficult to correlate our data with those of Stoerker and Glaser (1983), who essentially rescued transforming virus from a hybrid cell line that contained the DNA of the non-immortalising P3HR1 strain of EBV using a portion of the viral genome encompassed within *Bam*HI fragments W, Y and H (see Fig. 1). Theirs is a very complex experiment, and the structure of the DNA within their hybrid cell has not been characterised and may be very heterogeneous; also, their assay is immortalisation of cord blood lymphocytes. Our experiments, on the other hand, are more direct. Nonetheless, this apparent conflict of data exists. Among several projected explanations, the possibility that the mechanism by which EBV transforms B-lymphocytes and epithelial cells differs must receive strong consideration; that is, this large virus may produce the same phenotype in different types of cells by alternative routes. This is an important concept and needs further investigation. In the long run, the situation will undoubtedly prove even more complex than this; for example,

Åman, Ehlin-Henriksson and Klein (1984) have recently shown that among B-lymphocytes, only resting cells could apparently be infected and immortalised by EBV. Thus, not only the type of cell but the state of the cell may play an important role in determining whether it responds to the viral information in such a way as to become immortalised and ultimately malignant. EBV has set up an effective symbiotic relationship with its host(s) and it will obviously require a great deal of effort and insight to understand the relationship (see discussion in Rickinson, 1984). Our experiments should allow us to explore an important aspect (immortalisation) of this relationship.

REFERENCES

ÅMAN, P., EHLIN-HENRIKSSON, B. and KLEIN, G. (1984). Epstein–Barr virus susceptibility of normal human B lymphocyte populations. *Journal of Experimental Medicine*, **159**, 208–20.

ANDERSON, M. and LINDAHL, T. (1976). EBV DNA in human lymphoid cell lines: *in vitro* conversion. *Virology*, **73**, 96–105.

BAER, R., BANKIER, A. T., BIGGIN, M. D., DENINGER, P. L., FARRELL, P. J., GIBSON, T. J., HATFULL, G., HUDSON, G. S., SATCHWELL, S. C., SÉGUIN, C., TUFFNELL, P. S. and BARRELL, B. G. (1984). DNA sequence and expression of the B95-8 Epstein–Barr virus genome. *Nature*, **310**, 207–11.

BIREK, C., AUBIN, J. E., BHARGARA, U., BRUNETTE, D. M. and MELCHER, H. (1982). Dome formation in oral epithelial cells *in vitro*. *In vitro*, **18**, 382–92.

BORNKAMM, G. W., DELIUS, H., ZIMBER, U., HUDEWENTZ, J. and EPSTEIN, M. A. (1980). Comparison of Epstein–Barr virus strains of different origin by analysis of the viral DNAs. *Journal of Virology*, **35**, 603–18.

CEREIJIDO, M. (1984). Differentiation of epithelial cells. *Federation Proceedings*, **43**, 2207.

CHANG, S. E., KEEN, J., LANE, E. B. and TAYLOR-PAPADIMITRIOU, J. (1982). Establishment and characterization of SV40-transformed human breast epithelial cell lines. *Cancer Research*, **42**, 2040–53.

DALLA-FAVERA, R., GREGNI, M., ERIKSON, J., PATTERSON, D., GALLO, R. and CROCE, C. M. (1982). Human c-*myc* oncogene is located on the region of chromosome 8 that is translocated in Burkitt's lymphoma cells. *Proceedings of the National Academy of Sciences, USA*, **79**, 7824–7.

DE THÉ, G., GESER, A., DAY, N. E., TUKEI, P. M., WILLIAMS, E. H., BERI, D. P., SMITH, P. G., DEAN, A. G., BORNKAMM, G. W., FEORINO, P. and HENLE, W. (1978). Epidemiological evidence for causal relationship between Epstein–Barr virus and Burkitt's lymphoma: Results of an Ugandan prospective study. *Nature*, **274**, 756–61.

DE THÉ, G., HO, J. H. C., ABLASHI, D. V., DAY, N. E., MACARIO, A. J. L., MARTIN-BERTHELON, M. C., PEARSON, G. and SOHIER, G. (1975). Nasopharyngeal carcinoma: Antibodies to EBNA and correlation to other EBV antigens in Chinese patients. *International Journal of Cancer*, **16**, 713–21.

EPSTEIN, M. A. and ACHONG, B. G. (1979). Introduction. Discovery and general biology of the virus. In *The Epstein–Barr Virus*, ed. M. A. Epstein and B. G. Achong, pp. 1–22. Berlin, Springer.

FALK, L., WOLFE, L., DEINHARDT, F., PACIGA, J., DOMBOS, L., KLEIN, G., HENLE, W. and HENLE, G. (1974). Epstein–Barr virus: Transformation of non-human primate lymphocytes *in vitro*. *International Journal of Cancer*, **13**, 363–76.

FRANK, A., ANDIMAN, W. A. and MILLER, G. (1976). Epstein–Barr virus and non-human primates: Natural and experimental infection. *Advances in Cancer Research*, **23**, 171–201.

GALLOWAY, D. A. and McDOUGALL, J. K. (1983). The oncogenic potential of *Herpes simplex* viruses: evidence for a 'hit and run' mechanism. *Nature*, **302**, 21–4.

GRIFFIN, B. E. and KARRAN, L. (1984). Immortalisation of monkey epithelial cells by specific fragments of Epstein–Barr virus DNA. *Nature*, **309**, 78–82.

HELLER, M., DAMBAUGH, T. and KIEFF, E. (1981). Epstein–Barr virus DNA. IX. Variation among viral DNAs from producer and non-producer infected cells. *Journal of Virology*, **38**, 632–48.

HENLE, G., HENLE, W. and DIEHL, W. (1968). Relation of Burkitt's tumor-associated herpes-type virus to infectious mononucleosis. *Proceedings of the National Academy of Sciences, USA*, **59**, 94–101.

HESTON, L., RABSON, M., BROWN, N. and MILLER, G. (1982). New Epstein–Barr virus variants from cellular subclones of P3-HR-1 Burkitt's lymphoma. *Nature*, **295**, 160–3.

HOHN, B. and COLLINS, J. (1980). A small cosmid for efficient cloning of large DNA fragments. *Gene*, **11**, 291–8.

ITO, Y., OHIGASHI, H., KOSHIMIZU, K. and ZENG, Y. (1983). Epstein–Barr virus-activating principle in the ether extracts of soils collected from under plants which contain active diterpene esters. *Cancer Letters*, **19**, 113–17.

JEFFERSON, D. M., COBB, M. H., GENNARO, J. F. and SCOTT, W. N. (1980). Transporting renal epithelium: Culture in hormonally defined serum-free medium. *Science*, **210**, 912–14.

JONES, M. D., FOSTER, L., SHEEDY, T. and GRIFFIN, B. E. (1984). The EB virus genome in Daudi Burkitt's lymphoma cells has a deletion similar to that observed in a non-transforming strain (P3HR1) of the virus. *EMBO Journal*, **3**, 813–21.

KATAMINE, S., OTSU, M., TADA, K., TSUCHIYA, S., SATO, T., ISHIDA, N., HONJO, T. and ONO, Y. (1984). Epstein–Barr virus transforms precursor B cells even before immunoglobulin gene rearrangements. *Nature*, **309**, 369–72.

KLEIN, G. (1979). Lymphoma development in mice and human: Diversity of initiation is followed by convergent cytogenetic evolution. *Proceedings of the National Academy of Sciences, USA*, **76**, 2442–6.

KLEIN, G. (1981). The role of gene dosage and genetic transpositions in carcinogenesis. *Nature*, **294**, 313–18.

KLEIN, G. (1983). Specific chromosomal translocations and the genesis of B cell-derived tumors in mice and men. *Cell*, **32**, 311–15.

KLEIN, G., GIOVANELLA, B. C., LINDAHL, T., FIALKOW, P. J., SINGH, S. and STEHLIN, J. S. (1974). Direct evidence for the presence of Epstein–Barr virus DNA and nuclear antigen in malignant epithelial cells from patients with poorly differentiated carcinoma of the nasopharynx. *Proceedings of the National Academy of Sciences, USA*, **71**, 4737–41.

KLEIN, G. and KLEIN, E. (1984). The changing faces of EBV research. In *Progress in Medical Virology*, vol. 30, ed. J. L. Melnick, pp. 87–106. Basel, Karger.

LANE, E. B. (1982). Monoclonal antibodies provide specific intra-molecular markers for the study of epithelial tonofilament organization. *Journal of Cell Biology*, **92**, 665–73.

LEDER, A. and LEDER, P. (1975). Butyric acid, a potent inducer of erythroid differentiation in cultured erythroleukemia cells. *Cell*, 5, 319–22.

LEVER, J. E. (1979). Inducers of mammalian cell differentiation stimulate dome formation in a differentiated kidney epithelial cell line (MDCK). *Proceedings of the National Academy of Sciences, USA*, **76**, 1323–7.

LI, Y., WU, B., ZHAO, Z., YI, L., LIU, H., HUANG, M., WANG, X. and WU, M. (1984). Studies on mechanism of entry of EB virus into non-B lymphocytes. *Scientia Sinica*, (B), **27**, 284–93.

MENEZES, J., LEIBOLD, W. and KLEIN, G. (1975). Biological differences between Epstein-Barr virus (EBV) strains with regard to lymphocyte transforming ability, superinfection and antigen induction. *Experimental Cell Research*, **92**, 478–84.

MILLER, G. (1984). Regions of the EBV genome involved in latency and lymphocyte immortalization. In *Progress in Medical Virology*, vol. 30, ed. J. L. Melnick, pp. 107–28. Basel, Karger.

MILLER, G., ROBINSON, J., HESTON, L. and LIPMAN, M. (1974). Differences between laboratory strains of Epstein–Barr virus based on immortalization, abortive infection and interference. *Proceedings of the National Academy of Sciences, USA*, **71**, 4006–10.

MITTLEMAN, F., MARK-VENDEL, E., MINEUR, A., GIOVANELLA, B. and KLEIN, G. (1983). A 3f+ marker chromosome in EBV-carrying nasopharyngeal carcinomas. *International Journal of Cancer*, **32**, 651–5.

NILSSON, K., GIOVANELLA, B. C., STEHLIN, J. S. and KLEIN, G. (1977). Tumorigenicity of human hematopoeitic cell lines in athymic nude mice. *International Journal of Cancer*, **19**, 337–44.

RABSON, M., GRADOVILLE, L., HESTON, L. and MILLER, G. (1982). Non-immortalizing P3J-HR-1 Epstein–Barr virus: A deletion mutant of its transforming parent, Jijoye. *Journal of Virology*, **44**, 834–44.

REEDMAN, B. M. and KLEIN, G. (1973). Cellular localization of an Epstein–Barr virus (EBV) associated complement-fixing antigen in producer and non-producer lymphoblastoid cell lines. *International Journal of Cancer*, **11**, 499–520.

RICKINSON, A. (1984). Epstein–Barr virus in epithelium. *Nature, News and Views*, **310**, 99–100.

ROBINSON, J. and SMITH, D. (1981). Infection of human B lymphocytes with high multiplicities of Epstein–Barr virus: kinetics of EBNA expression, cellular DNA synthesis and mitosis. *Virology*, **109**, 336–43.

SIXBEY, J. W., VESTERINEN, E. H., NEDRUND, J. G., RAAB-TRAUB, N., WALTON, L. A. and PAGANO, J. S. (1983). Replication of Epstein–Barr virus in human epithelial cells infected *in vitro*. *Nature*, **306**, 480–3.

STOERKER, J. and GLASER, R. (1983). Rescue of transforming Epstein–Barr virus (EBV) from EBV-genome-positive epithelial hybrid cells transfected with subgenomic fragments of EBV DNA. *Proceedings of the National Academy of Sciences, USA*, **80**, 1726–9.

TAUB, R., KIRSCH, I., MORTON, C., LENOIR, G., SWAN, D., TRONICK, S., AARONSON, S. and LEDER, P. (1982). Translocation of the c-*myc* gene into the immunoglobulin heavy chain locus in human Burkitt's lymphoma and murine plasmacytoma cells. *Proceedings of the National Academy of Sciences, USA*, **79**, 7837–41.

VAN DER EB, A. J. and GRAHAM, F. L. (1980). Assay of transforming activity of tumor virus DNA. *Methods in Enzymology*, **65**, 826–39.

WEDDERBURN, N., EDWARDS, J. M. B., DESGRANGES, C., FONTAINE, C., COHEN, B. and DE THÉ, G. (1984). Infectious mononucleosis-like response in common marmosets infected with Epstein–Barr virus (EBV). *Journal of Infectious Diseases*, in press.

ZENG, Y., LIU, Y., LIU, C., CHEN, S., WEI, J., ZHU, J. and ZAI, H. (1980). Application of an immunoenzymatic method and an immunoautoradiographic method for a mass survey of nasopharyngeal carcinoma. *Intervirology*, **13,** 162–8.

ZENG, Y., ZHONG, J. M., MO, Y. K. and MIAO, X. C. (1983). Epstein–Barr virus early antigen induction in Raji cells by Chinese medicinal herbs. *Intervirology*, **19,** 201–4.

INTERVENTION AGAINST EB VIRUS-ASSOCIATED TUMOURS

M. A. EPSTEIN

*Department of Pathology, University of Bristol Medical School,
University Walk, Bristol BS8 1TD, UK*

Since the discovery of Epstein–Barr (EB) virus just over 20 years ago (Epstein, Achong and Barr, 1964) work from many laboratories has established remarkably close associations between this agent and endemic Burkitt's lymphoma (BL) (Burkitt, 1963), undifferentiated nasopharyngeal carcinoma (NPC) (Shanmugaratnam, 1971), and the lymphomas which occur in immunosuppressed human allograft recipients with an unusually high frequency (Penn, 1978; Kinlen *et al.*, 1979; Weintraub and Warnke, 1982). The basis of these associations is well known (for reviews see Epstein and Achong, 1979; de Thé, 1980; Klein and Purtilo, 1981) and work on cellular oncogene activation in BL now suggests possible explanations. Thus, the virus, together with certain co-factors, appears to be an essential link in a series of interlocking steps which includes characteristic chromosomal translocations (Lenoir *et al.*, 1982) leading to activation of the cellular *myc* oncogene (Dalla-Favera *et al.*, 1982; Taub *et al.*, 1982). This may be involved in progression to full malignancy; the B*lym*-1 oncogene also seems to be implicated (Diamond *et al.*, 1983).

But whatever the details of such oncogene activation in BL, and irrespective of whether they operate in the other EB virus-associated tumours, quite recent studies on the experimental induction of lymphomas by the virus *in vivo* demonstrate that it can potently, rapidly, and directly set in motion the chain of events which leads to the appearance of malignant tumours (Cleary *et al.*, 1984*b*). Such a direct role is not entirely surprising in the light of recent findings on transformation by EB virus *in vitro*; this has often been categorized as being merely a form of 'immortalization' (Miller, 1980), yet careful tests have shown that in addition to the latter phenomenon some cells are changed in such a way as to possess from the outset many of the attributes of malignant transformation (Zerbini and Ernberg, 1983).

The accumulation of information both on the general biological behaviour of EB virus and on its role in human malignancies has progressed at an ever increasing pace. But although this is unquestionably of scientific importance, it has seemed for some years that the value of such activities would be considerably enhanced if they could lead to intervention against infection by the virus which might in consequence reduce the incidence of the associated tumours. It was in this context that proposals were first put forward for a vaccine against EB virus (Epstein, 1976) and considerable progress in this direction has subsequently been achieved.

IS A VACCINE AGAINST EB VIRUS JUSTIFIABLE ECONOMICALLY?

The striking evidence implicating EB virus, together with co-factors, in the causation of endemic BL and undifferentiated NPC has already been mentioned. BL occurs frequently only in rather limited areas and even in these does not involve very large numbers (Burkitt, 1963); furthermore, the high incidence areas are just those with many more pressing medical and community health problems. In contrast, undifferentiated NPC is the commonest tumour of men and the second most common of women amongst Southern Chinese (Shanmugaratnam, 1971), has a high incidence amongst Eskimos (Lanier *et al.*, 1980), and there are moderately high incidence levels in North Africa (Cammoun, Hoerner and Mourali, 1974), East Africa (Clifford, 1970), and through most of South-East Asia (Shanmugaratnam, 1971). Thus, in world cancer terms, undifferentiated NPC is of very considerable significance and is thought to be responsible for more than 100 000 deaths a year; this figure alone justifies efforts to develop a vaccine against EB virus.

PRECEDENTS FOR ANTI-VIRAL VACCINATION IN CANCER

The control of a naturally occurring herpesvirus-induced lymphoma of chickens, Marek's disease (Marek, 1907; Payne, Frazier and Powell, 1976), by inoculation with apathogenic virus (Churchill, Payne and Chubb, 1969; Okazaki, Purchase and Burmester, 1970) provided the first example of anti-viral vaccination affecting the frequency of a cancer. Later work with the malignant lymphoma which can be induced experimentally by inoculation of *Herpesvirus saimiri* in South American subhuman primates (Meléndez *et al.*, 1969) has shown that animals given killed virus vaccine were protected against challenge infection and did not therefore get tumours (Laufs and Steinke, 1975). Furthermore, in the Marek's disease system, antigen-containing membranes from cells infected with Marek's disease herpesvirus markedly reduced lymphoma incidence when used as an experimental vaccine in chickens (Kaaden and Dietzschold, 1974) and even soluble viral antigens extracted from such cells protected in the same way (Lesnick and Ross, 1975). Similar approaches with EB virus, one of the five human herpesviruses, have long appeared worthy of investigation.

CHOICE OF VACCINE ANTIGEN

It has been known for many years that the virus-neutralizing antibodies developed by EB virus-infected individuals are those directed against the virus-determined cell surface membrane antigen (MA) (Pearson et al., 1970; Gergely, Klein and Ernberg, 1971; Pearson, Henle and Henle, 1971; de Schryver et al., 1974) and this information prompted the suggestion that MA be used as an anti-viral vaccine (Epstein, 1976).

Investigations into the molecular structure of MA have identified two high molecular weight glycoprotein components of 340 000 and 270 000 daltons (gp340 and gp270) (Qualtière and Pearson, 1979; Strnad et al., 1979; Thorley-Lawson and Edson, 1979; North, Morgan and Epstein, 1980; Qualtière and Pearson, 1980) and the concordance between human antibodies to MA and EB virus-neutralization has been formally explained by the demonstration of these same glycoproteins in both the viral envelope and the cell membrane MA (North et al., 1980). Not surprisingly, therefore, monoclonal antibodies which react with both MA components neutralize EB virus (Hoffman, Lazarowitz and Hayward, 1980; Thorley-Lawson and Geilinger, 1980) and gp340/270 can themselves elicit virus-neutralizing antibodies (North et al., 1982a). Most EB virus-producing lymphoid cell lines synthesize roughly equal amounts of gp340/270.

REQUIREMENTS FOR A VACCINE BASED ON EB VIRUS MA

To elaborate a vaccine against EB virus based on MA gp340/270, the essential prerequisites set out in Table 1 must be made available or devised.

Table 1. *Requirements for a vaccine based on EB virus MA gp340*

      (1) Susceptible test animals
      (2) A sensitive antigen-monitoring assay
      (3) An efficient antigen preparation method
      (4) A potently immunogenic product
      (5) A quantitative test for specific antibodies

Susceptible test animals

Only two kinds of animal are known to be fully susceptible to experimental infection with EB virus, the owl monkey (*Aotus*) (Epstein, Hunt and Rabin, 1973*a*; Epstein *et al.*, 1973*b*; Epstein *et al.*, 1975) and the cotton-top tamarin (*Saguinus oedipus oedipus*) (Shope, Dechairo and Miller, 1973; Miller *et al.*, 1977; Miller, 1979). However, the former 'species' has recently been found to be very heterogeneous, with at least nine different karyotypes (Ma *et al.*, 1976; Ma *et al.*, 1978; Ma, 1981), and shows considerable variation in susceptibility to certain infections.

The cotton-top tamarin is therefore the species of choice for experimental studies *in vivo* with EB virus even though it was placed on the endangered species list some years ago. For, although there was rather little information about this animal and the possibility of its successful propagation in captivity, the necessary management and husbandry conditions have recently been defined, and flourishing breeding colonies have been established (Brand, 1981; Kirkwood, 1983; Kirkwood, Epstein and Terlecki, 1983; Kirkwood and Epstein, 1984). Nevertheless, there are severe constraints on the numbers of the rare and costly tamarins which can be used in each experiment, similar to those operating in work with hepatitis B virus where biological tests require the use of chimpanzees. Because of these constraints, it is also necessary to test all methodologies with banal laboratory animals (which will make antibodies, for example, even though they cannot be infected with EB virus), before applying them to the tamarins.

An assay for MA gp340

In order to work out an efficient and reliable method for the preparation of antigen, it is essential that the product can be quantified and monitored at each step to permit modifications which maximize yields. Accordingly, a highly sensitive, quantitative radioim-munoassay (RIA) was developed for gp340. Small amounts of this molecule were prepared in extremely pure, radioiodinated form, were shown to be antigenic, and were thereafter used in a conventional competition RIA to quantify unlabelled samples of gp340 using a defined system of arbitrary units. A full account of the RIA has been published elsewhere (North et al., 1982b).

Preparation method for MA gp340

As mentioned above, EB virus-producing cell lines usually synthesize equal amounts of gp340/270, but the B95-8 marmoset line (Miller et al., 1972) is anomalous in that it expresses almost exclusively the larger component, thus providing an important advantage for molecular-mass-based purification. With the crucial help of the RIA, a preparative sodium dodecyl sulphate–polyacrylamide gel electro-phoresis (SDS–PAGE) procedure was worked out for gp340 from B95-8 cell membranes which included an important new technique for ensuring that the product was renatured and thus in an antigenic form. This was achieved by removing SDS in the presence of $4 \, mol \, l^{-1}$ urea to inhibit protein refolding, followed by dilution of the urea to permit protein renaturation; the details have already been given (Morgan, North and Epstein, 1983).

Enhancement of immunogenicity of MA gp340

gp340 made by the method just described proved only weakly immunogenic in mice and rabbits after repeated injection and the use of Freund's adjuvant. To eliminate the need for these two disadvantageous procedures, gp340 was incorporated in liposomes (Morein et al., 1978; Manesis, Cameron and Gregoriadis, 1979), sometimes with the addition of lipid A (Naylor et al., 1982), and comparative immunogenicity studies were undertaken to determine the best routes and methods of administration. Liposomes containing gp340, with or without lipid A, gave good titres of EB-virus-neutralizing antibodies in mice, rabbits, and cotton-top tamarins after rather few inoculations, and all the sera were specific in that they reacted only with MA gp340 and failed to recognize any other molecules from the surface or interior of B95-8 cells. These experiments have been reported in full (North et al., 1982a; Morgan, Epstein and North, 1984a).

A test for antibodies to MA gp340

In order to exploit immunogenicity studies to the full, a sensitive test to quantify antibody responses to gp340 was essential. A rapid enzyme-linked immunosorbent assay (ELISA) has therefore been developed based on gp340 purified by affinity chromatography using a monoclonal antibody immunoabsorbent (Randle et al., 1984). This ELISA has proved a thousand-fold more sensitive than conventional indirect immunofluorescence tests and has made it possible to follow accurately the sequential production of specific antibodies to gp340 during the immunization of animals. The ELISA is described in a recent publication (Randle and Epstein, 1984).

VACCINATION OF COTTON-TOP TAMARINS

To demonstrate vaccine protection of immunized tamarins a dose and mode of administration of challenge EB virus has been worked out which will ensure the induction of lesions in 100% of unprotected normal animals. The lesions have been extensively investigated and both on histological (Dorfman, Burke and Berard, 1982) and molecular biological (Arnold et al., 1983; Cleary et al., 1984a) grounds must clearly be regarded as malignant lymphomas with several interesting features (Cleary et al., 1984b).

Pilot experiment

When this challenge dose of virus was used in a small-scale preliminary experiment, a vaccinated animal whose serum had been shown to have potent virus-neutralizing capacity (Pearson et al., 1970; Moss and Pope, 1972) was found to be totally protected, whereas other animals with less neutralizing antibody were not (Epstein, 1984).

Confirmatory experiments

The demonstration that purified gp340 in liposomes can induce virus-neutralizing antibodies in cotton-top tamarins, and the preliminary indication that these protect against a highly pathogenic dose of challenge virus, provide a clear mandate for confirmatory tests with larger numbers of the expensive animals.

In this connection, evaluation is under way both of gp340 obtained by the molecular mass-based technique (SDS–PAGE; Morgan *et al.*, 1983) used from the outset and of gp340 purified by the more recent monoclonal antibody immunoaffinity chromatography method (Randle *et al.*, 1984). Comparison of the protection induced by inoculation of these two types of material should give valuable insights into the biological complexity of gp340, since the first of the two preparation procedures presumably yields all epitopes on molecules of the appropriate molecular weight, whereas the monoclonal antibody is known to bind only about 50 to 60% of the epitopes (Randle *et al.*, 1984). It will be interesting to see which immunogen is most efficacious.

FUTURE PROSPECTS

Once it has been confirmed that experimentally induced antibodies to EB virus-determined MA components indeed protect tamarins against infection by the virus, the situation will be exactly comparable to that long known for the Marek's disease herpesvirus and *Herpesvirus saimiri* systems (Kaaden and Dietzschold, 1974; Laufs and Steinke, 1975; Lesnick and Ross, 1975). Planning for a gp340-based vaccine for man should therefore be considered sooner, rather than later.

The most advantageous human context in which to test such a vaccine is in relation to infectious mononucleosis (IM). It is well known that, in Western countries, groups of young adults can be screened to detect those who have escaped primary EB virus infection in childhood and who are therefore at risk for delayed primary infection which is accompanied by the clinical manifestations of IM in 50% of cases (Niederman *et al.*, 1970; University Health Physicians *et al.*, 1971). Screening could therefore be applied to new students entering Universities or Colleges followed by a double-blind vaccine trial amongst informed, consenting volunteers in the 'at risk' category. The effectiveness of immunization in preventing infection and reducing the expected incidence of IM would rapidly be evident.

Thereafter, the effect of vaccination and consequential prevention of disease should be assessed in a high incidence region for endemic BL. This tumour has a peak incidence at about the age of seven (Burkitt, 1963) and the influence of vaccination on this should therefore be apparent within a decade. If this were successful there would then be inescapable reasons for tackling the far more difficult, but more important problem of intervention against undifferentiated NPC. Since this is a disease of middle and later life in high incidence areas (Shanmugaratnam, 1971), immunity would have to be maintained over many years.

MA components prepared in the ways discussed here (Morgan *et al.*, 1983; Randle *et al.*, 1984) have never been considered suitable for anything beyond the present experimental prototype vaccine (Epstein, 1984). It was therefore important to know something of the structure of gp340 and of the contribution, if any, of the sugar moiety to antigenicity. Experiments have been undertaken in which gp340 was analysed after treatment with a battery of glycosidases and V8 protease, with or without preliminary exposure during synthesis to tunicamycin. This work, reported by Morgan *et al.* (1984*b*), has shown that carbohydrate represents more than 50% of the total mass of gp340, that it is both O- and N-linked, that V8 protease digestion fragments are antigenic, and that specific antibody appears to bind the protein not the sugar.

The seemingly preponderant importance of the protein in the immunogenicity of gp340 means that, for use in man, the possibility of exploiting new procedures can be explored. The fragment of EB virus DNA carrying the gene coding for MA has already been identified (Hummel, Thorley-Lawson and Kieff, 1984) and the sequence probably relating to this gene is also known (Biggin, Farrell and Barrell, 1984). The potential for cloning the gene and seeking to make the product by expression in suitable prokaryotic or eukaryotic cells is thus very real. In addition, it can be readily envisaged that the practicability of using synthetic gp340 peptides as immunogens will soon be investigated. And, however the subunit vaccine molecule is ultimately obtained, yet further possibilities lie in the direction of greatly enhanced immunogenicity using powerful new adjuvants (Morein *et al.*, 1984).

Finally, there is an excellent chance that it may prove feasible to incorporate the EB virus MA gene into the genome of vaccinia virus and thus ensure its direct expression during vaccination in man (Smith, Mackett and Moss, 1983). Such an achievement could well solve many of the biological and logistic problems of an EB virus vaccine intended for intervention in relation to undifferentiated NPC.

This work was supported by the Medical Research Council, London (SPG978/32) and the Cancer Research Campaign, London (out of funds donated by the Bradbury Investment Company of Hong Kong).

REFERENCES

ARNOLD, A., COSSMAN, J., BAKHSHI, A., JAFFE, E. S., WALDMANN, T. A. and KORSMEYER, S. J. (1983). Immunoglobulin-gene rearrangements as unique clonal markers in human lymphoid neoplasms. *New England Journal of Medicine*, **309**, 1593–9.

BIGGIN, M., FARRELL, P. J. and BARRELL, B. G. (1984). Transcription and DNA sequence of the *Bam*HI L fragment of B95-8 Epstein–Barr virus. *EMBO Journal*, **3**, 1083–90.

BRAND, H. M. (1981). Husbandry and breeding of a newly established colony of cotton-topped tamarins (*Sanguinus oedipus*). *Laboratory Animals*, **15**, 7–11.

BURKITT, D. (1963). A lymphoma syndrome in tropical Africa. In *International Review of Experimental Pathology*, vol. 2, ed. G. W. Richter and M. A. Epstein, pp. 67–138. New York and London, Academic Press Inc.

CAMMOUN, M., HOERNER, G. V. and MOURALI, N. (1974). Tumors of the naso-pharynx in Tunisia: an anatomic and clinical study based on 143 cases. *Cancer*, **33**, 184–92.

CHURCHILL, A. E., PAYNE, L. N. and CHUBB, R. C. (1969). Immunization against Marek's disease using a live attenuated virus. *Nature*, **221**, 744–7.

CLEARY, M. L., CHAO, J., WARNKE, R. and SKLAR, J. (1984a). Immunoglobulin gene rearrangement as a diagnostic criterion of B cell lymphoma. *Proceedings of the National Academy of Sciences, USA*, **81**, 593–7.

CLEARY, M. L., EPSTEIN, M. A., FINERTY, S., DORFMAN, R. F., BORNKAMM, G. W., KIRKWOOD, J. K., MORGAN, A. J. and SKLAR, J. (1984b). Individual tumours of multifocal EB virus-induced malignant lymphomas in tamarins arise from different B cell clones. *Science*, in press.

CLIFFORD, P. (1970). A review: On the epidemiology of nasopharyngeal carcinoma. *International Journal of Cancer*, **5**, 287–309.

DALLA-FAVERA, R., BREGNI, M., ERIKSON, J., PATTERSON, D., GALLO, R. C. and CROCE, C. M. (1982). Human c-*myc onc* gene is located on the region of chromosome 8 that is translocated in Burkitt lymphoma cells. *Proceedings of the National Academy of Sciences, USA*, **79**, 7824–7.

DE SCHRYVER, A., KLEIN, G., HENLE, W. and HENLE, G. (1974). EB virus-associated antibodies in Caucasian patients with carcinoma of the nasopharynx and in long-term survivors after treatment. *International Journal of Cancer*, **13**, 319–25.

DE THÉ, G. (1980). Role of Epstein–Barr virus in human diseases: infectious mono-nucleosis, Burkitt's lymphoma, and nasopharyngeal carcinoma. In *Viral Oncology*, ed. G. Klein, pp. 769–97. New York, Raven Press.

DIAMOND, A., COOPER, G. M., RITZ, J. and LANE, M-A. (1983). Identification and molecular cloning of the Human Blym transforming gene activated in Burkitt's lymphomas. *Nature*, **305**, 112–16.

DORFMAN, R. F., BURKE, J. S. and BERARD, C. (1982). A working formulation of non-Hodgkin's lymphomas: background recommendations, histological criteria, and relationship to other classifications. In *Malignant Lymphomas*, ed. S. Rosenberg and H. Kaplan, pp. 351–68. New York, Academic Press.

EPSTEIN, M. A. (1976). Epstein–Barr virus – is it time to develop a vaccine program? *Journal of the National Cancer Institute*, **56**, 697–700.

EPSTEIN, M. A. (1984). A prototype vaccine to prevent Epstein–Barr (EB) virus-associated tumours. *Proceedings of the Royal Society of London, B*, **221**, 1–20.

EPSTEIN, M. A. and ACHONG, B. G. (1979). The relationship of the virus to Burkitt's lymphoma. In *The Epstein–Barr Virus*, ed. M. A. Epstein and B. G. Achong, pp. 321–37. Berlin, Heidelberg and New York, Springer.

EPSTEIN, M. A., ACHONG, B. G. and BARR, Y. M. (1964). Virus particles in cultured lymphoblasts from Burkitt's lymphoma. *Lancet*, **i**, 702–3.

EPSTEIN, M. A., HUNT, R. D. and RABIN, H. (1973*a*). Pilot experiments with EB virus in owl monkeys (*Aotus trivirgatus*). I. Reticuloproliferative disease in an inoculated animal. *International Journal of Cancer*, **12**, 309–18.

EPSTEIN, M. A., RABIN, H., BALL, G., RICKINSON, A. B., JARVIS, J. and MELÉNDEZ, L. V. (1973*b*). Pilot experiments with EB virus in owl monkeys (*Aotus trivirgatus*). II. EB virus in a cell line from an animal with reticuloproliferative disease. *International Journal of Cancer*, **12**, 319–32.

EPSTEIN, M. A., ZUR HAUSEN, H., BALL, G. and RABIN, H. (1975). Pilot experiments with EB virus in owl monkeys (*Aotus trivirgatus*). III. Serological and biochemical findings in an animal with reticuloproliferative disease. *International Journal of Cancer*, **15**, 17–22.

GERGELY, L., KLEIN, G. and ERNBERG, I. (1971). Appearance of Epstein–Barr virus-associated antigens in infected Raji cells. *Virology*, **45**, 10–21.

HOFFMAN, G. J., LAZAROWITZ, S. G. and HAYWARD, S. D. (1980). Monoclonal antibody against a 250 000-dalton glycoprotein of Epstein–Barr virus identifies a membrane antigen and a neutralizing antigen. *Proceedings of the National Academy of Sciences, USA*, **77**, 2979–83.

HUMMEL, M., THORLEY-LAWSON, D. A. and KIEFF, E. (1984). An Epstein–Barr virus DNA fragment encodes messages for the two major envelope glycoproteins (gp350/300 and gp220/200). *Journal of Virology*, **49**, 413–17.

KAADEN, O. R. and DIETZSCHOLD, B. (1974). Alterations of the immunological specificity of plasma membranes of cells infected with Marek's disease and turkey herpesviruses. *Journal of General Virology*, **25**, 1–10.

KINLEN, L. J., SHEIL, A. G. R., PETO, J. and DOLL, R. (1979). Collaborative United Kingdom–Australasian study of cancer in patients treated with immunosuppressive drugs. *British Medical Journal*, **2**, 1461–6.

KIRKWOOD, J. K. (1983). Effects of diet on health, weight and litter size in captive cotton-top tamarins *Saguinus oedipus oedipus*. *Primates*, **24**, 515–20.

KIRKWOOD, J. K. and EPSTEIN, M. A. (1984). Rearing a second generation of cotton-top tamarins *Saguinus oedipus oedipus*. Submitted.

KIRKWOOD, J. K., EPSTEIN, M. A. and TERLECKI, A. J. (1983). Factors influencing population growth of a colony of cotton-top tamarins. *Laboratory Animals*, **17**, 35–41.

KLEIN, G. and PURTILO, D. T. (1981). Epstein–Barr virus-induced lymphoproliferative diseases in immunodeficient patients, ed. G. Klein and D. T. Purtilo, pp. 4209–304. *Cancer Research*, **41** (Suppl.).

LANIER, A., BENDER, T., TALBOT, M., WILMETH, S., TSCHOPP, C., HENLE, W., HENLE, G., RITTER, D. and TERASAKI, P. (1980). Nasopharyngeal carcinoma in Alaskan Eskimos, Indians and Aleuts: A review of cases and study of Epstein–Barr virus, HLA and environmental risk factors. *Cancer*, **46**, 2100–6.

LAUFS, R. and STEINKE, H. (1975). Vaccination of non-human primates against malignant lymphoma. *Nature*, **253**, 71–2.

LENOIR, G. M., PREUD'HOMME, J. L., BERNHEIM, A. and BERGER, R. (1982). Correlation between immunoglobulin light chain expression and variant translocation in Burkitt's lymphoma. *Nature*, **298**, 474–6.

LESNICK, F. and ROSS, L. J. N. (1975). Immunization against Marek's disease using Marek's disease virus-specific antigens free from infectious virus. *International Journal of Cancer*, **16**, 153–63.

MA, N. S. F. (1981). Chromosome evolution in the owl monkey, *Aotus*. *American Journal of Physical Anthropology*, **54**, 293–303.

MA, N. S. F., JONES, T. C., MILLER, A. C., MORGAN, L. M. and ADAMS, E. A. (1976). Chromosome polymorphism and banding patterns in the owl monkey (*Aotus*). *Laboratory Animal Science*, **26**, 1022–36.

MA, N. S. F., ROSSAN, R. N., KELLEY, S. T., HARPER, J. S., BEDARD, M. T. and

JONES, T. C. (1978). Banding patterns of the chromosomes of two new karyotypes of the owl monkey, *Aotus*, captured in Panama. *Journal of Medical Primatology*, **7**, 146–55.

MANESIS, E. K., CAMERON, C. H. and GREGORIADIS, G. (1979). Hepatitis B surface antigen-containing liposomes enhance humoral and cell-mediated immunity to the antigen. *FEBS Letters*, **102**, 107–11.

MAREK, J. (1907). Multiple Nervenentzündung (Polyneuritis) bei Hühnern. *Deutsche Tierärztliche Wochenschrift*, **15**, 417–21.

MELÉNDEZ, L. V., HUNT, R. D., DANIEL, M. D., GARCIA, P. G. and FRASER, C. E. O. (1969). Herpesvirus saimiri. II. An experimentally induced primate disease resembling reticulum cell sarcoma. *Laboratory Animal Care*, **19**, 378–86.

MILLER, G. (1979). Experimental carcinogenicity by the virus *in vivo*. In *The Epstein–Barr Virus*, ed. M. A. Epstein and B. G. ACHONG, pp. 351–72. Berlin, Heidelberg and New York, Springer.

MILLER, G. (1980). Biology of the Epstein–Barr virus. In *Viral Oncology*, ed. G. Klein, pp. 713–38. New York, Raven Press.

MILLER, G., SHOPE, T., COOPE, D., WATERS, L., PAGANO, J., BORNKAMM, G. W. and HENLE, W. (1977). Lymphoma in cotton-top marmosets after inoculation with Epstein–Barr virus: tumor incidence, histologic spectrum, antibody responses, demonstration of viral DNA, and characterization of viruses. *Journal of Experimental Medicine*, **145**, 948–67.

MILLER, G., SHOPE, T., LISCO, H., STITT, D. and LIPMAN, M. (1972). Epstein–Barr virus: transformation, cytopathic changes, and viral antigens in squirrel monkey and marmoset leukocytes. *Proceedings of the National Academy of Sciences, USA*, **69**, 383–7.

MOREIN, B., HELENIUS, A., SIMONS, K., PETTERSSON, R., KÄÄRIÄINEN, L. and SCHIRRMACHER, V. (1978). Effective subunit vaccines against an enveloped animal virus. *Nature*, **276**, 715–18.

MOREIN, B., SUNDQUIST, B., HÖGLUND, S., DALSGAARD, K. and OSTERHAUS, A. (1984). Iscom, a novel structure for antigenic presentation of membrane proteins from enveloped viruses. *Nature*, **308**, 457–60.

MORGAN, A. J., EPSTEIN, M. A. and NORTH, J. R. (1984a). Comparative immunogenicity studies on Epstein–Barr (EB) virus membrane antigen (MA) with novel adjuvants in mice, rabbits and cotton-top tamarins. *Journal of Medical Virology*, **13**, 281–92.

MORGAN, A. J., NORTH, J. R. and EPSTEIN, M. A. (1983). Purification and properties of the gp340 component of Epstein–Barr (EB) virus membrane antigen (MA) in an immunogenic form. *Journal of General Virology*, **64**, 455–60.

MORGAN, A. J., SMITH, A. R., BARKER, R. N. and EPSTEIN, M. A. (1984b). A structural investigation of the Epstein–Barr (EB) virus membrane antigen glycoprotein, gp340. *Journal of General Virology*, **69**, 397–404.

MOSS, D. J. and POPE, J. H. (1972). Assay of the infectivity of Epstein–Barr virus by transformation of human leucocytes *in vitro*. *Journal of General Virology*, **17**, 233–6.

NAYLOR, P. T., LARSEN, H. L., HUANG, L. and ROUSE, B. T. (1982). *In vivo* induction of anti-*Herpes simplex* virus immune response by Type I antigens and Lipid A incorporated into liposomes. *Infection and Immunity*, **36**, 1209–16.

NIEDERMAN, J. C., EVANS, A. S., SUBRAHMANYAN, L. and McCOLLUM, R. W. (1970). Prevalence, incidence and persistence of EB virus antibody in young adults. *New England Journal of Medicine*, **282**, 361–5.

NORTH, J. R., MORGAN, A. J. and EPSTEIN, M. A. (1980). Observations on the EB virus envelope and virus-determined membrane antigen (MA) polypeptides. *International Journal of Cancer*, **26**, 231–40.

NORTH, J. R., MORGAN, A. J., THOMPSON, J. L. and EPSTEIN, M. A. (1982a). Purified

EB virus gp340 induces potent virus-neutralizing antibodies when incorporated in liposomes. *Proceedings of the National Academy of Sciences, USA*, **79**, 7504–8.

NORTH, J. R., MORGAN, A. J., THOMPSON, J. L. and EPSTEIN, M. A. (1982*b*). Quantification of an EB virus-associated membrane antigen (MA) component. *Journal of Virological Methods*, **5**, 55–65.

OKAZAKI, W., PURCHASE, H. G. and BURMESTER, B. R. (1970). Protection against Marek's disease by vaccination with a herpesvirus of turkeys. *Avian Diseases*, **14**, 413–29.

PAYNE, L. N., FRAZIER, J. A. and POWELL, P. C. (1976). Pathogenesis of Marek's disease. In *International Review of Experimental Pathology*, vol. 16, ed. G. W. Richter and M. A. Epstein, pp. 59–154. New York, San Francisco, London, Academic Press.

PEARSON, G., DEWEY, F., KLEIN, G., HENLE, G. and HENLE, W. (1970). Relation between neutralization of Epstein–Barr virus and antibodies to cell-membrane antigens induced by the virus. *Journal of the National Cancer Institute*, **45**, 989–95.

PEARSON, G., HENLE, G. and HENLE, W. (1971). Production of antigens associated with Epstein–Barr virus in experimentally infected lymphoblastoid cell lines. *Journal of the National Cancer Institute*, **46**, 1243–50.

PENN, I. (1978). Malignancies associated with immunosuppressive or cytotoxic therapy. *Surgery*, **83**, 492–502.

QUALTIÈRE, L. F. and PEARSON, G. R. (1979). Epstein–Barr virus-induced membrane antigens: Immunochemical characterization of Triton X100 solubilized viral membrane antigens from EBV-superinfected Raji cells. *International Journal of Cancer*, **23**, 808–17.

QUALTIÈRE, L. F. and PEARSON, G. R. (1980). Radioimmune precipitation study comparing the Epstein–Barr virus membrane antigens expressed on P_3HR-1 virus-superinfected Raji cells to those expressed on cells in a B95-8 virus-transformed producer culture activated with tumor-promoting agent (TPA). *Virology*, **102**, 360–9.

RANDLE, B. J. and EPSTEIN, M. A. (1984). A highly sensitive enzyme-linked immunosorbent assay to quantify antibodies to Epstein–Barr virus membrane antigen gp340. *Journal of Virological Methods*, in press.

RANDLE, B. J., MORGAN, A. J., STRIPP, S. A. and EPSTEIN, M. A. (1984). Large-scale purification of Epstein–Barr virus membrane antigen gp340 using a monoclonal immunoabsorbent. Submitted.

SHANMUGARATNAM, K. (1971). Studies on the etiology of nasopharyngeal carcinoma. In *International Review of Experimental Pathology*, vol. 10, ed. G. W. Richter and M. A. Epstein, pp. 361–413. New York and London, Academic Press.

SHOPE, T., DECHAIRO, D. and MILLER, G. (1973). Malignant lymphoma in cotton-top marmosets after inoculation with Epstein–Barr virus. *Proceedings of the National Academy of Sciences, USA*, **70**, 2487–91.

SMITH, G. L., MACKETT, M. and Moss, B. (1983). Infectious vaccinia virus recombinants that express hepatitis B virus surface antigen. *Nature*, **302**, 490–5.

STRNAD, B. C., NEUBAUER, R. H., RABIN, H. and MAZUR, R. A. (1979). Correlation between Epstein–Barr virus membrane antigen and three large cell surface glycoproteins. *Journal of Virology*, **32**, 885–94.

TAUB, R., KIRSCH, I., MORTON, C., LENOIR, G., SWAN, D., TRONICK, S., AARONSON, S. and LEDER, P. (1982). Translocation of the c-*myc* gene into the immunoglobulin heavy chain locus in human Burkitt's lymphoma and murine plasmacytoma cells. *Proceedings of the National Academy of Sciences, USA*, **79**, 7837–41.

THORLEY-LAWSON, D. A. and EDSON, C. M. (1979). The polypeptides of the Epstein–Barr virus membrane antigen complex. *Journal of Virology*, **32**, 458–67.

THORLEY-LAWSON, D. A. and GEILINGER, K. (1980). Monoclonal antibodies against the major glycoprotein (gp350/220) of Epstein–Barr virus neutralize infectivity. *Proceedings of the National Academy of Sciences, USA*, **77**, 5307–11.

UNIVERSITY HEALTH PHYSICIANS and PHLS LABORATORIES (1971). Infectious mononucleosis and its relationship to EB virus antibody. *British Medical Journal*, **iv**, 643–6.

WEINTRAUB, J. and WARNKE, R. A. (1982). Lymphoma in cardiac allotransplant recipients: clinical and histological features and immunological phenotype. *Transplantation*, 33, 347–51.

ZERBINI, M. and ERNBERG, I. (1983). Can Epstein–Barr virus infect and transform all the B-lymphocytes of human cord blood? *Journal of General Virology*, **64**, 539–47.

ADENOVIRUS GENES INVOLVED IN TRANSFORMATION. WHAT DETERMINES THE ONCOGENIC PHENOTYPE?

PHILLIP H. GALLIMORE, PHILIP J. BYRD and ROGER J. A. GRAND

Department of Cancer Studies, University of Birmingham, The Medical School, Birmingham B15 2TJ, UK

The major impetus to adenovirus research came in 1962 with the discovery that hamsters, inoculated as newborns with adenovirus type 12 (Ad 12), developed tumours (Trentin, Yabe and Taylor, 1962). Subsequently, other human adenovirus serotypes (Huebner, Rowe and Lane, 1962; Girardi, Hilleman and Zwickey, 1964; Pereira, Pereira and Clarke, 1965; Trentin, Van Hoosier and Samper, 1968) and adenoviruses isolated from a number of different animal species (Hull *et al.*, 1965; Sarma, Huebner and Lane, 1965; Darbyshire, 1966; Sarma *et al.*, 1967) were shown to be tumourigenic viruses. A considerable literature concerned with the human adenoviruses has built up over the past 23 years and because of this we have a better understanding of these viruses than of animal isolates. This chapter will therefore be restricted to studies with human adenoviruses.

Highly oncogenic Ad 12 has been shown to have a broad host range for tumour induction; rodents are particularly susceptible hosts (Trentin *et al.*, 1962; Huebner *et al.*, 1963; Yabe *et al.*, 1964; Rabson and Kirschstein, 1964). Adenovirus tumours are mainly primitive sarcomas (Berman, 1967), although Ad 12 has recently been shown to produce 'retinoblastoma-like' tumours when inoculated intraocularly in baboons (Mukai *et al.*, 1980). Not all adenoviruses are oncogenic (Trentin *et al.*, 1962) and some only produce tumours after a long latent period (Girardi *et al.*, 1964) (see classifications, Table 1).

The complexity of studying tumourigenesis led to the development of tissue culture transformation systems (McBride and Wiener, 1964; Freeman *et al.*, 1967; Graham and van der Eb, 1973). Adenoviruses, whether of oncogenic or non-oncogenic serotypes, or adenovirus transforming DNA, produce phenotypic changes in rare cells in the culture (1 in 10 000 to 1 in 100 000) that are manifested as a visible

Table 1. *Properties of human adenoviruses*

Group	Serotype	DNA homology	DNA[a] (%C+C)	Oncogenicity in newborn: Hamsters	Rats	Mice	Cell types transformed	Tumourigenicity of transformed cell line in: Animals of same species	Athymic nude mice	Pathogenicity and epidemiology
A	12, 18, 31	48–69% within group, 8–20% with other types	Low (47–49)	+ Highly oncogenic; short latent period 2–4 months	+	+/−	Hamster Rabbit Rat Mouse Human[b]	+ NR + + ND	+ NR + + +	Associated with upper respiratory tract infections and diarrhoea; also inapparent infections.
B	3, 7, 11, 14, 16, 21, 34, 35	89–94% within group, 9–20% with other types	Intermediate (49–52)	+ Weakly oncogenic; long latent period; low tumour incidence	NR	NR	Rat	+	NR	Acute respiratory disease (ARD), pharyngealconjunctival fever (PCF), febrile pneumonia, acute haemorrhagic cystitis (Ad 11 and occasionally 21) and diarrhoea

C	1, 2, 5, 6	99–100% within group, 10–16% with other types	High (57–59)	–	Hamster[b,c]; Rat; Mouse, Human[b]	+ (approx. 50% of cell lines); ± (2 + out of 80 lines tested); NR; ND	+; +; NR; +	Mild to severe infections of the upper respiratory tract of infants and children, latent infections of lymphoid tissue
D	8, 9, 10, 13, 15, 17, 19, 20, 22, 23, 24, 25, 26, 27, 28, 29, 30, 32, 33, 36, 37, 38, 39	94–99% within group, 4–17% with other types	High (57–59)	+[d]	Hamster; Rat	± (some tumourigenic cell lines in newborn and immuno-suppressed)	NR	Common cause of epidemic kerato-conjunctivitis (EKC); mainly Ad 8 and 19
E	4	4–23% with others	High	–	NR	NR	NR	ARD, PCF and EKC
F	40	NR	NR	–	NR	NR	NR	Infantile and young-adult gastro-enteritis
G	41	NR	NR	–	NR	NR	NR	enteritis

ND, not done; NR, not reported.
[a] Taken from Green et al. (1979).
[b] Transformation by DNA.
[c] Transformation by temperature-sensitive mutants or UV-irradiated virus.
[d] Female rat mammary adenomas.

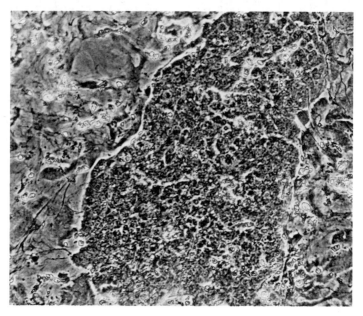

Fig. 1. An adenovirus type 12 transformed focus developing 14 days after infection of 21-day-old mouse kidney cells.

transformed foci 3–5 weeks after infection. These foci (Fig. 1) can be isolated individually and developed into immortal cell lines. Adenoviruses can induce transformation in tissue culture systems in which the virus interacts either abortively (Ad 12 with hamster cells: Doerfler, 1969) or semi-permissively (Ad 2 with rat cells: Gallimore, 1974). Transformation of permissive cells has been achieved by using mutant virus with a temperature-sensitive lesion in a structural gene (Williams, 1973), by transfection of sheared virus DNA (Graham *et al.*, 1977) and by DNA transfection (Byrd, Brown and Gallimore, 1982*a*) or microinjection (Whittaker *et al.*, 1984) of molecularly cloned adenovirus transforming genes. Adenovirus transformed human permissive cells have proved to be important for the isolation of mutants with lesions in the transforming region (Graham, Harrison and Williams, 1978; Gallimore *et al.*, 1984).

The aim of this chapter is to familiarize the reader with the adenovirus gene regions influencing transformation and tumourigenicity, to summarize our current understanding of the properties of adenovirus transforming proteins and to discuss the current concepts of what factors determine the 'malignant phenotype'.

ADENOVIRUS GENES INVOLVED IN TRANSFORMATION

Integration and the extent of viral genome representation and expression in transformed cells

The transformation process involves the integration and expression of viral DNA sequences. Viral DNA appears to integrate randomly in the cellular genome (Doerfler *et al.*, 1982), although in some cell lines patch-like regions of homology between the terminal adenovirus sequences and the flanking host sequences have been identified (Deuring *et al.*, 1981; Westin *et al.*, 1982; Gahlmann and Doerfler, 1983). Integration may proceed through circular intermediates (Ruben, Bacchetti and Graham, 1983) since several cell lines have been found to contain linked left and right ends that were inverted relative to each other and joined either directly or through short tracts of cellular DNA (Visser *et al.*, 1979, 1982; Vardimon and Doerfler, 1981). At the viral/cellular DNA junction cellular sequences may be completely conserved, whereas deletion of terminal viral nucleotides is a common feature (Gahlmann *et al.*, 1982; Gahlmann and Doerfler, 1983).

The amount of the adenovirus genome that is retained in transformed cells is dependent on the permissivity of the cells. Complete, unrearranged viral genomes have been found frequently in non-permissive cells (Doerfler *et al.*, 1982; Fisher *et al.*, 1982; Ruben, Bacchetti and Graham, 1982) whereas extensive fragmentation is the general rule in semi-permissive cells (Gallimore, Sharp and Sambrook, 1974; Flint *et al.*, 1976; Johansson *et al.*, 1978; Green, Wold and Buttner, 1981). Early studies of the Ad 2 DNA integrated in semi-permissive rat cell lines showed that only the left-end 14% of the viral genome was common to all transformed lines examined (Gallimore *et al.*, 1974). Cells containing intact genes derived from other parts of the viral genome express those genes that are normally active before viral DNA replication in the lytic cycle (early genes) but not those that are transcribed after replication (late genes: Flint *et al.*, 1976; Doerfler *et al.*, 1982; Rijnders *et al.*, 1983). This differential gene expression may be partly due to extensive methylation of late gene blocks in cells that have been passaged *in vitro* for long periods (Doerfler *et al.*, 1982). However, additional undefined factors may be operating because the integrated viral genes in cells from Ad 12 induced hamster tumours may be silent in spite of minimal methylation (Kuhlmann and Doerfler, 1983).

Organization of early region 1 (E1)

The left-end 14% of the adenovirus genome contains the E1 region, which comprises two transcriptional units, E1a and E1b (Berk and Sharp, 1978; Chow, Broker and Lewis, 1979; Perricaudet *et al.*, 1979; Wilson, Fraser and Darnell, 1979; Baker and Ziff, 1981). The DNA sequences of the E1 regions of representative non-oncogenic (Ad 2 and Ad 5), weakly oncogenic (Ad 7) and highly oncogenic (Ad 12) serotypes have been determined and the structures of their E1a and E1b mRNA species defined (Dijkema, Dekker and van Ormondt, 1980*a*, 1982; Sugisaki *et al.*, 1980; van Ormondt, Maat and van Beveren, 1980*b*; Bos *et al.*, 1981; Kimura *et al.*, 1981; Gingeras *et al.*, 1982; Virtanen *et al.*, 1982) (see Fig. 2).

The E1a region

The E1a region is the first transcriptional unit to be expressed following infection; its transcription is constitutive in that it is independent of other viral gene products and a product of this region modulates viral transcription (Berk *et al.*, 1979; Jones and Shenk, 1979*b*; Lewis and Mathews, 1980; Nevins, 1981).

Fig. 2.(a) Important features of the adenovirus transcription map. The adenovirus genome is a double-stranded DNA molecule of 34–36 kb that is transcribed in both the rightward (r-stand) and leftward (l-strand) directions. The conventional adenovirus map is divided into 100 equal units. The locations and directions of transcription of the early (E) and late (L) transcription units are shown as large unfilled arrows. Transcripts from the major late promoter (MLP) at 16.5 map units initially extend almost to the right end of the genome but differential splicing and poly(A) site selection generates families of mRNAs (L1 to L5). The MLP is also active at early times but only the L1 mRNAs are made and, in addition to the leader segments (shown as short dashes) common to all late L1 to L5 messages, they have an extra leader sequence. Species of L5 mRNAs may have one or more of three additional leader segments. The E2a and E2b regions are transcribed from the same promoter, the bodies of the messages being spliced to the leader segments. The spliced structures of E1a, E1b, E3, E4 and IVa$_2$ mRNAs are not shown. The transcription units for the structural proteins IX and IVa$_2$ are transcribed at predominantly late times.

(*b*) and (*c*) Structures of the E1 region and mRNAs of Ad 5 and Ad 12. The locations of promoters (TATA) and poly(A) addition signals are shown. The coordinates of the 5' caps, 3'-poly(A) addition sites and splice donor and acceptor sites are indicated. Carets denote sequences that are spliced out of the mRNAs. The relative sizes of the mRNAs and the proteins (in terms of numbers of amino acids) they encode are shown to the left of the E1a mRNAs and to the right of the E1b mRNAs. The locations of the open reading frames from which the proteins are translated are shown as open boxes. Initiation codons and termination codons are designated ♀ and | respectively. Dashed lines connect the coding regions in different frames that are joined in the mRNAs. The E1b mRNAs (other than the protein IX message) have the potential to direct the synthesis of more than one protein. The predicted coding regions for all these proteins (identified and hypothetical) are shown. The splice acceptor site at nucleotide 3276 for Ad 5 was identified by homology with Ad 2 which contains a splice site at 3270. These diagrams were drawn from data presented by Sugisaki *et al.* (1980), Perricaudet *et al.* (1980), van Ormondt *et al.* (1980*a*), Bos *et al.* (1981), Kimura *et al.* (1981), Virtanen *et al.* (1982), van Ormondt and Hesper (1983) and van Ormondt and Galibert (1984).

(a)

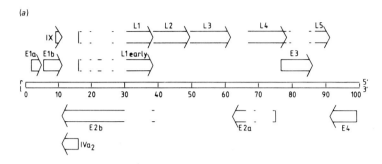

(b) Ad 5

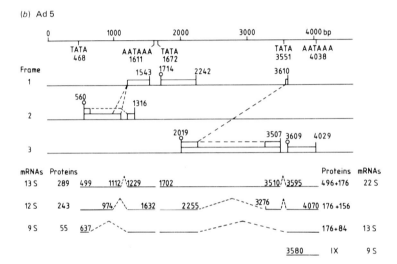

(c) Ad 12

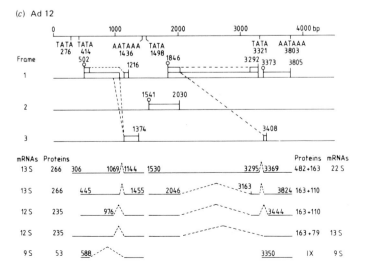

The E1a regions of Ad 2, Ad 5 and Ad 7 are transcribed off the r-strand into 13S and 12S early mRNAs which are 5′ and 3′ coterminal and differ only in the sizes of their introns; the 13S and 12S mRNAs have different 5′ donor splice sites but use the same 3′ acceptor (Spector, McGrogan and Raskas, 1978; van Ormondt *et al.*, 1978; Perricaudet *et al.*, 1979; Dijkema *et al.*, 1980*b*; van Ormondt, Maat and Dijkema, 1980*a*; Yoshida and Fujinaga, 1980; Lupker *et al.*, 1981). The E1a region of Ad 12 has two Goldberg–Hogness boxes (TATAATA) which direct the synthesis of two pairs of mRNAs that are analogous to the 13S and 12S E1a species of Ad 2, Ad 5 and Ad 7 (Sugisaki *et al.*, 1980; Kimura *et al.*, 1981). Splicing of the RNA transcripts joins the two major open reading frames for protein synthesis, removing A + T-rich regions that contain nonsense codons in all three frames. The proteins translated from the 13S and 12S mRNAs contain 289 and 243 amino acids respectively for Ad 5, 261 and 230 amino acids for Ad 7 and 266 and 235 amino acids for Ad 12; different groups have assigned different molecular weights to these and the E1b proteins, presumably because of differences in the gel electrophoresis conditions used to determine their size, so the E1 proteins will be referred to in terms of the number of amino acids in them to avoid confusion. The proteins translated from the 13S mRNAs contain 46 (Ad 2 and Ad 5), 31 (Ad 7) and 31 (Ad 12) amino acids that are not found in those translated from the 12S mRNAs. A 9S E1a mRNA species which is preferentially synthesized at late times in infection is translated in a different frame to the larger transcripts downstream of the splice junction (Spector *et al.*, 1978; Chow *et al.*, 1979; Wilson and Darnell, 1981; Virtanen and Pettersson, 1983); as yet no 9S E1a analogue has been found for Ad 12 but a potential splice donor site at nucleotide 588 that could be used to synthesize a message of this size and structure has been identified (van Ormondt and Galibert, 1984). Although the E1a mRNAs overlap, each species appears to be generated by a separate splicing event from unspliced nuclear precursor RNA, i.e. the 13S and 12S species are not used as splicing intermediates (Svensson, Pettersson and Akusjarvi, 1983). The 3′ ends of the E1a mRNAs contain the hexanucleotide AAUAAA polyadenylation signal which is required for efficient cleavage of the transcripts before polyadenylation; in the absence of the signal, spliced polyadenylated transcripts spanning the E1a and E1b regions are generated (Montell *et al.*, 1983*b*). The E1a Goldberg–Hogness box(es) were found to be important for specifying precise transcriptional start sites, but mutants lacking this controlling element grow as well as the wild-type in HeLa cells (Osborne, Gaynor and Berk, 1982; Hearing and Shenk, 1983a). Sequences upstream of the E1a promoter enhance expression of this region and function independently of position and perhaps orientation to stimulate transcription from the promoters of a variety of genes to which they have been linked (Hearing and Shenk, 1983b; Hen et al., 1983; Imperiale, Feldman and Nevins, 1983). The enhancer elements located between nucleotides 195 and 353 (Hearing and Shenk, 1983b) are probably most relevant to control of adenovirus E1a expression since they were identified in experiments with mutant viruses that contained E1a 5′ flanking region deletions.

The E1b region

The E1b regions of Ad 2, Ad 5 and Ad 12 contain two promoters, one of which directs the synthesis of an unspliced, predominantly late mRNA species encoding the structural polypeptide protein IX (Spector et al., 1978; Chow et al., 1979; Alestrom et al., 1980; Bos et al., 1981; Kimura et al., 1981; van Ormondt and Hesper, 1983); this protein is not considered to contribute to transformation although it is produced in a minority of transformed cell lines (Lewis and Mathews, 1982). Ad 7 is unusual because there is no Goldberg–Hogness box upstream of the gene for protein IX (Dijkema et al., 1982).

A family of 5′ and 3′ coterminal spliced mRNAs are transcribed off the r-strand from the other E1b promoter. Until recently Ad 2 and Ad 5 were thought to be transcribed into only two mRNA species of 22S and 13S that differed only in the sizes of the intervening sequences that were removed by splicing; both mRNA species use the same 3′ acceptor splice junction. However, Anderson et al. (1984) found that in a cell-free translation system programmed with E1b-selected mRNA, a novel polypeptide of 155 amino acids was synthesized in addition to the 495 and 175 amino acid proteins expected. Sequence analysis of tryptic peptides of the E1b 155 amino acid protein showed that it shared carboxy and amino termini with the 495 amino acid protein and also that an internal peptide could only have been generated by splicing of nucleotide 2249 to nucleotide 3270. Thus it appears that this Ad 2 E1b mRNA species contains a second exon and that the protein it encodes is an internally deleted version of the 495 amino acid protein. In an earlier report Virtanen et al. (1982) sequenced cDNA clones of Ad 12 E1b mRNAs with just this structure. The E1b region of Ad 7 was found to direct the transcription of an analogous mRNA (Dijkema et al., 1982) the coordinates of which were inferred from a minor Ad 2 E1b mRNA that had been detected by electron microscopical analysis of adenovirus DNA–RNA r-loops (L. T. Chow, cited in Dijkema *et al.*, 1982). The major 495 and minor 155 amino acid Ad 2 E1b proteins are translated from the second AUG downstream of the promoter whereas the 13S mRNA encodes a polypeptide of 175 amino acids that is translated from the first, promoter-proximal AUG. The 175 amino acid protein is translated from a frame that is different from and overlaps with that used for the translation of the 495 and 155 amino acid proteins. The 22S mRNA contains the AUG initiation triplets for both major E1b proteins and in translation systems *in vitro* it directs the synthesis of both the 495 and 175 amino acid proteins (Esche, Mathews and Lewis, 1980; Lupker *et al.*, 1981). A mutant defective in the splice donor site of the Ad 5 E1b 13S species encoding the 176 amino acid protein (the Ad 5 E1b proteins are one amino acid larger than those of Ad 2: Gingeras *et al.*, 1982) makes both major E1b proteins (C. Montell, cited in Anderson *et al.*, 1984). In some transformed cell lines the 22S mRNA is the predominant E1b transcript, the 13S mRNA specifying the 175 amino acid protein being undetectable, but both major E1b proteins are synthesized (Sawada and Fujinaga, 1980; Saito *et al.*, 1981; Saito, Shiroki and Shimojo, 1983; van den Elsen *et al.*, 1983c). The 495, 175 and 155 amino acid proteins are all entirely encoded in the sequences upstream of the splice junction that joins to the common 3′ acceptor splice point at nucleotide 3589 in Ad 2. The possibility that translation could initiate on the 13S mRNA at the second AUG, i.e. the initiation codon for the 496 and 156 amino acid proteins, and continue downstream of the 3′ acceptor splice site has been considered (Bos *et al.*, 1981; Kimura *et al.*, 1981; see Fig. 2), but convincing evidence that this happens *in vivo* has not been presented.

All aspects of the organization of the E1b region considered above primarily for Ad 2 and Ad 5 are also shared by Ad 7 and Ad 12, with the exception that no Ad 7 or Ad 12 equivalent of the Ad 2 E1b 155 amino acid protein has been identified. While the coding information of the E1b region appears to lie exclusively on the r-strand and little consideration has been given to l-strand transcripts, a leftward early transcript across the E1a and E1b regions has been identified (Katze, Persson and Philipson, 1982); the structure of this putative gene and its contribution to transformation are still unknown. There is no open reading frame on the l-strand spanning the E1a and E1b regions of Ad 12 (van Ormondt and Galibert, 1984).

There is currently no evidence of enhancer elements immediately

5′ of the E1b promoter, but it has been suggested that the expression of the E1b region might be modulated to a degree by the E1a enhancer (Spector, 1982; Hearing and Shenk, 1983b) or by E1a structural sequences (Sassone-Corsi et al., 1983). The phenotypes of certain Ad 5 E1a mutants indicate that the E1b region is subject to positive control by the E1a region (see later). Sequences involved in this regulation have been mapped between positions −135 and +11 relative to the E1b mRNA cap site (Bos and ten Wolde-Kraamwinkel, 1983).

Transformation mediated by subgenomic DNA fragments

Confirmation that the E1 region is the transforming region was obtained from DNA transfection studies performed on cultures of normal rat cells, normal hamster cells, normal human cells and continuous lines of rat cells (Graham and van der Eb, 1973; Shiroki et al., 1977; van der Eb et al., 1977, 1979; Yano et al., 1977; Sekikawa et al., 1978; Dijkema et al., 1979; Mak et al., 1979; Byrd et al., 1982a, b; Jochemsen et al., 1982; Rowe et al., 1984). Recent investigations have focused on the roles of the E1a and E1b regions in the initiation and maintenance of transformation, the morphological changes accompanying cell transformation and the roles of specific E1 products in determining tumourigenicity.

Transformation of primary baby rat kidney (BRK) cells has been achieved with the E1a region of Ad 5 (van der Eb et al., 1979; Houweling, van den Elsen and van der Eb, 1980) and Ad 12 (Gallimore et al., 1984), at reduced frequencies (4–6-fold and 30-fold respectively) relative to the complete E1 region. In contrast the Ad 12 E1a region transforms a continuous rat cell line (3Y1) as efficiently as the complete E1 region (Shiroki et al., 1979b). Fragments of Ad 5 DNA containing the E1a region and the E1b sequences encoding the 176 amino acid protein transformed BRK cells as efficiently as the complete E1 region (Houweling et al., 1980) whereas the transforming activity of the analogous Ad 12 DNA fragment was 4–7-fold lower than a complete E1 region fragment (Jochemsen et al., 1982; Gallimore et al., 1984). BRK cell foci transformed by Ad12 and Ad 5 E1a regions are more difficult to establish as continuous cell lines than are foci generated by the complete E1 region (Gallimore et al., 1984; A. J. van der Eb, personal communication). Comparisons of the properties of E1a transformants with those of cells expressing E1a and E1b proteins have identified E1a lines as 'incomplete transformants' exhibiting morphological differences, less vigorous growth on plastic or in soft agar, and reduced or no tumourigenic capacity (Shiroki et al., 1979b; Houweling et al., 1980). These observations and analysis of the relative contribution of E1a and E1b to the morphological alterations associated with transformation have led to the suggestion that the E1a region initiates transformation, confers immortality and has a potential to alter cellular morphology – a potential that is apparently only realized in the presence of E1b proteins (Shiroki et al., 1979b, van der Eb et al., 1979; Houweling et al., 1980; van den Elsen, Houweling and van der Eb, 1983a). While an immortalizing function for E1a is an attractive hypothesis the fact that E1a transformants are so difficult to establish suggests that such transformants may be switched into a finite phase of self-renewal and that during this period rare genetic, immortalizing alterations are sustained by a small proportion of cells (Gallimore et al., 1984; P. H. Gallimore, P. J. Byrd, J. L. Whittaker and R. J. A. Grand, unpublished).

Analysis of the role in transformation of the E1b region in isolation from the E1a region was only possible when the regulatory elements upstream of the coding sequences, through which E1a exerts its influence, were replaced by the early promoter of SV40. Transfection of BRK cells or 3Y1 cells with this construct alone failed to elicit any evidence of transformation in spite of the fact that the E1b proteins were synthesized (van den Elsen, Houweling and van der Eb, 1983b). The transfection experiments clearly show that the E1b region has a role in stabilizing or increasing the frequency of initiation events. Both major Ad 12 E1b proteins appear to be required for the complete response whereas this role is entirely fulfilled by the Ad 5 E1b 496 or 176 amino acid protein (van der Eb et al., 1977, 1979; Houweling et al., 1980; Jochemsen et al., 1982; Gallimore et al., 1984; van den Elsen et al., 1983a). Analysis of the effects of specific mutations in either of the Ad 12 E1b major proteins, i.e. the 482 amino acid protein (analogous to the Ad 5 E1b protein of 496 residues) or the 163 amino acid protein (analogous to the Ad 5 E1b protein of 176 residues), indicated that a functionally intact 482 amino acid protein is critical for the enhancement of transformation to the level of the wild-type E1 region (Bernards et al., 1983a). A role in the modulation of expression of Ad 5 E1a by the E1b region has been intimated by van den Elsen et al. (1983a), who also proposed that the abnormal phenotype of Ad 5 E1a transformants may be the result of low concentrations of E1a mRNA. The results (van den

Elsen et al., 1983a) further suggested that this function would be a property of both the 496 and 176 amino acid proteins of Ad 5 E1b and, extending this hypothesis to Ad 12, would be expected to be the property of the 482 amino acid protein solely (on the basis of phenotypes of the mutants constructed by Bernards et al., 1983a). However, at the protein level there is no difference in the level of expression of the Ad 12 E1a region in BRK cells transformed by E1a alone, E1a plus the E1b 163 amino acid protein or the complete E1 region (P. H. Gallimore, P. J. Byrd, J. L. Whittaker and R. J. A. Grand, unpublished).

The use of mutants in determining the roles of the E1a and E1b regions in transformation

The isolation and characterization of adenovirus E1 mutants has provided additional information on the roles of the individual E1 proteins in transformation and oncogenicity. The isolation of many of these mutants depended on the availability of permissive cell lines containing and expressing the viral E1 region. The Ad 5 transformed human embryo kidney cell line 293 (Graham et al., 1977) has been widely used for this purpose, but human KB tumour lines expressing the Ad 5 E1 region have also been developed (Babiss et al., 1983b; Shiroki et al., 1983). A series of Ad 12 transformed human embryo retinal cell lines have been used successfully in the isolation of Ad 12 E1 mutants (Byrd et al., 1982a).

Mutants with lesions in the E1 region have been generated by classical mutagenesis, selection of pre-existent mutants in viral DNA preparations, or by mutation in vitro (using chemical mutagens or molecular biological approaches) of E1 fragments that are subsequently rescued by ligation to the rest of the viral genome and by transfection of permissive normal or transformed cells (Harrison, Graham and Williams, 1977; Jones and Shenk, 1979a; Stow, 1981). The majority of Ad 5 E1 mutants were found to fall into two groups according to their host range (hr). Group I mutants only grow in line 293 cells, while group II mutants grow in both 293 cells and human embryo kidney cells but not in HeLa cells (Harrison et al., 1977). A proportion of Ad 5 E1 mutants were found to contain conditional lethal mutations that prevented their growth in HeLa cells at temperatures below 38.5 °C, i.e. they are host range, cold-sensitive (hr^{cs}) mutants (Ho, Galos and Williams, 1982). A series of classical hr mutants (Harrison et al., 1977) was mapped by both marker rescue and recombinational mapping with a separate series of host range deletion (*dl*) mutants (Jones and Shenk, 1979*a*; Shenk *et al.*, 1979) the coordinates of which within the E1 region were known. Complementation group I mutants, i.e. *hr*1, *hr*2, *hr*3, *hr*4, *hr*5, *hr*cs11, *hr*cs12, *dl*311 and *dl*312, belong to the E1a region whilst complementation group II mutants, i.e. *hr*6 (recently identified as *hr*cs: Ho *et al.*, 1982) *hr*7, *hr*cs13, *dl*313 and *sub*315 (*sub* = substitution), have been assigned to the E1b region (Harrison *et al.*, 1977; Frost and Williams, 1978; Jones and Shenk, 1978, 1979*a*; Galos *et al.*, 1980; Ross *et al.*, 1980; Ho *et al.*, 1982). At temperatures of 37°C or lower all these mutants, with the exception of *dl*311 which contains a 75 bp deletion downstream of the E1a 3′ acceptor splice point, are completely transformation-defective on rat embryo or rat embryo brain cells (Graham *et al.*, 1978; Shenk *et al.*, 1979). However, group I *hr* mutants induce semi-abortive or abnormal transformation events in BRK cells, such that cell lines can only be obtained by co-cultivation of 20 or more foci (Graham *et al.*, 1978; Ruben *et al.*, 1982). Unexpectedly *hr*1 and *hr*2 which are not cold-sensitive for lytic infection in HeLa cells were found to transform rat embryo cells at 38.5°C but not at lower temperatures; this is the transformation phenotype exhibited by the group I *hr*cs mutants (Ho *et al.*, 1982). Temperature shift experiments indicated that *hr*1 can initiate transformation at 32.5°C but that the event is only manifested at 38.5°C because the lesion affects a gene which is required to maintain the transformed phenotype (Ho *et al.*, 1982). Essentially the same observations were made after *hr*1 infection of a cloned rat embryo fibroblast line (CREF; Babiss, Ginsberg and Fisher, 1983*a*).

The mutation in *hr*1 was identified as a single base pair (bp) deletion at nucleotide 1055 in the region transcribed uniquely into the E1a 13S mRNA, such that the 289 amino acid protein terminates prematurely with ten aberrant C-terminal amino acids (Ricciardi *et al.*, 1981). Babiss, Fisher and Ginsberg (1984) constructed insertion and deletion mutants in or around the *Xma*I/*Sma*I site at 1007 in the 13S unique region and found evidence which indicated that the sequence between nucleotides 1001 to 1009 (coding for glutamine-histidine-proline, residues 148 to 150) must be intact for preservation of some transforming capability, albeit cold-sensitive. A restriction enzyme linker insertion mutant (H5 *in*500) at the *Sma*I site which was made independently by Carlock and Jones (1981) was found to be transformation-defective, but there is no information as to whether this mutant is cold-sensitive for transformation.

Montell and co-workers have used oligonucleotide-directed mutagenesis techniques to construct mutant viruses which have E1 alterations that allow expression of only the 289 (Ad 2/5 *pm*975) or 243 (*dl*1500) amino acid proteins from the E1a region (Montell *et al.*, 1983*a*, 1984). From the efficient but cold-sensitive transforming ability of *dl*1500 and the inefficient, temperature-independent transforming activity of *pm*975 it would seem that the 243 amino acid protein stimulates initiation events which are subsequently maintained by the 289 amino acid protein. That this is almost certainly an oversimplification is suggested by the inability of wild-type virus to transform as efficiently as *dl*1500 and by the observation that the ability of the 243 amino acid protein to stimulate the growth of transformed cells to form macroscopic colonies in soft agar is also cold-sensitive (Ho *et al.*, 1982; Montell *et al.*, 1984). Although the cold-sensitive transformed state induced by the 243 amino acid protein argues against the hypothesis of Babiss *et al.* (1984), mutant *dl*105 (Babiss *et al.*, 1984), which has an in-frame 13S unique region deletion of 69 bp and should therefore make a complete 243 amino acid protein and an internally deleted 289 amino acid protein, is absolutely defective for transformation. This is a further indication of the complexity of the roles of the E1a proteins in transformation.

In common with the results from DNA transfection experiments, transformation studies with E1 mutants have shown that BRK cells and CREF cells that express incomplete E1a proteins or which do not appear to express E1a proteins are fibroblastic (Ruben *et al.*, 1982; Babiss *et al.*, 1984), although *hr* transformed rat embryo cells show the typical epithelioid morphology of adenovirus transformed cells (Ho *et al.*, 1982).

Classical mutants and E1 mutants of Ad 12 constructed *in vitro* have been isolated and tested for their transformation and tumourigenic potentials (Gallimore *et al.*, 1984). Mutants H12 *hr*700 and H12 *hr*701 appear to map in the E1a region along with H12 *in*600, which is a linker insertion mutant at nucleotide 1004 in the unique region of the 13S message. Although H12 *hr*700 and H12 *hr*701 are only partially transformation-defective, H12 *in*600 is absolutely transformation-defective. The phenotype of H12 *in*600 might indicate that the functions of the Ad 12 and Ad 5 E1a regions in transformation are subtly different. Alternatively this mutant might be the Ad 12 equivalent of *dl*105, which is also completely transformation-defective. Further mutants in this segment of the E1a region are required to resolve this issue.

The classical group II Ad 5 E1b host range mutants exhibit two phenotypes: cold-sensitive for lytic infection and absolutely defective for lytic infection in HeLa cells. Unlike the group I mutants there is no further distinction between them because they all transform extremely poorly at any temperature and regardless of temperature shifts following infection (Ho *et al.*, 1982). A number of these mutants have been broadly mapped to those sequences of the E1b 22S message which are spliced out of the 13S mRNA. The mutant *hr*6 is now known to contain two transversions in this region at nucleotides 2347 and 2947, both of which substitute phenylalanine for cysteine, while *hr*cs13 contains a transversion at nucleotide 2469 that introduces a stop codon in the frame of the 496 amino acid protein (J. Williams, personal communication). Mutants of Ad 12 (H12 *hr*703 and H12 *hr*704) the phenotypes of which parallel those of the Ad 5 group II *hr* mutants have also been tentatively mapped to the sequences encoding the largest E1b protein (Gallimore *et al.*, 1984). The disparity between these virus transformation studies and the DNA transfection studies which indicate that the 496 amino acid protein is dispensable for transformation was partly resolved by Rowe and Graham (1983) who showed that *hr*6 and *hr*50 viral DNA transformed BRK cells as efficiently as wild-type DNA. The best interpretation of these observations is that initiation of virion-mediated transformation requires a functional E1b 496 amino acid protein but that transformation by transfected DNA operates through a different mechanism independent of this protein (Rowe and Graham, 1983). Once again, however, the use of different cells in the transformation assay presents a different picture because Babiss and Ginsberg (1984) reported that Ad 5 mutants with deletions or insertions around nucleotides 2805 and 3329 generate completely transformed foci at reduced frequency during virion-mediated transformation of CREF cells and that the efficiency drops progressively as the mutations approach the N-terminus of the 496 amino acid protein. Overall it appears that the phenotypes of the group II mutants define a function for the E1b 496 amino acid protein that is required in initiation of transformation, but the simple hypothesis that only this protein is required for initiation whilst the E1a proteins function exclusively in maintenance is probably wrong, because rare clonal *hr*cs13 transformants show a cold-sensitive phenotype similar to that of E1a *hr*cs transformants (Ho *et al.*, 1982) and the contribution of the E1b 176 amino acid protein is not considered.

Mutants of Ad 12 with reduced oncogenicity were isolated as early as 1968 (Takemori, Riggs and Aldrich, 1968, 1969). These cytolytic (*cyt*) mutants, so called because they elicit extensive cellular destruction and produce large clear plaques (wild-type plaques have fuzzy edges), are heterogeneous for growth in KB cells. The replication defect seems to be caused by lesions either in a viral endonuclease gene (*cyt*61) or in a viral antinuclease gene (*cyt*52, 62, 68 and 70: Ezoe, LaiFatt and Mak, 1981). The mutants *cyt*52, 62, 68 and 70 map to the E1b 163 amino acid protein (LaiFatt and Mak, 1982) and have been found to be partially or severely transformation-defective (Mak and Mak, 1983). Fukui *et al.* (1984) found that mutants with insertions or deletions at or around nucleotide 1597 in the Ad 12 E1b 163 amino acid protein could be isolated in normal human embryo kidney cells, i.e. this protein is dispensable for lytic infection, but these mutants were found to have much reduced transformation capacity and clonal transformed isolates failed to form colonies in soft agar. A series of large plaque (*lp*), i.e. cytocidal, mutants of Ad 2 have also been mapped to the E1b 175 amino acid protein and these too exhibit transformation defectiveness of varying degree (Chinnadurai, Chinnadurai and Brusca, 1979; Chinnadurai, 1983). Mutations affecting the E1b 175 amino acid protein or perhaps more specifically its N-terminus reduce transformation efficiency and impair colony formation in soft agar, leading to the supposition that this protein, in common with the E1a 289 amino acid protein, has a maintenance function in transformation.

In spite of the elegant approaches (Montell *et al.*, 1983a, 1984) being adopted in the analysis of the adenovirus genes involved in transformation, and the extraordinary insights afforded by these studies, a complete understanding of the roles of the E1 proteins in the initiation and maintenance of transformation does not look imminent. Complex interactions between the E1a and E1b regions and perhaps other viral genes or functions appear increasingly likely in transformation. In this respect it is worth noting that mutations in two genes outside the E1 region, i.e. *ts*36 and *ts*149 (in the viral DNA polymerase gene) and *ts*125 (in the single-strand DNA binding protein gene), can affect (negatively and positively respectively) the level of transformation (Ginsberg *et al.*, 1974; Williams, Young and Austin, 1974). The group I *hr* mutants fail to replicate their genomes in HeLa cells (Harrison *et al.*, 1977; Lassam, Bayley and Graham, 1978; Jones and Shenk, 1979b; Carlock and Jones, 1981; Solnick and Anderson, 1982; Ho *et al.*, 1982), in contrast to the group II mutants which are replication-proficient, but the significance of this deficiency to their defective transformation phenotype is unclear. Additionally, analysis of early mRNA synthesis in HeLa cells infected with *hr* mutants has shown that the E1a 289 amino acid protein regulates the expression of the other early viral regions, i.e. E1b, E2, E3 and E4 (Berk *et al.*, 1979; Jones and Shenk, 1979*a*, *b*; Montell *et al.*, 1984). While *hr*1 fails to express these regions and Montell *et al.* (1984) have ascribed this regulatory function to the 46 amino acids unique to the E1a 289 amino acid protein, other mutants such as H5 *in*500 (Carlock and Jones, 1981) and in particular *hr*440 (Solnick and Anderson, 1982), which makes only a truncated E1a protein representing the N-terminal 140 amino acids, are only partially impaired in this function, suggesting that the functional domain is within the common N-terminal residues. There is ample evidence now that E1a can exert its regulatory capacity on non-viral genes, and the evidence for this and the basis of the control is discussed in the next section. So far the regulatory aspects of only the E1a region have been considered, but E1b is required to turn off host protein synthesis in lytic infection (Babiss and Ginsberg, 1984); whether this could operate selectively in transformed cells to affect the expression of specific cellular genes is not known.

E1 gene functions and oncogenicity

The ability of Ad 12 to complement Ad 5 E1a and E1b mutants in cells in which their growth is normally restricted is evidence of the functional relatedness of the products encoded by the transforming sequences of these viruses (Williams, Ho and Galos, 1981; Rowe and Graham, 1981; Brusca and Chinnadurai, 1981; Shiroki *et al.*, 1982). The DNA sequences of the Ad 5 and Ad 12 E1 regions and the amino acid sequences of their E1 proteins show striking homologies (Sugisaki *et al.*, 1980; van Ormondt *et al.*, 1980*a*; Bos *et al.*, 1981; Kimura *et al.*, 1981; van Ormondt and Hesper, 1983). Despite their similarities the divergence of the E1 proteins must be largely sufficient to account for the differential oncogenicity of the viruses and the differences in the tumourigenic capacities of Ad 12 as opposed to Ad 5 (or Ad 2) transformed cell lines.

In experiments to define the reasons for these differences in oncogenicity Bernards *et al.* (1982) transfected BRK cells with hybrid plasmids containing either Ad 12 E1a/Ad 5 E1b or Ad 5 E1a/Ad 12 E1b. Tumourigenicity experiments conducted in athymic nude mice showed that cells transformed with the Ad 5 E1a/Ad 12 E1b construct were as oncogenic as Ad 12 E1a/Ad 12 E1b transformants, suggesting that the E1b region determines the neoplastic phenotype. Following up this result Bernards *et al.* (1983*a*) tested the tumour-igenicity in nude mice of cells transformed by plasmids which contained the E1a regions from both Ad 5 and Ad 12 but which carried either the Ad 12 E1b 163 amino acid protein and the Ad 5 E1b 496 amino acid protein or the Ad 12 E1b 482 amino acid protein and the Ad 5 E1b 176 amino acid protein. The conclusion from this study was that the difference in oncogenic potential between Ad 5 and Ad 12 was determined by the large E1b protein but that expression of the small E1b protein from either virus was required for the manifestation of oncogenicity. A role for the small E1b protein in tumour production has been identified by the reduced oncogenicity of the Ad 12 *cyt* mutants (Takemori *et al.*, 1968, 1969; LaiFatt and Mak, 1982; Mak and Mak, 1983) and more recently by the completely non-tumourigenic Ad 12 E1b 163 amino acid protein mutants constructed by Fukui *et al.* (1984). Although Jochemsen *et al.* (1982) found that BRK cells transformed by Ad 12 DNA fragments containing the E1a region and the coding sequences for the E1b 163 amino acid protein were not tumourigenic in nude mice, Gallimore *et al.* (1984) found that even Ad 12 E1a transformed BRK cell lines (four out of seven) were tumourigenic in the syngeneic host. They also concluded that while the Ad 12 E1b 482 amino acid protein is dispensable for transformation of primary rat cells by naked DNA, it must enhance the survival or growth rate of the transformed cells *in vivo*.

Similar results were obtained by Rowe and Graham (1983) and Rowe *et al.* (1984), but they also demonstrated that the same spectrum of tumourigenicity was exhibited in the syngeneic host by Ad 5 transformed hamster cells expressing only the E1a proteins, the E1a proteins and the E1b 176 amino acid protein, or all E1 proteins. In contrast to the rat system, therefore, it seems that the E1b proteins do not contribute to the survival and growth of transformed cells in the hamster. Although Ad 12 E1b deletion mutants would not be expected to produce tumours in newborn hamsters, since the E1b 482 amino acid protein is required for virus-mediated transformation, a hybrid Ad 12 virus the E1a region of which had been replaced by the Ad 5 E1a region would be expected to be oncogenic. Viruses have been constructed in which the E1b region or the entire E1 region of Ad 5 has been replaced by the analogous Ad 12 region (Bernards *et al.*, 1983*c*, 1984), but the failure of these hybrids to induce tumours in newborn hamsters may be attributable to the fact that hamster cells are permissive for Ad 5.

A critical role for the E1a region in tumourigenicity was shown by the inability of BRK cells transformed by plasmids lacking the E1a region but containing the SV40 enhancer to produce tumours in nude mice, in spite of high-level expression of normal E1b mRNAs (Bos *et al.*, 1983). Similarly *hr*1 transformed BRK cells, which surprisingly expressed normal levels of E1b proteins, were found to be unable to produce tumours in athymic nude mice (Ruben *et al.*, 1982), demonstrating that expression of the E1b region *per se* is not sufficient for tumourigenicity. The fact that Ad 12 transformed rat cells are tumourigenic in athymic nude mice and the syngeneic host, while Ad 5 E1 or Ad 5 E1a/Ad 12 E1b transformed rat cells are only oncogenic in immunodeficient nude mice has been linked to escape of T-cell-mediated cytolysis through reduction of major histocompatibility complex (MHC) class I gene expression by Ad 12 (Bernards *et al.*, 1983*b*; Schrier *et al.*, 1983). The observations of Bos *et al.* (1983) described above may identify another role for the E1a region in tumourigenicity because the level of MHC antigen expression should be immaterial to nude mice which are deficient in T-cell-mediated cytolysis.

PRODUCTS OF THE Ad 12 TRANSFORMING REGION

This section will concentrate mainly on the Ad 12 E1 proteins but will include some discussion of the E1 proteins of other serotypes for additional information. Since a number of groups have presented data to confirm that the only region of the Ad 12 genome required for cellular transformation is the E1 region, it is reasonable to suppose that the limited number of proteins encoded by this region of DNA are able to generate all of those changes in cellular morphology and biochemistry which are associated with transformation. Some of the properties of the Ad 12 E1 proteins are summarized in Table 2.

Initial investigations into the polypeptides encoded by the E1 region of Ad 12 DNA relied upon the fact that the viral proteins could be immunoprecipitated from lysates of infected or transformed cells with antisera from syngeneic rats or hamsters bearing tumours induced either with Ad 12 or with Ad 12 transformed cells. The immunoprecipitated proteins were then electrophoresed on polyacrylamide gels (PAG) run in the presence of SDS. Because the level of protein expression is so low (see below) it was essential to label the cells with radioactive amino acid (generally [^{35}S]methionine) to increase the sensitivity of the procedure. The first set of results obtained by this technique was rather confusing, probably because the various groups of investigators used different SDS-PAGE systems which gave rather different molecular weights for the viral proteins. Shiroki *et al.* (1979*a*, 1980) demonstrated, using partially transformed rat cells, that the *Acc*IH fragment (0–4.7 map units) of Ad 12 DNA coded for a group of acidic proteins (pI 5–5.5) of molecular weights 35 000–40 000. Using different cell lines they also showed that the region of the Ad 12 genome from 4.7 to 6.8 map units coded for a small neutral protein (molecular weight 10 kD, pI 6.4). All of these polypeptides were also detected in Ad 12 infected cells. In a more extensive study van der Eb *et al.* (1979) found that the three major Ad 12 proteins, expressed in rat cells transformed with the *Eco*RIC DNA fragment, had molecular weights of 41, 19 and 60 kD, whilst in cells transformed with the *Hind*IIIG fragment (0–6.8 map units) only the 41 and 19 kD proteins could be detected. They concluded that the 41 kD family of proteins was encoded by the left-hand end of the Ad 12 genome (E1a) whilst the 60 kD protein was encoded by the right-hand end of the E1 region (E1b). The 19 kD proteins also came from the E1b region. Later studies (for example Byrd *et al.*, 1982*a*; Paraskeva *et al.*, 1982) have confirmed these observations. Similar proteins are expressed in cell lines transformed with fragments of Ad 2 and Ad 5 DNA, although the precise molecular weights vary somewhat. For example, van der Eb *et al.* (1979) found the major Ad 5 E1a protein to be of molecular weight 52 kD, and the major E1b proteins to be 19 and 65 kD.

An alternative approach to the problem of the identification of the proteins coded by Ad 12 E1 DNA has been to examine the proteins synthesized after translation of viral mRNA in a cell-free translation system (Esche *et al.*, 1980; Jochemsen *et al.*, 1982; Esche and Siegmann, 1982; Saito *et al.*, 1983). Although the results are slightly equivocal due to the multitude of protein bands detected, these data tend to support the view that E1a codes for a family of proteins of molecular weight about 35–40 kD and E1b codes for three major proteins of molecular weights 17, 19 and 60 kD.

As the nucleotide sequence of Ad 12 E1a DNA has been determined (see above) a comparison can be made between the predicted polypeptides and those observed after immunoprecipitation and cell-free translation of Ad 12 mRNAs. These comprise 266 and 235 amino acids and correspond to the 13S and 12S mRNAs respectively (Perricaudet *et al.*, 1980). On the basis of the nucleotide

Table 2. *Some properties of the Ad 12 E1 proteins*

	Protein		
	E1a proteins	Smaller E1b protein	Larger E1b protein
Molecular weight (a) By SDS-PAGE (b) Amino acid number	Two major proteins, 40 kD 235 and 266	19 kD 163	52 kD 482
Post-translational modification	Phosphorylated	Acylated	None identified in Ad 12 but the 496 amino acid protein is phosphorylated in Ad 5
Protein–protein interactions	Histones H1 and H4 (*in vitro*);cytoskeletal components	–	Dimerizes; binds to p53. In Ad 5 the 496 amino acid polypeptide interacts with an E4 protein
Subcellular localization	Nucleus, cytoplasm, cytoskeleton	All membranes (nucleus, mitochondrion, lysosomes, and microsomes). Also a cell surface protein	Nucleus, cytoplasm

Enzymic activity			
(a) Protein kinase	—	—	Probably negative but positive in Ad 5
(b) ATPase	—	—	—
Requirement in trans-formation			
(a) By DNA	Yes	No	No
(b) By virus	Yes	Yes	Yes
Other properties	Earliest proteins expressed in lytic infection. Controls expression of other early regions. Induces heat-shock protein (HSP70). Represses Class I MHC. Activates transcription of some eukaryotic genes in plasmids *in vivo* and *in vitro*	Possibly responsible for tumour-specific transplantation immunity	—

sequences it is predicted that both the 266 and 235 amino acid E1a proteins should be very acidic, which is in accord with the observations of Shiroki et al. (1980) and Grand and Gallimore (1984). However, most of the acidic residues are grouped in the N-terminal halves of the proteins – for example for the 13S message the net charge for the N-terminal 133 amino acids is about -31 whereas for residues 134–266 it is only -5; this may be of significance in interactions with other polypeptides or in some other function.

The molecular weights of the E1a proteins calculated from the nucleotide sequences are considerably less than those observed on SDS-PAGE after immunoprecipitation. Sugisaki et al. (1980) have calculated values of 26 021 and 29 658 for the proteins encoded by the 12S and 13S messages respectively whereas the major E1a proteins have observed molecular weights of about 40 kD. It has been suggested that this discrepancy is due to the high proline content of the proteins, but it seems more likely that it is indicative of incomplete denaturation under the conditions of SDS-PAGE. It has previously been noted that another small acidic protein, calmodulin, runs anomalously on polyacrylamide gels due to the extreme stability of its tertiary structure in the presence of calcium.

Amino acid sequences have been predicted for the E1 proteins of a number of adenovirus serotypes on the basis of nucleotide sequences (see above). Ad 2 and Ad 5 E1 polypeptides are very similar indeed (van Ormondt et al., 1980a; Bos et al., 1981; Gingeras et al., 1982), with variations in the amino acid sequences of the E1 proteins being limited to six conservative substitutions in the E1a polypeptides and a limited number of differences at the N termini of the E1b proteins. Differences between Ad 5 (or Ad 2), Ad 7 and Ad 12 sequences, however, are much more pronounced. Van Ormondt et al. (1980b) have calculated the overall homology for the E1a regions of these three viruses to be about 50% for both the coding and non-coding regions. The authors did note, however, that in some areas of the proteins the similarities were much more marked. A comparison of the predicted E1b protein sequences of Ad 5, Ad 7 and Ad 12 (van Ormondt and Hesper, 1983) revealed a level of homology of 55–60% when two serotypes were compared and 45% when the three serotypes were compared. Areas of non-coding DNA diverged much more radically, as did those sequences coding for the C terminus of the smaller E1b protein and amino acids 23–120 of the larger E1b protein.

Not surprisingly, the nucleotide sequences of E1a DNA from Ad 2,

5, 7 and 12 show considerable homology; however, recently it has been reported that the proteins encoded by the v-*myc*, v-*myb* and human c-*myc* and chicken c-*myb* oncogenes have similarities to the proposed sequences of the adenovirus E1a proteins (Ralston and Bishop, 1983). Although the overall homology is not very high (15–21% of identical amino acids in paired comparisons between proteins from different families there are short runs of amino acids where the similarities are quite striking: for example amino acid residues 115–126 in Ad 12 E1a correspond very closely to an area of v-*myc*. Ralston and Bishop (1983) have concluded that these homologies between the nucleotide sequences of E1a and *myc* suggest a common ancestral sequence, indicating either a common ancestry for adenovirus and eukaryotic cells or, probably more convincingly, recombination between ancestral viral sequences and cellular *myc* sequences.

Functions of the Ad 12 tumour (T) antigens

Despite the expenditure of considerable time and effort by a large number of workers our knowledge of how the three major adenovirus E1 proteins transform cells in culture or cause tumours in rodents is still extremely limited.

Experiments in which attempts have been made to study proteins directly have been severely hampered by the very low levels of expression in transformed or even infected cells (probably 0.01–0.001% of the total cellular protein). Recently attempts to pinpoint the subcellular localization of the E1 proteins have met with some success (see below), but a rather more profitable area of study has been the investigation of the properties of mutant and reconstructed viruses in which specific changes have been introduced into the E1 region DNA. Although many of these experiments have been performed with Ad 5 it should be possible to extrapolate from these results to the Ad 12 system. As already mentioned a number of studies have shown that the products of the Ad 5 E1a region exert a controlling influence over the expression of other adenovirus mRNAs (E1b, E2, E3, E4 and L1) (Berk *et al.*, 1979; Jones and Shenk, 1979b; Nevins, 1981; Svensson and Akusjarvi, 1984). The mechanism by which this regulation occurs is still far from clear, but it has been suggested that the E1a protein is able to increase the level of expression of other adenovirus genes through its interaction with, and inactivation of, 'cellular negative-acting factors' (Nevins *et al.*, 1984). It has also been observed that after infection of HeLa cells with

Ad 5 there is a substantially increased synthesis of the 70 kD human heat-shock protein (Nevins, 1982). Nevins (1982) and Nevins *et al.* (1984) have argued that activation of this heat-shock gene and of adenovirus E1b region therefore operate through a similar mechanism; whilst this is possible it seems more likely, at present, that the severe stress administered to the cell by the onset of viral infection would be sufficient to trigger expression of the heat-shock gene. It is now generally accepted that production of these heat-shock proteins does not occur only in response to elevated temperature but to many other sorts of cellular upheaval. Recent data of Svensson and Akusjarvi (1984) have, however, suggested that the E1a 13s mRNA product is capable of stimulating expression of the rabbit β-globin gene as well as adenovirus genes in transfected cells.

A somewhat different mechanism of E1a control of viral gene expression has been proposed by Persson *et al.* (1981) and Katze *et al.* (1982) on the basis of a study of the treatment of Ad 5 infected cells with various drugs which inhibit protein synthesis. They concluded that a viral protein encoded by E1a (or E1b) 'inactivates a cellular gene product which de-stabilizes or inhibits transport of early viral mRNAs'.

The observations by Schrier *et al.* (1983) and Bernards *et al.* (1983*b*) that expression of the heavy chain of the class I MHC complex was very much reduced in rat cells transformed with Ad 12 but not with Ad 5 may be more important and have relevance to malignancy in this system. The suppression of the rat class I MHC was at the level of mRNA synthesis and was strictly correlated with the presence of the Ad 12 E1a genes (see below for a more detailed discussion of these results).

Rossini (1983), using microinjection of cloned segments of viral DNA into hamster cells, has observed that expression of E1a stimulates the production of the viral DNA binding protein transcribed from the early promoter but inhibits synthesis from the late promoter of the E2a region. It is suggested that these two opposing functions are encoded by the different E1a proteins (Rossini, 1983). Further research is obviously required to resolve all of these contradictory views, but an additional problem arises when considering cellular transformation by adenovirus because it has certainly not been proven that the viral proteins perform similar functions in the transformed cell to those necessary for lytic infection.

Adenovirus infection induces a whole spectrum of changes in cellular metabolism that contribute to the uncoupling of the cell cycle from normal control (Braithwaite, Murray and Bellett, 1981; Cheetham and Bellett, 1982; Braithwaite *et al.*, 1983) and expression of the E1a region has been implicated as the main culprit in this subversion (Braithwaite *et al.*, 1983). At present it is not clear what significance these changes and others such as chromosomal alterations (McDougall, Dunn and Gallimore, 1974; Murray *et al.*, 1982) have to transformation and oncogenicity.

As described above Montell *et al.* (1983*a*, 1984) and Svensson and Akusjarvi (1984) have constructed Ad 2 mutants which express only one of the E1a proteins. The mutant expressing only the 13S message had a phenotype indistinguishable from wild-type, but both E1a proteins were required to produce a fully transformed phenotype (Montell *et al.*, 1984). Svensson and Akusjarvi (1984) found that the product of the 12S message was incapable of stimulating expression of other genes. Although these results suggest that all of the replication functions of E1a reside in the 13S message it is obvious that the smaller E1a protein must play an important, although as yet undetermined, role in infection as the structure of E1a (i.e. the two spliced mRNAs with most of their sequences in common) has been retained in a number of different adenovirus serotypes which diverged, evolutionarily, a considerable time ago.

Direct studies on adenovirus E1 proteins have not, so far, led to any significant advances in our knowledge of their mode of action in infection or transformation. However, a number of observations have been made which are of interest. The levels of expression of the T antigens in Ad 12 transformed cell lines appear to be within strictly defined limits. It has been observed that in a large number of rat embryo brain and BRK lines transformed with the *Acc*IH or *Eco*RIC fragments of DNA that the amounts of E1a proteins expressed were virtually identical; similarly in all of the *Eco*RIC transformed lines examined the level of expression of the larger E1b protein was constant (R. J. A. Grand and P. H. Gallimore, unpublished). A large number of human embryo retina and human embryo kidney cell lines were also found to contain the same amounts of Ad 12 E1a and E1b proteins (R. J. A. Grand and P. H. Gallimore, unpublished). In many of these cells, however, the number of Ad 12 gene copies varied considerably (for example from 1 to over 100). It is reasonable to suppose, therefore, that for a transformed cell line to become established a low level of expression of the adenovirus T-antigens is essential. Presumably if this is not the case metabolism of the cell is so disrupted that cell death occurs.

van der Eb *et al.* (1979) reported that in Ad 5 transformed cells the E1a protein was phosphorylated and Paraskeva *et al.* (1982) observed that the E1a protein was also phosphorylated in Ad 12 infected cells. We have recently demonstrated that the products of both the Ad 5 and Ad 12 12S and 13S mRNAs are phosphorylated in human and rat cells transformed with fragments of adenovirus DNA as well as in Ad 5 and Ad 12 infected human embryo kidney cells (R. J. A. Grand, C. Fennell and P. H. Gallimore, unpublished). Spindler, Rosser and Berk (1984) have suggested, on the basis of studies using protein produced by mutant Ad 2 virus, that the major phosphorylation sites for the E1a proteins occur in the C-terminal 70 amino acids. There is no evidence in the literature that any of the other Ad 12 E1 proteins are phosphorylated, but it is now firmly established that the Ad 5 E1b 496 amino acid polypeptide is a phosphoprotein (Levinson and Levine, 1977; Sarnow *et al.*, 1982*a*; Malette, Yee and Branton, 1983), with at least three phosphorylation sites – two at serine residues and one at a threonine residue (Malette *et al.*, 1983). It is not clear whether the phosphate groups on the Ad 12 E1a proteins and the Ad 5 E1a and larger E1b proteins turn over rapidly in response to cellular changes or whether phosphorylation occurs immediately after protein synthesis and remains unchanged. Obviously if the former case applies phosphorylation is likely to be of considerably more significance for the functioning of the protein. Preliminary data suggest that both of the Ad 5 and Ad 12 major E1a polypeptides are present with a constant level of phosphorylation in infected and transformed cells (R. J. A. Grand, C. Fennell and P. H. Gallimore, unpublished), perhaps implying that the phosphate moieties do not turn over rapidly. At present there is no evidence to suggest what enzymes are implicated in the phosphorylation process, but it is probable that there is no involvement of cAMP-dependent protein kinase in the modification of either the Ad 5 E1b 496 amino acid protein (Malette *et al.*, 1983) or the Ad 12 E1a proteins (R. J. A. Grand and P. H. Gallimore, unpublished).

The Ad 12 E1b 163 amino acid protein, which is not essential for replication but is required for transformation (Fukui *et al.*, 1984), is also post-translationally modified in both infected cells and in rat cells transformed with the E1 region of Ad 12 DNA, as it contains lipid which is covalently bound to either a serine or threonine residue (R. J. A. Grand, C. Roberts and P. H. Gallimore, unpublished). This acylation is probably of importance in anchoring the protein in the cell membrane (see below) and has been shown to occur in other membrane-bound viral proteins: for example, the pp60$^{v\text{-}src}$ transforming protein of Rous sarcoma virus (Sefton *et al.*, 1982), that proportion of SV40 large T-antigen which is membrane bound (Klockmann and Deppert, 1983*a*, *b*), and a number of other viral structural proteins.

Several studies have recently been undertaken to examine the subcellular localization of the adenovirus E1 proteins in transformed and infected cells. In Ad 12 transformed rat and human cells the E1a proteins and the larger E1b protein are present in both the nuclei and the cytoplasm in approximately equal proportions whereas the smaller E1b protein is largely membrane-bound, being present mainly in the microsomal fraction but also in the nuclear, mitochondrial and lysosomal fractions (Grand and Gallimore, 1984). Similarly in Ad 5 infected cells the E1a proteins and the E1b 496 amino acid protein are located in both nucleus and cytoplasm (Feldman and Nevins, 1983; Rowe, Graham and Branton, 1983*a*; Yee *et al* ., 1983) and the 176 amino acid E1b protein is membrane associated (Rowe *et al.*, 1983*a*). Persson, Katze and Philipson (1982) have also purified the smaller Ad 2 E1b protein to homogeneity from a membrane fraction of Ad 2 infected HeLa cells.

Labelling of proteins on the surface of Ad 12 transformed cells leads to incorporation of radioactivity into the 163 amino acid E1b polypeptide. It has therefore been concluded that this protein is responsible for cell surface T-antigen activity in transformed cells, as none of the other E1 proteins was labelled (R. J. A. Grand, C. Roberts and P. H. Gallimore, unpublished). This observation confirms the findings of Fohring *et al.* (1983) who came to a similar conclusion on the basis of a study of cytolytic T-cell killing of a number of different Ad 12 transformed cell lines. These workers found that the only Ad 12 E1 protein shared by all cell lines killed in cytolytic assays was the 163 amino acid polypeptide.

Many proteins exert a regulatory influence on biochemical processes by interacting with (or binding to) controlling enzymes. In view of this some preliminary attempts have been made to examine whether the adenovirus E1 proteins form complexes with cellular components. Sarnow *et al.* (1982*a*) observed that when the 496 amino acid E1b protein was immunoprecipitated from Ad 5 transformed mouse cells a 53 kD polypeptide was co-precipitated. This protein was shown to be identical to the p53 present in cells transformed by SV40 (Lane and Crawford, 1979; Linzer and Levine, 1979; Sarnow *et al.*, 1982*a*). It is now apparent that p53 levels vary considerably in non-transformed cells at different times during the cell cycle, reaching a maximum in late G1 (Milner and Milner, 1981; Reich and Levine, 1984). In transformed cells, however, regulation of p53 expression appears to be altered and the steady-state levels are much higher. This is probably a consequence of the fact that the half-life of p53 is very different in SV40 transformed and untransformed cells, being about 24 hours and 0.5 hours respectively (Oren, Maltzman and Levine, 1981; Reich and Levine, 1984). It appears that the cellular protein is stabilized by its interaction with SV40 large T-antigen (Oren *et al.*, 1981). Reich and Levine (1984) have hypothesized that the stimulation of DNA synthesis observed after microinjection of SV40 large T-antigen (Tjian, Fey and Graessman, 1978) may be due to an increase in p53 stability caused by its binding to the viral protein. They concluded that 'the altered control of p53 expression would have a key role in the altered control of cell division in the transformed state'. Presumably a similar mechanism could apply in adenovirus transformed cell lines where the 496 amino acid E1b protein in Ad 5 (Sarnow *et al.*, 1982a) and the 482 amino acid E1b protein in Ad 12 (Brown, 1983; R. J. A. Grand and P. H. Gallimore, unpublished) could play similar stabilizing roles. As the p53 is located in the cell nucleus it has been suggested that the nuclear fraction of the E1b protein fulfils at least part of its function in transformation by binding to p53 whilst the cytoplasmic fraction plays an entirely different role (Grand and Gallimore, 1984).

Sarnow, Sullivan and Levine (1982b) observed that immuno-precipitation of the Ad 5 E1b 496 amino acid protein led to the co-precipitation of a 25 kD polypeptide besides the p53. This was later identified as an Ad 5 E4 protein and it has now been firmly established that the two adenovirus proteins form a complex *in vivo* (unpublished data quoted in Winchester (1983) and Babiss and Ginsberg (1984)). This complex may play an important role in the shut-off of host protein synthesis during lytic infection with Ad 5 (Babiss and Ginsberg, 1984), either by direct action or through an intermediary polypeptide. Obviously such a mechanism is not going to be of relevance in transformation events which are attributable to E1 proteins alone, but it does help to confirm that the viral T-antigens exert their influence through a wide variety of protein–protein interactions. The Ad 12 E1b 482 amino acid protein also forms complexes with itself, producing 100 kD dimers stabilized by intermolecular disulphide bonds (Grand and Gallimore, 1984). It appears that this dimerization occurs soon after protein synthesis in infected cells as well as in transformed cells, but at present it is not clear whether the monomer or dimer is the active form of the protein.

The Ad 12 E1a proteins have been shown to form very stable complexes with histones H1 and H4 *in vitro* (Grand and Gallimore, 1984) and rather weak complexes with H2A, H2B and H3. Whether any of these interactions occur *in vivo* is not clear at present, but it is certainly an appealing possibility that the adenovirus proteins could influence cellular events during both infection and transformation by interacting with the chromatin. The E1a proteins probably also inter-act with one or more cytoskeletal components, as there is an enhancement of these polypeptides in microtubules and 10 nm fila-ments in Ad 12 transformed rat cells (Grand and Gallimore, 1984). Rowe *et al.* (1983*a*) have also observed an enhanced level of E1a protein in a cytoskeletal fraction of Ad 5 infected cells. It is possible that this binding to the cytoskeleton could be one way in which the viral T-antigens could cause the alterations in cellular morphology which have been associated with transformation. In *in vitro* studies, however, no binding of viral proteins to actin, tropomyosin or calmodulin could be detected (Grand and Gallimore, 1984).

Collett and Erikson (1978) found that the tumour antigen (pp $60^{v\text{-}src}$) produced by Rous sarcoma virus was a membrane-associated protein kinase of high activity and specificity which was particularly efficient in the phosphorylation of tyrosine residues (Hunter and Sef-ton, 1980). Since that time attempts have been made to detect similar enzymic activities in T-antigens encoded by other viruses. It was observed that when immunoprecipitates from Ad 5 infected cells were incubated in the presence of ATP, phosphate groups were introduced onto serine residues of the immunoglobulin heavy chains or onto added histones (Lassam *et al.*, 1979; Branton *et al.*, 1981). This kinase activity appears to co-purify with the E1b 496 amino acid protein in a quite specific manner (Yee and Branton, 1983).

The results obtained with Ad 12 infected and transformed cells are, however, rather more equivocal. Raska, Geis and Fohring (1979) found that kinase activity directed against immunoglobulin heavy chains could be immunoprecipitated with Ad 12 tumour bearer serum, but also with some normal rat sera. Grand and Gallimore (1984) could detect little or no protein kinase activity associated with the E1 proteins in a large number of Ad 12 transformed cell lines. It is to be hoped that reasons for these contradictions will become apparent after further investigation.

A second enzymic activity which has been associated with DNA tumour virus T-antigens is the ability to hydrolyse ATP. Tjian and Robbins (1979) and Giacherio and Hager (1979) demonstrated an ATPase activity inherent to SV40 large T-antigen and Gaudray, Clertant and Cuzin (1980) showed a similar activity associated with polyoma large T-antigen. However, no ATPase activity could be detected in Ad 12 E1 proteins (Grand and Gallimore, 1984). It has been suggested that the larger Ad 5 E1b protein is ribosylated in infected cells (Goding, Shaw, Blair and Russell, 1983) as is the case for SV40 large T-antigen (Goldman, Brown and Khoury, 1981). However, using intact Ad 12 infected or transformed cells we have found that ribosylation does not occur in the Ad 12 system (R. J. A. Grand and P. H. Gallimore, unpublished). It is probable that in the Ad 5 study (Goding *et al.*, 1983) ribosylation only occurred after severe cellular disruption and therefore the process may not occur *in vivo*.

Recently, Ruley (1983) has shown that whilst BRK cells cannot be transformed with either the polyoma middle T-antigen gene or the T24 Ha-*ras*-1 gene, they can be transformed quite readily if Ad 2 E1a genes are co-transfected. It was concluded that in this situation the E1a genes or the proteins translated from them immortalized the cells in some way before transformation was accomplished by the Ha-*ras*-1 or middle T-antigen genes. Ruley (1983) concluded that the adenovirus E1 proteins express functions similar to oncogenes and act in such a way as to immortalize cells grown in culture. No evidence that the cells were immortal was presented in this paper. Whilst these results might be consistent with those studies which have shown that cells transfected with the *Acc*IH fragment of Ad 12 DNA show only an incomplete transformed phenotype (Shiroki *et al.*, 1979*b*; Gallimore *et al.*, 1983), definite conclusions will have to await further investigation.

FACTORS DETERMINING ADENOVIRUS TUMOURIGENICITY

The observations that adenovirus oncogenesis was restricted to the two subgroups A and B (Table 1) and that many serotypes were non-oncogenic, provided early evidence that one factor determining tumour induction was the genetic constitution of the adenovirus. Over the years a number of host-determined parameters which also influence adenovirus oncogenesis have been defined. These include:

(a) *The genetic constitution of the experimental animal.* Within a species, e.g. the mouse, inbred strains may be susceptible or resistant to adenovirus tumour induction (Yabe *et al.*, 1964; Allison, Berman and Levey, 1967). Yohn and his colleagues (Yohn *et al.*, 1965; Yohn, Funk and Grace, 1967) showed that female hamsters had an increased incidence of Ad 12 tumours compared with males and that the tumours in female hamsters developed with a shorter latent period and progressed more rapidly.

(b) *The host's age at the time of virus inoculation.* Tumours only appear in animals inoculated as newborns or within a few days of birth and the animals become refractory to adenovirus oncogenesis by 3–4 weeks of age (Yabe, Trentin and Taylor, 1962).

(c) *The immune status of the host.* Thymectomy (Kirschtein, Rabson and Peters, 1964; Yohn *et al.*, 1965) and immunosuppression induced by either anti-lymphocyte serum (Allison *et al.*, 1967) or steroids (Yohn, Funk and Grace, 1968) have been shown to elevate tumour incidence. Similarly, animals which were either naturally immunodeficient (Gallimore, McDougall and Chen, 1977; van den Elsen *et al.*, 1982) or immunosuppressed with anti-thymocyte serum (Gallimore, 1972; Harwood and Gallimore, 1975; Gallimore and Paraskeva, 1979) were found to be susceptible to tumour formation when inoculated with cell lines transformed *in vitro* by non-oncogenic adenoviruses. Newborn and very young animals must also be considered as immunologically immature hosts. Other factors influencing adenovirus oncogenesis include the virus dose inoculated (Yohn *et al.*, 1967) and the route of inoculation (Yabe *et al.*, 1963).

A number of approaches both *in vitro* and *in vivo* have been used to try to define a cellular phenotype that determines malignancy and to identify what host defence systems oncogenic viruses and malignant cell lines evade as the tumour develops. So far no *in vitro* growth parameter has been shown to correlate with a malignant phenotype (Gallimore *et al.*, 1977, 1979; Gallimore and Paraskeva, 1979). The recently reported observation that Ad 12 E1a but not Ad 5 E1a down-regulates Class I MHC antigens on transformed cells (Schrier *et al.*, 1983; Bernards *et al.*, 1983b) led these authors to hypothesize that it is through this mechanism that Ad 12 transformed cells escape immunological detection. Bernards *et al.* (1983b) reported that Ad 12 transformed rat cell lines were significantly more resistant to *in vitro* cytolysis induced by allogeneic cytotoxic T-cells than comparable Ad 5 transformed lines. Confirmation that Ad 12 transformed rat cell lines had reduced levels of Class I MHC has been provided by

Mellow *et al.* (1984). However, these authors found no correlation between sensitivity to allogeneic cytotoxic T-cells and tumour rejection. It seems likely that cytotoxic T-lymphocytes, which are MHC restricted (i.e. they recognize a virus transplantation antigen in the context of Class I MHC antigens: Zinkernagel and Doherty, 1974; McMichael *et al.*, 1977), are functional in the tumour rejection observed in tumour transplantation immunity experiments (Sjogren, Minowada and Ankerst, 1967; Ankerst and Sjogren, 1970). To date, experiments *in vivo* have failed to show any correlation between the levels of transplantation immunity induced by adenovirus transformed cell lines of widely differing tumourigenic potential (Gallimore and Paraskeva, 1979; Lewis and Cook, 1982). Similar findings have been reported from studies in which cytotoxic T-lymphocytes, generated as a secondary response to adenovirus tumour antigens, were used as effector cells *in vitro* (Raska and Gallimore, 1982). Compelling evidence has been presented from two animal species, the hamster and rat, that malignancy in the adenovirus system is defined by resistance to natural defence mechanisms, i.e. macrophages (Cook, Hibbs and Lewis, 1982) and natural killer cells (Cook *et al.*, 1982; Raska and Gallimore, 1982; Cook and Lewis, 1984; Sheil *et al.*, 1984). This spontaneous cell-mediated cytotoxicity was independent of Class I MHC antigens (Raska and Gallimore, 1982) and has been shown to develop at an early stage of immune ontogeny in both the rat (Raska and Gallimore, 1982) and hamster (Cook and Lewis, 1984). It remains to be resolved whether the mechanisms that inhibit tumour formation by non-oncogenic adenoviruses are the same as the responses in the syngeneic host which eliminate cell lines produced in tissue culture, by infection with non-oncogenic adenoviruses.

CONCLUSION

The expression of virtually the entire adenovirus genome (34–36 kb) in a permissive cell leads to virus replication and cell death. The expression of a limited number of adenovirus early genes, notably those contained within the E1 region, can bring about morphological transformation of non-permissive, semi-permissive and permissive cell types. Our present understanding of the structure of the E1 region has been refined by molecular and genetic studies. These approaches have so far failed to elucidate the precise biological/biochemical activity of individual E1 proteins. It is hoped that new developments, in particular the use of more specific antisera – i.e. monoclonal antibodies (Sarnow *et al.*, 1982*b*; Shiroki *et al.*, 1984) and antipeptide antibodies (Green *et al.*, 1983; Rowe *et al.*, 1983*b*; Yee *et al.*, 1983) – with expression vector systems (Fukui *et al.*, 1983; Scott *et al.*, 1984) will provide highly purified adenovirus transforming proteins that may be examined for protein/protein interactions (both cellular and viral) and for their intrinsic activity.

We are presently left with a number of very important questions unanswered:

1. Do adenovirus transforming genes immortalize mammalian cells or is immortalization a feature of subtle genetic changes occurring after the stabilization of the transformed phenotype?
2. How do adenovirus genes outside the E1 region influence transformation and malignancy?
3. Does the observed reduction of Class I MHC molecules in Ad 12 transformed cell lines represent the mechanism by which the oncogenic serotypes escape detection *in vivo*, or is this due to resistance to 'natural, unprimed defence mechanisms'?
4. Do adenovirus transforming proteins have similar or related functions to the transforming proteins of other DNA transforming viruses such as polyomavirus and SV40, or are they highly evolved cellular proto-oncogenes?

The authors are indebted to Miss Debbie Williams for her patience and secretarial expertise and to the Cancer Research Campaign (CRC) for supporting adenovirus research.

REFERENCES

ALESTROM, P., AKUSJARVI, G., PERRICAUDET, M., MATHEWS, M. B., KLESSIG, D. F. and PETTERSSON, U. (1980). The gene for polypeptide IX of adenovirus type 5 and its unspliced messenger RNA. *Cell*, **19**, 671–81.

ALLISON, A. C., BERMAN, L. D. and LEVEY, R. H. (1967). Increased tumour induction by adenovirus type 12 in thymectomized mice and mice treated with antilymphocyte serum. *Nature, London*, **215**, 185–7.

ANDERSON, C. W., SCHMITT, R. C., SMART, J. E. and LEWIS, J. B. (1984). Early region 1b of adenovirus 2 encodes two coterminal proteins of 495 and 155 amino acid residues. *Journal of Virology*, **50**, 387–96.

ANKERST, J. and SJOGREN, J. J. O. (1970). Cross-reacting tumour-specific transplantation antigens in tumours induced by adenoviruses 3, 14 and 12. *Cancer Research*, **30**, 1499–505.

BABISS, L. E., FISHER, P. B. and GINSBERG, H. S. (1984). Deletion and insertion mutations in early region 1a of type 5 adenovirus that produce cold-sensitive or defective phenotypes for transformation. *Journal of Virology*, **49**, 731–40.

BABISS, L. E. and GINSBERG, H. S. (1984). Adenovirus type 5 early region 1b gene product is required for efficient shut off of host protein synthesis. *Journal of Virology*, **50**, 202–12.

BABISS, L. E., GINSBERG, H. S. and FISHER, P. B. (1983*a*). Cold sensitive expression of transformation by a host range mutant of type 5 adenovirus. *Proceedings of the National Academy of Sciences, USA*, **80**, 1352–6.

BABISS, L. E., YOUNG, C. S. H., FISHER, P. B. and GINSBERG, H. S. (1983*b*). Expression of adenovirus E1a and E1b gene products and the *Escherichia coli XGPRT* gene in KB cells. *Journal of Virology*, **46**, 454–65.

BAKER, C. W. and ZIFF, E. B. (1981). Promoters and heterogeneous 5′ termini of the messenger RNAs of adenovirus serotype 2. *Journal of Molecular Biology*, **149**, 189–221.

BERK, A. J., LEE, F., HARRISON, T., WILLIAMS, J. and SHARP, P. A. (1979). Phenotypes of adenovirus 5 host range mutants for early mRNA synthesis. *Cold Spring Harbor Symposia on Quantitative Biology*, **44**, 429–36.

BERK, A. J. and SHARP, P. A. (1978). Structure of the adenovirus 2 early mRNAs. *Cell*, **14**, 695–711.

BERMAN, L. D. (1967). Comparative morphologic study of the virus induced solid tumours of Syrian hamsters. *Journal of the National Cancer Institute*, **39**, 847–901.

BERNARDS, R., DE LEEUW, M. G. W., VAESSEN, M.-J., HOUWELING, A. and VAN DER EB, A. J. (1984). Oncogenicity by adenovirus is not determined by the transforming region only. *Journal of Virology*, **50**, 847–53.

BERNARDS, R., HOUWELING, A., SCHRIER, P. I., BOS, J. L. and VAN DER EB, A. J. (1982). Characterization of cells transformed by Ad 5/Ad 12 hybrid early region 1 plasmids. *Virology*, **120**, 422–32.

BERNARDS, R., SCHRIER, P. I., BOS, J. L. and VAN DER EB, A. J. (1983*a*). Role of adenovirus types 5 and 12 early region 1b tumour antigens in oncogenic transformation. *Virology*, **127**, 45–53.

BERNARDS, R., SCHRIER, P. I., HOUWELING, A., BOS, J. L., VAN DER EB, A. J., ZIJLSTRA, M. and MELIEF, C. J. M. (1983*b*). Tumourigenicity of cells transformed by adenovirus type 12 by evasion of T-cell immunity. *Nature, London*, **305**, 776–79.

BERNARDS, R., VAESSEN, M. J., VAN DER EB, A. J. and SUSSENBACH, J. S. (1983*c*). Construction and characterization of an adenovirus type 5/adenovirus type 12 recombinant virus. *Virology*, **131**, 30–8.

BOS, J. L., JOCHEMSEN, A. G., BERNARDS, R., SCHRIER, P. I., VAN ORMONDT, H. and VAN DER EB, A. J. (1983). Deletion mutants of region E1a of Ad 12 E1 plasmids: effect on oncogenic transformation. *Virology*, **129**, 393–400.

BOS, J. L., POLDER, L. J., BERNARDS, R., SCHRIER, P. I., VAN DEN ELSEN, P. J., VAN DER EB, A. J. and VAN ORMONDT, H. (1981). The 2.2 kb E1b mRNA of human Ad 12 and Ad 5 codes for two tumour antigens starting at different AUG triplets. *Cell*, **27**, 121–31.

BOS, J. L. and TEN WOLDE-KRAAMWINKEL, H. C. (1983). The E1b promoter of Ad 12 in mouse Ltk⁻ cells is activated by adenovirus region E1a. *EMBO Journal*, **2**, 73–6.

BRAITHWAITE, A. W., CHEETHAM, B. F., LI, P., PARISH, C. R., WALDRON-STEVENS, L. K. and BELLETT, A. J. D. (1983). Adenovirus-induced alterations of the cell growth cycle: a requirement for expression of E1a but not of E1b. *Journal of Virology*, **45**, 192–9.

BRAITHWAITE, A. W., MURRAY, J. D. and BELLETT, A. J. D. (1981). Alterations to controls of cellular DNA synthesis by adenovirus infection. *Journal of Virology*, **39**, 331–40.

BRANTON, P. E., LASSAM, N. J., DOWNEY, J. F., YEE, S.-P., GRAHAM, F. L., MAK, S. and BAYLEY, S. T. (1981). Protein kinase activity immunoprecipitated from adenovirus infected cells by sera from tumour-bearing hamsters. *Journal of Virology*, **37**, 601–8.

BROWN, K. W. (1983). Protein changes associated with epithelial cell transformation. PhD Thesis, Birmingham University.

BRUSCA, J. S. and CHINNADURAI, G. (1981). Transforming genes among three different oncogenic subgroups of human adenoviruses have similar replicative functions. *Journal of Virology*, **39**, 300–5.

BYRD, P. J., BROWN, K. W. and GALLIMORE, P. H. (1982*a*). Malignant transformation of human embryo retinoblasts by cloned adenovirus 12 DNA. *Nature, London*, **298**, 69–71.

BYRD, P. J., CHIA, W., RIGBY, P. W. J. and GALLIMORE, P. H. (1982*b*). Cloning of DNA fragments from the left end of the adenovirus type 12 genome: transformation by cloned early region 1. *Journal of General Virology*, **60**, 279–93.

CARLOCK, L. R. and JONES, N. C. (1981). Transformation defective mutant of adenovirus type 5 containing a single altered E1a mRNA species. *Journal of Virology*, **40**, 657–64.

CHEETHAM, B. F. and BELLETT, A. J. D. (1982). A biochemical investigation of the adenovirus induced G1 to S phase progression: thymidine kinase, ornithine decarboxylase and inhibitors of polyamine biosynthesis. *Journal of Cellular Physiology*, **110**, 114–22.

CHINNADURAI, G. (1983). Adenovirus 2 lp^+ locus codes for a 19 kd tumour antigen that plays an essential role in cell transformation. *Cell*, **33**, 759–66.

CHINNADURAI, G., CHINNADURAI, S. and BRUSCA, J. (1979). Physical mapping of a large plaque mutation of adenovirus type 2. *Journal of Virology*, **32**, 623–8.

CHOW, L. T., BROKER, T. R. and LEWIS, J. B. (1979). Complex splicing patterns of RNAs from the early regions of adenovirus 2. *Journal of Molecular Biology*, **134**, 265–303.

COLLETT, M. S. and ERIKSON, R. L. (1978). Protein kinase activity associated with the Avian Sarcoma *src* gene product. *Proceedings of the National Academy of Sciences, USA*, **78**, 1695–9.

COOK, J. L., HIBBS, J. B. and LEWIS, A. M. JR (1982). DNA virus transformed hamster cell–host effector cell interactions: level of resistance to cytolysis is correlated with tumourigenicity. *International Journal of Cancer*, **30**, 795–803.

COOK, J. L. and LEWIS, A. M. JR (1984). Differential NK cell and macrophage killing of hamster cells infected with non-oncogenic and oncogenic adenoviruses. *Science*, **224**, 612–14.

DARBYSHIRE, J. H. (1966). Oncogenicity of bovine adenovirus type 3 in hamsters. *Nature, London*, **211**, 102.

DEURING, R., WINTERHOFF, U., TAMANOI, F., STABEL, S. and DOERFLER, W. (1981). Site of linkage between adenovirus type 12 and cell DNAs in hamster tumour line CLAC 3. *Nature, London*, **293**, 81–4.

DIJKEMA, R., DEKKER, B. M. M., VAN DER FELTZ, M. J. M. and VAN DER EB, A. J. (1979). Transformation of primary rat kidney cells by DNA fragments of weakly oncogenic adenoviruses. *Journal of Virology*, **32**, 943–50.

DIJKEMA, R., DEKKER, B. M. M. and VAN ORMONDT, H. (1980*a*). The nucleotide sequence of the transforming Bgl II-H fragment of adenovirus type 7 DNA. *Gene*, **9**, 141–56.

DIJKEMA, R., DEKKER, B. M. M. and VAN ORMONDT, H. (1982). Gene organisation of the transforming region of adenovirus 7 DNA. *Gene*, **18**, 143–56.

DIJKEMA, R., DEKKER, B. M. M., VAN ORMONDT, H., DE WAARD, A., MAAT, J. and
   BOYER, H. W. (1980b). Gene organization of the transforming region of weakly
   oncogenic adenovirus type 7: the E1a region. *Gene*, **12**, 287–99.
DOERFLER, W. (1969). Non-productive infection of baby hamster kidney cells
   (BHK-21) with adenovirus type 12. *Virology*, **38**, 587–606.
DOERFLER, W., KUHLMANN, I., WINTERHOFF, U., NEUMANN, R., STABEL, S., SCHIRM,
   S. and EICK, D. (1982). Integration, methylation and expression of adenovirus
   type 12 DNA in transformed and tumour cells. In *Genes and Tumour Genes*, ed.
   E. Winnacker and H.-H. Schoene, pp. 25–37. New York, Raven Press.
ESCHE, H., MATHEWS, M. B. and LEWIS, J. B. (1980). Proteins and messenger RNAs
   of the transforming region of wild type and mutant adenoviruses. *Journal of
   Molecular Biology*, **42**, 399–417.
ESCHE, H. and SIEGMANN, B. (1982). Expression of early viral gene products in
   adenovirus type 12 infected and transformed cells. *Journal of General Virology*,
   60, 99–113.
EZOE, H., LAIFATT, R. B. and MAK, S. (1981). Degradation of intracellular DNA in
   KB cells infected with *cyt* mutants of human adenovirus 12. *Journal of Virology*,
   40, 20–7.
FELDMAN, L. T. and NEVINS, J. R. (1983). Localisation of the adenovirus E1a pro-
   tein, a positive-acting transcriptional factor in infected cells. *Molecular and Cellular Biology*, **3**, 829–38.
FISHER, P. B., BABISS, L. E., WEINSTEIN, I. B. and GINSBERG, H. S. (1982). Analysis
   of type 5 adenovirus transformation with a cloned rat embryo cell line (CREF).
   Proceedings of the National Academy of Sciences, USA, **79**, 3527–31.
FLINT, S. J., SAMBROOK, J., WILLIAMS, J. F. and SHARP, P. A. (1976). Viral nucleic
   acid sequences in transformed cells. IV. A study of the sequences of adenovirus
   type 5 DNA and RNA in four lines of adenovirus 5 transformed rodent cells using
   specific fragments of the viral genome. *Virology*, **72**, 456–70.
FOHRING, B., GALLIMORE, P. H., MELLOW, G. H. and RASKA, K. (1983). Adenovirus
   type 12 specific cell surface antigen in transformed cells is a product of the E1b
   early region. *Virology*, **131**, 463–72.
FREEMAN, A. E., BLACK, P. H., WOLFORD, R. and HUEBNER, R. J. (1967). The
   adenovirus type 12/rat embryo transformation system. *Journal of Virology*, **1**,
   362–7.
FROST, E. and WILLIAMS, J. (1978). Mapping temperature sensitive and host-range
   mutations of adenovirus type 5 by marker rescue. *Virology*, **91**, 39–50.
FUKUI, Y., SAITO, I., SHIROKI, K. and SHIMOJO, H. (1984). Isolation of transformation-defective, replication-nondefective early region 1b mutants of adenovirus 12.
   Journal of Virology, **49**, 154–61.
FUKUI, Y., SAITO, I., SHIROKI, K., SHIMOJO, H., TAKEBE, Y. and KAZIRO, Y. (1983).
   The 19 kDal protein encoded by early region 1b of adenovirus type 12 is
   synthesized efficiently in *Escherichia coli* only as a fused protein. *Gene*, **23**,
   1–13.
GAHLMANN, R. and DOERFLER, W. (1983). Integration of viral DNA into the genome
   of the adenovirus type 2-transformed hamster cell line HE5 without loss or alteration of cellular nucleotides. *Nucleic Acids Research*, **11**, 7347–61.
GAHLMANN, R., LEISTEN, R., VARDIMON, L. and DOERFLER, W. (1982). Patch homologies and the integration of adenovirus DNA in mammalian cells. *EMBO Journal*, **1**, 1101–4.
GALLIMORE, P. H. (1972). Tumour production in immunosuppressed rats with cells
   transformed *in vitro* by adenovirus type 2. *Journal of General Virology*, **16**,
   99–102.
GALLIMORE, P. H. (1974). Interactions of adenovirus type 2 with rat embryo cells.

Permissiveness, transformation and *in vitro* characteristics of adenovirus transformed rat embryo cells. *Journal of General Virology*, **25**, 263–73.

GALLIMORE, P. H., BYRD, P., GRAND, R., BREIDING, D., WHITTAKER, J. and WILLIAMS, J. (1984). An examination of the transforming and tumour-inducing capacity of a number of adenovirus type 12 early region 1, host-range mutants and cells transformed by subgenomic fragments of Ad 12 E1 region. In *Cancer Cells 2, Oncogenes and Viral Genes*, ed. G. F. Vande Woude *et al.*, pp. 519–26. New York, Cold Spring Harbor Laboratory.

GALLIMORE, P. H., McDOUGALL, J. K. and CHEN, L. B. (1977). *In vitro* traits of adenovirus-transformed cell lines and their relevance to tumourigenicity in nude mice. *Cell*, **10**, 669–78.

GALLIMORE, P. H., McDOUGALL, J. K. and CHEN, L. B. (1979). Malignant behaviour of three adenovirus 2 transformed rat embryo brain cell lines and their methyl cellulose selected sub-clones. *International Journal of Cancer*, **24**, 477–84.

GALLIMORE, P. H. and PARASKEVA, C. (1979). A study to determine the reasons for differences in the tumourigenicity of rat cell lines transformed by adenovirus 2 and adenovirus 12. *Cold Spring Harbor Symposia on Quantitative Biology*, **44**, 703–13.

GALLIMORE, P. H., SHARP, P. A. and SAMBROOK, J. (1974). Viral DNA in transformed cells. III. A study of the sequences of adenovirus 2 DNA in nine lines of transformed rat cells using specific fragments of the viral genome. *Journal of Molecular Biology*, **89**, 49–72.

GALOS, R. S., WILLIAMS, J., SHENK, T. and JONES, N. (1980). Physical location of host range mutations of adenovirus type 5; deletion and marker rescue mapping. *Virology*, **104**, 510–13.

GAUDRAY, P., CLERTANT, P. and CUZIN, F. (1980). ATP phosphohydrolase (ATPase) activity of a polyoma virus T antigen. *European Journal of Biochemistry*, **109**, 553–60.

GIACHERIO, D. and HAGER, L. P. (1979). A poly(dT)-stimulated ATPase activity associated with simian virus 40 large T-antigen. *Journal of Biological Chemistry*, **254**, 8113–16.

GINGERAS, T. F., SCIAKY, D., GELINAS, R. E., BING-DONG, J., YEN, C. E., KELLY, M. M., BULLOCK, P. A., PARSONS, B. L., O'NEILL, K. E. and ROBERTS, R. J. (1982). Nucleotide sequences from the adenovirus-2 genome. *Journal of Biological Chemistry*, **257**, 13475–91.

GINSBERG, H. S., ENSINGER, M. J., KAUFFMAN, R. S., MAYER, A. J. and LUNDHOLM, U. (1974). Cell transformation: a study of regulation with types 5 and 12 adenovirus temperature-sensitive mutants. *Cold Spring Harbor Symposia on Quantitative Biology*, **39**, 419–26.

GIRARDI, A. J., HILLEMAN, M. R. and ZWICKEY, R. E. (1964). Tests in hamsters for oncogenic quality of ordinary viruses including adenovirus type 7. *Proceedings of the Society for Experimental Biology and Medicine*, **115**, 1141–50.

GODING, C. R., SHAW, C. H., BLAIR, G. E. and RUSSELL, W. C. (1983). ADP-ribosylation in *in vitro* systems synthesizing adenovirus DNA. *Journal of General Virology*, **64**, 477–83.

GOLDMAN, N., BROWN, M. and KHOURY, G. (1981). Modification of SV40 T antigen by poly ADP-ribosylation. *Cell*, **24**, 567–72.

GRAHAM, F. L., HARRISON, T. and WILLIAMS, J. (1978). Defective transforming capacity of adenovirus type 5 host range mutants. *Virology*, **86**, 10–21.

GRAHAM, F. L., SMILEY, J., RUSSELL, W. C. and NAIRN, R. (1977). Characteristics of a human cell line transformed by DNA from human adenovirus type 5. *Journal of General Virology*, **36**, 59–72.

GRAHAM, F. L. and VAN DER EB, A. J. (1973). Transformation of rat cells by DNA of human adenovirus 5. *Virology*, **54**, 536–9.

GRAND, R. J. A. and GALLIMORE, P. H. (1984). Adenovirus 12 early region 1 proteins: a study of their subcellular localisation and protein–protein interactions. *Journal of General Virology*, in press.

GREEN, M., BRACKMANN, K. W., LUCHER, L. A. and SYMINGTON, J. S. (1983). Antibodies to synthetic peptides targeted to the transforming genes of human adenoviruses: an approach to understanding early viral gene function. In *The Molecular Biology of Adenoviruses 2*. Current Topics in Microbiology and Immunology no. 110, ed. W. Doerfler, pp. 167–92. Berlin, Heidelberg, New York and Tokyo, Springer-Verlag.

GREEN, M., MACKEY, J. K., WOLD, W. S. M. and RIGDEN, P. (1979). Thirty-one human adenovirus serotypes (Ad 1–Ad 31) form five groups (A–E) based upon DNA genome homologies. *Virology*, **93**, 481–92.

GREEN, M., WOLD, W. S. M. and BUTTNER, W. (1981). Integration and transcription of group C human adenovirus sequences in the DNA of five lines of transformed rat cells. *Journal of Molecular Biology*, **151**, 337–66.

HARRISON, T., GRAHAM, F. L. and WILLIAMS, J. (1977). Host range mutants of adenovirus type 5 defective for growth in Hela cells. *Virology*, **77**, 319–29.

HARWOOD, L. M. J. and GALLIMORE, P. H. (1975). A study of the oncogenicity of adenovirus type 2 transformed rat embryo cells. *International Journal of Cancer*, **16**, 498–508.

HEARING, P. and SHENK, T. (1983a). Functional analysis of the nucleotide sequence surrounding the cap site for adenovirus type 5 region E1a messenger RNAs. *Journal of Molecular Biology*, **167**, 809–22.

HEARING, P. and SHENK, T. (1983b). The adenovirus type 5 E1a transcriptional control region contains a duplicated enhancer element. *Cell*, **33**, 695–703.

HEN, R., BORRELLI, E., SASSONE-CORSI, P. and CHAMBON, P. (1983). An enhancer element is located 340 base pairs upstream from the adenovirus-2 E1a cap site. *Nucleic Acids Research*, **11**, 8747–60.

HO, Y.-S., GALOS, R. and WILLIAMS, J. (1982). Isolation of type 5 adenovirus mutants with a cold-sensitive host range phenotype: genetic evidence of an adenovirus transformation maintenance function. *Virology*, **122**, 109–24.

HOUWELING, A., VAN DEN ELSEN, P. J. and VAN DER EB, A. J. (1980). Partial transformation of primary rat cells by the left-most 4.5% fragment of adenovirus 5 DNA. *Virology*, **105**, 537–50.

HUEBNER, R. J., ROWE, W. P. and LANE, W. T. (1962). Oncogenic effects in hamsters of human adenovirus types 12 and 18. *Proceedings of the National Academy of Sciences, USA*, **48**, 2051–8.

HUEBNER, R. J., ROWE, W. P., TURNER, H. C. and LANE, W. T. (1963). Specific adenovirus complement-fixing antigens in virus-free hamster and rat tumours. *Proceedings of the National Academy of Sciences, USA*, **50**, 379–89.

HULL, R. N., JOHNSON, I. S., CULBERTSON, C. G., REINER, C. B. and WRIGHT, H. F. (1965). Oncogenicity of the simian adenoviruses. *Science*, **150**, 1044–6.

HUNTER, T. and SEFTON, B. M. (1980). Transforming gene product of Rous sarcoma virus phosphorylates tyrosines. *Proceedings of the National Academy of Sciences, USA*, **77**, 1311–15.

IMPERIALE, M. J., FELDMAN, L. T. and NEVINS, J. R. (1983). Activation of gene expression by adenovirus and herpesvirus regulatory genes acting in *trans* and by a *cis*-acting adenovirus enhancer element. *Cell*, **35**, 127–36.

JOCHEMSEN, H., DANIELS, G. S. G., HERTOGHS, J. J. L., SCHRIER, P. I., VAN DEN ELSEN, P. J. and VAN DER EB, A. J. (1982). Identification of adenovirus-type 12 gene products involved in transformation and oncogenesis. *Virology*, **122**, 15–28.

JOHANSSON, K., PERSSON, H., LEWIS, A. M., PETTERSSON, U., TIBBETTS, C. and PHILIPSON, L. (1978). Viral DNA sequences and gene products in hamster cells transformed by adenovirus type 2. *Journal of Virology*, **27**, 628–39.

JONES, N. and SHENK, T. (1978). Isolation of deletion and substitution mutants of adenovirus type 5. *Cell*, **13**, 181–8.

JONES, N. and SHENK, T. (1979*a*). Isolation of adenovirus type 5 host range deletion mutants defective for transformation of rat embryo cells. *Cell*, **17**, 683–9.

JONES, N. and SHENK, T. (1979*b*). An adenovirus type 5 early gene function regulates expression of the other early viral genes. *Proceedings of the National Academy of Sciences, USA*, **76**, 3665–9.

KATZE, M. G., PERSSON, H. and PHILIPSON, L. (1982). A novel mRNA and a low molecular weight polypeptide encoded in the transforming region of adenovirus DNA. *EMBO Journal*, **1**, 783–90.

KIMURA, T., SAWADA, Y., SHINAGAWA, M., SHIMIZU, Y., SHIROKI, K., SHIMOJO, H., SUGISAKI, H., TAKANAMI, M., UEMIZU, Y. and FUJINAGA, K. (1981). Nucleotide sequence of the transforming early region E1b of adenovirus type 12 DNA: structure and gene organisation and comparison with those of adenovirus type 5 DNA. *Nucleic Acids Research*, **9**, 6571–89.

KIRSCHSTEIN, R. L., RABSON, A. S. and PETERS, E. A. (1964). Oncogenic activity of adenovirus 12 in thymectomized BALB/c and C3H/HeN mice. *Proceedings of the Society for Experimental Biology and Medicine*, **117**, 198–200.

KLOCKMANN, U. and DEPPERT, W. (1983*a*). Acylated simian virus 40 large T-antigen: a new subclass associated with a detergent resistant lamina of the plasma membrane. *EMBO Journal*, **2**, 1151–7.

KLOCKMANN, U. and DEPPERT, W. (1983*b*). Acylation: a new post-translational modification specific for plasma-membrane-associated simian virus 40 large T-antigen. *FEBS Letters*, **151**, 257–9.

KUHLMANN, I. and DOERFLER, W. (1983). Loss of viral genomes from hamster tumour cells and nonrandom alterations in patterns of methylation of integrated adenovirus type 12 DNA. *Journal of Virology*, **47**, 631–6.

LAIFATT, R. B. and MAK, S. (1982). Mapping of an adenovirus function involved in the inhibition of DNA degradation. *Journal of Virology*, **42**, 969–77.

LANE, D. P. and CRAWFORD, L. V. (1979). T-antigen is bound to a host protein in SV40 transformed cells. *Nature, London*, **278**, 261–3.

LARSON, W. M., CONRAD, P. A., CLARK, W. R. and HILLEMAN, M. R. (1971). Tests for oncogenicity of viruses under conditions of altered host and virus. *Proceedings of the Society for Experimental Biology and Medicine*, **136**, 1304–13.

LASSAM, N. J., BAYLEY, S. T. and GRAHAM, F. L. (1978). Synthesis of DNA, late polypeptides and infectious virus by host range mutants of adenovirus 5 in nonpermissive cells. *Virology*, **87**, 463–7.

LASSAM, N. J., BAYLEY, S. T., GRAHAM, F. L. and BRANTON, P. E. (1979). Immunoprecipitation of protein kinase activity from adenovirus 5-infected cells using antiserum directed against tumour antigens. *Nature, London*, **277**, 241–3.

LEVINSON, A. D. and LEVINE, A. J. (1977). The isolation and identification of the adenovirus group C tumour antigens. *Virology*, **76**, 1–11.

LEWIS, A. M. JR and COOK, J. L. (1982). Spectrum of tumourigenic phenotypes among adenovirus 2-, adenovirus 12- and simian virus 40-transformed Syrian hamster cells defined by host cellular immune–tumour cell interactions. *Cancer Research*, **42**, 939–44.

LEWIS, J. B. and MATHEWS, M. B. (1980). Control of adenovirus early gene expression, a class of immediate early products. *Cell*, **21**, 303–13.

LEWIS, J. B. and MATHEWS, M. B. (1982). Viral messenger RNAs in six lines of adenovirus-transformed cells. *Journal of Virology*, **115**, 345–60.

LINZER, D. I. H. and LEVINE, A. J. (1979). Characterisation of 54 kD cellular SV40 tumour antigen present in transformed cells and uninfected embryonal carcinoma cells. *Cell*, **17**, 43–72.

LUPKER, J. H., DAVIS, A., JOCHEMSEN, H. and VAN DER EB, A. J. (1981). *In vitro* synthesis of adenovirus type 5 T-antigens. I. Translation of early region 1-specific RNA from lytically infected cells. *Journal of Virology*, **37**, 524–9.

MCBRIDE, W. D. and WIENER, A. (1964). *In vitro* transformation of hamster kidney cells by human adenovirus type 12. *Proceedings of the Society for Experimental Biology and Medicine*, **115**, 870–4.

MCDOUGALL, J. K., DUNN, A. R. and GALLIMORE, P. H. (1974). Recent studies on the characteristics of adenovirus infected and transformed cells. *Cold Spring Harbor Symposia on Quantitative Biology*, **39**, 591–600.

MCMICHAEL, A. J., TING, A., ZWEERINK, H. J. and ASKONAS, B. A. (1977). HLA restriction of cell-mediated lysis of influenza virus-infected human cells. *Nature, London*, **270**, 524–6.

MAK, I. and MAK, S. (1983). Transformation of rat cells by *cyt* mutants of adenovirus type 12 and mutants of adenovirus type 5. *Journal of Virology*, **45**, 1107–17.

MAK, S., MAK, I., SMILEY, J. R. and GRAHAM, F. L. (1979). Tumourigenicity and viral gene expression in rat cells transformed by Ad 12 virions or by the Eco RI-C fragment of Ad 12 DNA. *Virology*, **98**, 456–60.

MALETTE, P. L., YEE, S.-P. and BRANTON, P. E. (1983). Studies on the phosphory-lation of the 58 000 dalton early region 1b protein of human adenovirus type 5. *Journal of General Virology*, **64**, 1069–78.

MELLOW, G. H., FOHRING, B., DOUGHERTY, T., GALLIMORE, P. H. and RASKA, K. (1984). Tumourigenicity of adenovirus-transformed rat cells and expression of Class I major histocompatibility antigen. *Virology*, **134**, 460–5.

MILNER, J. and MILNER, S. (1981). SV40 53K antigen: a possible role for 53K in normal cells. *Virology*, **112**, 785–8.

MONTELL, C., COURTOIS, G., ENG, C. and BERK, A. J. (1984). Complete transforma-tion by adenovirus 2 requires both E1A proteins. *Cell*, **36**, 951–61.

MONTELL, C., FISHER, E. F., CARUTHERS, M. H. and BERK, A. J. (1983*a*). Resolving the function of overlapping viral genes by site-specific mutagenesis at a mRNA splice site. *Nature, London*, **295**, 380–4.

MONTELL, C., FISHER, E. F., CARUTHERS, M. H. and BERK, A. J. (1983*b*). Inhibition of RNA cleavage but not polyadenylation by a point mutation in RNA 3′ consen-sus sequence AAUAAA. *Nature, London*, **305**, 600–5.

MUKAI, N., KALTER, S. S., CUMMINS, L. B., MATHEWS, V. A., NISHIDA, T. and NAKAJIMA, T. (1980). Retinal tumor induction in the baboon by human adenovirus 12. *Science*, **210**, 1023–5.

MURRAY, N. E., BELLETT, A. J. D., BRAITHWAITE, A. W., WALDRON, L. K. and TAYLOR, I. W. (1982). Altered cell cycle progression and aberrant mitosis in adenovirus infected rodent cells. *Journal of Cellular Physiology*, **111**, 89–96.

NEVINS, J. R. (1981). Mechanism of activation of early viral transcription by the adenovirus E1a gene product. *Cell*, **26**, 213–20.

NEVINS, J. R. (1982). Induction of the synthesis of a 70 000 dalton mammalian heat shock protein by the adenovirus E1A gene product. *Cell*, **29**, 913–19.

NEVINS, J. R., IMPERIALE, M. J., FELDMAN, L. T. and KAO, H.-T. (1984). Role of adenovirus transforming gene (E1A) in the general control of gene expression. *Transplantation Proceedings*, **16**, 438–40.

OREN, M., MALTZMAN, W. and LEVINE, A. J. (1981). Post-translation regulation of the 54K cellular tumour antigen in normal and transformed cells. *Molecular and Cellular Biology*, **1**, 101–10.

OSBORNE, T. F., GAYNOR, R. B. and BERK, A. J. (1982). The TATA homology and the mRNA 5' untranslated sequence are not required for expression of essential adenovirus E1a functions. *Cell*, **29**, 139–48.

PARASKEVA, C., BROWN, K. W., DUNN, A. R. and GALLIMORE, P. H. (1982). Adenovirus type 12-transformed rat embryo brain and rat liver epithelial cell lines: adenovirus type 12 genome content and viral protein expression. *Journal of Virology*, **44**, 759–64.

PEREIRA, M. S., PEREIRA, H. G. and CLARKE, S. K. R. (1965). Human adenovirus type 31. A new serotype with oncogenic properties. *Lancet*, **i**, 21–3.

PERRICAUDET, M., AKUSJARVI, G., VIRTANEN, A. and PETTERSSON, U. (1979). Structure of two spliced mRNAs from the transforming region of human subgroup C adenoviruses. *Nature, London*, **281**, 694–6.

PERRICAUDET, M., LE MOULLEC, J.-M., TIOLLAIS, P. and PETTERSSON, U. (1980). Structure of two adenovirus type 12 transforming polypeptides and their evolutionary implications. *Nature, London*, **288**, 174–6.

PERSSON, H., KATZE, M. G. and PHILIPSON, L. (1982). Purification of a native membrane-associated adenovirus tumour antigen. *Journal of Virology*, **42**, 905–17.

PERSSON, H., MONSTEIN, H.-J., AKUSJARVI, G. and PHILIPSON, L. (1981). Adenovirus early gene products may control viral mRNA accumulation and translation *in vivo*. *Cell*, **23**, 485–96.

RABSON, A. S. and KIRSCHSTEIN, R. L. (1964). Tumours produced by adenovirus 12 in *Mastomys* and mice. *Journal of the National Cancer Institute*, **32**, 77–87.

RALSTON, R. and BISHOP, J. M. (1983). The protein products of the *myc* and *myb* oncogenes and adenovirus E1a are structurally related. *Nature, London*, **306**, 803–6.

RASKA, K. JR and GALLIMORE, P. H. (1982). An inverse relation of the oncogenic potential of adenovirus-transformed cells and their sensitivity to killing by syngeneic natural killer cells. *Virology*, **123**, 8–18.

RASKA, K., GEIS, A. and FOHRING, B. (1979). Adenovirus type 12 tumour antigen. II. Immunoprecipitation of protein kinase from infected and transformed cells by antisera to T antigen and some normal rat sera. *Virology*, **99**, 174–8.

REICH, N. C. and LEVINE, A. J. (1984). Growth regulation of a cellular tumour antigen, p53, in nontransformed cells. *Nature, London*, **308**, 199–201.

RICCIARDI, R. P., JONES, R. L., CEPKO, C. L., SHARP, P. A. and ROBERTS, B. E. (1981). Expression of early adenovirus genes requires a viral encoded acidic polypeptide. *Proceedings of the National Academy of Sciences, USA*, **78**, 6121–5.

RIJNDERS, A. W. M., VAN MAARSCHALKERWEERD, M. W., VISSER, L., REEMST, A. M. C. B., SUSSENBACH, J. S. and ROZIJN, T. H. (1983). Expression of integrated viral DNA sequences outside the transforming region in eight adenovirus-transformed cell lines. *Biochimica et Biophysica Acta*, **739**, 48–56.

ROSS, S. R., LEVINE, A. J., GALOS, R. S., WILLIAMS, J. and SHENK, T. (1980). Early viral proteins in Hela cells infected with adenovirus type 5 host range mutants. *Virology*, **103**, 475–92.

ROSSINI, M. (1983). The role of adenovirus early region 1A in the regulation of early regions 2A and 1B expression. *Virology*, **131**, 49–58.

ROWE, D. T., BRANTON, P. E., YEE, S.-P., BACCHETTI, S. and GRAHAM, F. L. (1984). Establishment and characterization of hamster cell lines transformed by restriction endonuclease fragments of adenovirus 5. *Journal of Virology*, **49**, 162–70.

ROWE, D. T. and GRAHAM, F. L. (1981). Complementation of adenovirus type 5 host range mutants by adenovirus type 12. *Journal of Virology*, **38**, 191–7.

ROWE, D. T. and GRAHAM, F. L. (1983). Transformation of rodent cells by DNA extracted from transformation-defective adenovirus mutants. *Journal of Virology*, **46**, 1039–44.

ROWE, D. T., GRAHAM, F. L. and BRANTON, P. E. (1983*a*). Intracellular localisation of adenovirus type 5 tumour antigens in productively infected cells. *Virology*, **129**, 456–68.

ROWE, D. T., YEE, S., OTIS, J., GRAHAM, F. L. and BRANTON, P. E. (1983*b*). Characterization of human adenovirus type 5 early region 1A polypeptide using antitumour sera and an antiserum specific for the carboxy terminus. *Virology*, **127**, 253–71.

RUBEN, M., BACCHETTI, S. and GRAHAM, F. L. (1982). Integration and expression of viral DNA in cells transformed by host range mutants of adenovirus type 5. *Journal of Virology*, **41**, 674–85.

RUBEN, M., BACCHETTI, S. and GRAHAM, F. (1983). Covalently closed circles of adenovirus 5 DNA. *Nature, London*, **301**, 172–4.

RULEY, H. E. (1983). Adenovirus early region 1a enables viral and cellular transforming genes to transform primary cells in culture. *Nature, London*, **304**, 602–6.

SAITO, I., SATO, J.-I., HANDA, H., SHIROKI, K. and SHIMOJO, H. (1981). Mapping of RNAs transcribed from adenovirus type 12 early and VA RNA regions. *Virology*, **114**, 379–98.

SAITO, I., SHIROKI, K. and SHIMOJO, H. (1983). mRNA species and proteins of adenovirus type 12 transforming regions: identification of proteins translated from multiple coding stretches in 2.2 kb region 1B mRNA *in vitro* and *in vivo*. *Virology*, **127**, 272–89.

SARMA, P. S., HUEBNER, R. J. and LANE, W. T. (1965). Induction of tumours in hamsters with an avian adenovirus (CELO). *Science*, **149**, 1108.

SARMA, P. S., VASS, W., HUEBNER, R. J., IGEL, H., LANE, W. T. and TURNER, H. C. (1967). Induction of tumours in hamsters with infectious canine hepatitis virus. *Nature, London*, **215**, 293–4.

SARNOW, P., HO, Y. S., WILLIAMS, J. and LEVINE, A. J. (1982*a*). Adenovirus E1b 58 kd tumour antigen and SV40 large tumour antigen are physically associated with the same 54 kd cellular protein in transformed cells. *Cell*, **28**, 387–94.

SARNOW, P., SULLIVAN, C. A. and LEVINE, A. J. (1982*b*). A monoclonal antibody detecting the adenovirus type 5 E1b 58 kd tumour antigen: characterisation of the E1b 58 kd tumour antigen in adenovirus-infected and transformed cells. *Virology*, **120**, 510–17.

SASSONE-CORSI, P., HEN, R., BORRELLI, E., LEFF, T. and CHAMBON, P. (1983). Far upstream sequences are required for efficient transcription from the adenovirus-2 E1a transcription unit. *Nucleic Acids Research*, **11**, 8735–45.

SAWADA, Y. and FUJINAGA, K. (1980). Mapping of adenovirus 12 mRNAs transcribed from the transforming region. *Journal of Virology*, **36**, 639–51.

SCHRIER, P. I., BERNARDS, R., VAESSEN, R. T. M. J., HOUWELING, A. and VAN DER EB, A. J. (1983). Expression of Class I major histocompatibility antigens switched off by highly oncogenic adenovirus in transformed rat cells. *Nature, London*, **305**, 771–5.

SCOTT, M. O., KIMELMAN, D., NORRIS, D. and RICCIARDI, R. P. (1984). Production of a monospecific antiserum against the early region 1A proteins of adenovirus 12 and adenovirus 5 by an adenovirus 12 early region 1A-β-galactosidase fusion antigen expressed in bacteria. *Journal of Virology*, **50**, 895–903.

SEFTON, B. M., TROWBRIDGE, I. S., COOPER, J. A. and SCOLNICK, E. M. (1982). The transforming proteins of Rous sarcoma virus, Harvey sarcoma virus and Abelson virus contain tightly bound lipid. *Cell*, **31**, 465–74.

SEKIKAWA, K., SHIROKI, K., SHIMOJO, H., OJIMA, S. and FUJINAGA, K. (1978). Transformation of a rat cell line by an adenovirus 7 DNA fragment. *Virology*, **88**, 1–7.

SHEIL, J. M., GALLIMORE, P. H., ZIMMER, S. G. and SOPORI, M. L. (1984). Susceptibility of adenovirus 2-transformed rat cell lines to natural killer (NK) cells: direct correlation between NK resistance and *in vivo* tumourigenesis. *Journal of Immunology*, **132**, 1578–82.

SHENK, T., JONES, N., COLBY, W. and FOWLKES, D. (1979). Functional analysis of adenovirus-5 host-range deletion mutants defective for transformation of rat embryo cells. *Cold Spring Harbor Symposia on Quantitative Biology*, **44**, 367–75.

SHIROKI, K., HANDA, H., SHIMOJO, H., YANO, S., OJIMA, S. and FUJINAGA, K. (1977). Establishment and characterization of rat cell lines transformed by restriction endonuclease fragments of adenovirus 12 DNA. *Virology*, **82**, 462–71.

SHIROKI, K., HASHIMOTO, S., SAITO, I., FUKUI, Y., FUKUI, Y., KATO, H. and SHIMOJO, H. (1984). Expression of the E4 gene is required for establishment of soft-agar colony-forming rat cell lines transformed by the adenovirus 12 E1 gene. *Journal of Virology*, **50**, 854–63.

SHIROKI, K., MARUYAMA, K., SAITO, I., FUKUI, Y., YAZAKI, K. and SHIMOJO, H. (1982). Dependence of tumour-forming capacities of cells transformed by recombinants between adenovirus type 5 and 12 on expression of early region 1. *Journal of Virology*, **42**, 708–18.

SHIROKI, K., SAITO, I., MARUYAMA, K., FUKUI, Y., IMATANI, Y., ODA, K.-I. and SHIMOJO, H. (1983). Expression of adenovirus type 12 early region 1 in KB cells transformed by recombinants containing the gene. *Journal of Virology*, **45**, 1074–82.

SHIROKI, K., SEGAWA, K., SAITO, I., SHIMOJO, H. and FUJINAGA, K. (1979*a*). Products of the adenovirus-12 transforming genes and their functions. *Cold Spring Harbor Symposia on Quantitative Biology*, **44**, 533–40.

SHIROKI, K., SEGAWA, K. and SHIMOJO, H. (1980). Two tumour antigens and their polypeptides in adenovirus type 12-infected and transformed cells. *Proceedings of the National Academy of Sciences, USA*, **77**, 2274–8.

SHIROKI, K., SHIMOJO, H., SAWADA, Y., UEMIZU, Y. and FUJINAGA, K. (1979*b*). Incomplete transformation of rat cells by a small fragment of adenovirus 12 DNA. *Virology*, **95**, 127–36.

SJOGREN, H. O., MINOWADA, J. and ANKERST, J. (1967). Specific transplantation antigens of mouse sarcomas induced by adenovirus type 12. *Journal of Experimental Medicine*, **125**, 689–701.

SOLNICK, D. and ANDERSON, M. A. (1982). Transformation-deficient adenovirus mutant defective in expression of region 1a but not region 1b. *Journal of Virology*, **42**, 106–13.

SPECTOR, D. J. (1982). Transcription of adenovirus 5 early region 1b is elevated in permissive cells infected by a mutant with an upstream deletion. *Journal of Virology*, **44**, 544–54.

SPECTOR, D. J., MCGROGAN, M. and RASKAS, H. J. (1978). Regulation of the appearance of cytoplasmic RNAs from region 1 of the adenovirus 2 genome. *Journal of Molecular Biology*, **126**, 395–414.

SPINDLER, K. R., ROSSER, D. S. E. and BERK, A. J. (1984). Analysis of adenovirus transforming proteins from early regions 1A and 1B with antisera to inducible fusion antigens produced in *Escherichia coli*. *Journal of Virology*, **49**, 132–41.

STOW, N. D. (1981). Cloning of a DNA fragment from the left hand terminus of the adenovirus type 2 genome and its use in site-directed mutagenesis. *Journal of Virology*, **37**, 171–80.

SUGISAKI, H., SUGIMOTO, K., TAKANAMI, M., SHIROKI, K., SAITO, I., SHIMOJO, H., SAWADA, Y., UEMIZU, Y., UESUGI, S.-I. and FUJINAGA, K. (1980). Structure and gene organisation in the transforming HindIII-G fragment of Ad 12. *Cell*, **20**, 777–86.

SVENSSON, C. and AKUSJARVI, G. (1984). Adenovirus 2 early region 1A stimulates expression of both viral and cellular genes. *EMBO Journal*, **3**, 789–94.

SVENSSON, C., PETTERSSON, U. and AKUSJARVI, G. (1983). Splicing of adenovirus 2 early region 1A mRNAs is non-sequential. *Journal of Molecular Biology*, **165**, 475–95.

TAKEMORI, N., RIGGS, J. L. and ALDRICH, C. D. (1968). Genetic studies with tumourigenic adenoviruses. I. Isolation of cytocidal (*cyt*) mutants of adenovirus type 12. *Virology*, **36**, 575–86.

TAKEMORI, N., RIGGS, J. L. and ALDRICH, C. D. (1969). Genetic studies with tumourigenic adenoviruses. II. Heterogeneity of *cyt* mutants of adenovirus type 12. *Virology*, **38**, 8–15.

TJIAN, R., FEY, G. and GRAESSMAN, A. (1978). Biological activity of purified SV40 T-antigen proteins. *Proceedings of the National Academy of Sciences*, USA, **75**, 1279–83.

TJIAN, R. and ROBBINS, A. K. (1979). Enzymatic activities associated with a purified simian virus 40 T antigen-related protein. *Proceedings of the National Academy of Sciences, USA*, **76**, 610–14.

TRENTIN, J. J., VAN HOOSIER, G. L. and SAMPER, L. (1968). The oncogenicity of human adenoviruses in hamsters. *Proceedings of the Society for Experimental Biology and Medicine*, **127**, 683–9.

TRENTIN, J. J., YABE, Y and TAYLOR, G. (1962). The quest for human cancer viruses. *Science*, **137**, 835–41.

VAN DEN ELSEN, P. J., DE PATER, S., HOUWELING, A., VAN DER VEER, J. and VAN DER EB, A. J. (1982). The relationship between region E1a and E1b of human adenoviruses in cell transformation. *Gene*, **18**, 175–85.

VAN DEN ELSEN, P. J., HOUWELING, A. and VAN DER EB, A. J. (1983*a*). Morphological transformation of human adenoviruses is determined to a large extent by gene products of region E1a. *Virology*, **131**, 241–6.

VAN DEN ELSEN, P. J., HOUWELING, A. and VAN DER EB, A. J. (1983*b*). Expression of region E1b of human adenoviruses in the absence of region E1a is not sufficient for complete transformation. *Virology*, **128**, 377–90.

VAN DEN ELSEN, P., KLEIN, B., DEKKER, B., VAN ORMONDT, H. and VAN DER EB, A. J. (1983*c*). Analysis of virus-specific mRNAs present in cells transformed with re-striction fragments of adenovirus type 5 DNA. *Journal of General Virology*, **64**, 1079–90.

VAN DER EB, A. J., MULDER, C., GRAHAM, F. L. and HOUWELING, A. (1977). Trans-formation with specific fragments of adenovirus DNAs. I. Isolation of specific fragments with transformation activity of adenovirus 2 and 5 DNA. *Gene*, **2**, 115–32.

VAN DER EB, A. J., VAN ORMONDT, H., SCHRIER, P. I., LUPKER, J. H., JOCHEMSEN, H., VAN DEN ELSEN, P. J., DE LEYS, J., MAAT, J., VAN BEVEREN, C. P., DIJKEMA, R. and DE WAARD, A. (1979). Structure and function of the transforming genes of human adenoviruses and SV40. *Cold Spring Harbor Symposia on Quantitative Biology*, **44**, 383–98.

VAN ORMONDT, H. and GALIBERT, F. (1984). Nucleotide sequences of adenovirus DNAs. In *The Molecular Biology of Adenoviruses 2*, Current Topics in Micro-biology and Immunology no. 110, ed. W. Doerfler, pp. 73–142. Berlin, Heidel-berg, New York and Tokyo, Springer-Verlag.

VAN ORMONDT, H. and HESPER, B. (1983). Comparison of the nucleotide sequences of early region E1b DNA of human adenovirus types 12, 7 and 5 (subgroups A, B and C). *Gene*, **21**, 217–26.

VAN ORMONDT, H., MAAT, J., DE WAARD, A. and VAN DER EB, A. J. (1978). The nucleotide sequence of the transforming HpaI-E fragment of adenovirus type 5 DNA. *Gene*, **4**, 309–28.

VAN ORMONDT, H., MAAT, J. and DIJKEMA, R. (1980*a*). Comparison of nucleotide sequences of the early E1a regions for subgroups A, B and C of human adenoviruses. *Gene*, **12**, 63–76.

VAN ORMONDT, H., MAAT, J. and VAN BEVEREN, C. P. (1980*b*). The nucleotide sequence of the transforming early region E1 of adenovirus type 5 DNA. *Gene*, 11, 299–309.

VARDIMON, L. and DOERFLER, W. (1981). Patterns of integration of viral DNA in adenovirus type 2-transformed hamster cells. *Journal of Molecular Biology*, 147, 227–46.

VIRTANEN, A. and PETTERSSON, U. (1983). The molecular structure of the 9S mRNA from early region 1A of adenovirus serotype 2. *Journal of Molecular Biology*, 165, 496–9.

VIRTANEN, A., PETTERSSON, U., LE MOULLEC, J. M., TIOLLAIS, P. and PERRICAUDET, M. (1982). Different mRNAs from the transforming region of highly oncogenic and non-oncogenic human adenoviruses. *Nature, London*, 295, 705–7.

VISSER, L., REEMST, A. C. M. B., VAN MANSFELD, A. D. M. and ROZIJN, T. H. (1982). Nucleotide sequence of the linked left and right hand terminal regions of adenovirus type 5 DNA present in the transformed rat cell line 5RK20. *Nucleic Acids Research*, 10, 2189–98.

VISSER, L., VAN MAARSCHALKERWEERD, M. W., ROZIJN, T. H., WASSENAAR, A. D. C., REEMST, A. M. C. B. and SUSSENBACH, J. S. (1979). Viral DNA sequences in adenovirus-transformed cells. *Cold Spring Harbour Symposia on Quantitative Biology*, 44, 541–50.

WESTIN, G., VISSER, L., ZABIELSKI, J., VAN MANSFELD, A. D. M., PETTERSSON, U. and ROZIJN, T. H. (1982). Sequence organization of a viral DNA insertion present in the adenovirus type 5 transformed hamster cell BHK 268–C31. *Gene*, 17, 263–70.

WHITTAKER, J. L., BYRD, P. J., GRAND, R. J. A. and GALLIMORE, P. H. (1984). The isolation and characterisation of four adenovirus type 12 transformed human embryo kidney cell lines. *Molecular and Cellular Biology*, 4, 110–16.

WILLIAMS, J. F. (1973). Oncogenic transformation of hamster embryo cells *in vitro* by adenovirus type 5. *Nature, London*, 243, 162–3.

WILLIAMS, J. F., HO, Y.-S. and GALOS, R. (1981). Evidence for functional relatedness of products encoded by the transforming sequences of human adenovirus types 5 and 12. *Virology*, 110, 208–12.

WILLIAMS, J. F., YOUNG, C. S. H. and AUSTIN, P. E. (1974). Genetic analysis of human adenovirus type 5 in permissive and non-permissive cells. *Cold Spring Harbor Symposia on Quantitative Biology*, 39, 427–37.

WILSON, M. C. and DARNELL, J. E. JR (1981). Control of messenger RNA concentration by differential cytoplasmic half life: adenovirus messenger RNAs from transcription units 1A and 1B. *Journal of Molecular Biology*, 148, 231–51.

WILSON, M. C., FRASER, N. W. and DARNELL, J. E. (1979). Mapping of RNA initiation sites by high doses of UV irradiation: evidence for three independent promoters within the left 11% of the Ad 2 genome. Virology, 94, 175–84.

WINCHESTER, G. (1983). p53 protein and control of growth. *Nature, London*, 303, 660–1.

YABE, Y., SAMPER, L., BRYAN, E., TAYLOR, G. and TRENTIN, J. J. (1964). Oncogenic effect of human adenovirus type 12 in mice. *Science*, 143, 46–7.

YABE, Y., SAMPER, L., TAYLOR, G. and TRENTIN, J. J. (1963). Cancer induction in hamsters by human type 12 adenovirus. Effect by route of injection. *Proceedings of the Society for Experimental Biology and Medicine*, 113, 221–4.

YABE, Y., TRENTIN, J. J. and TAYLOR, G. (1962). Cancer induction in hamsters by human type 12 adenovirus. Effect of age and of virus dose. *Proceedings of the Society for Experimental Biology and Medicine*, 111, 343–4.

YANO, S., OJIMA, S., FUJINAGA, K., SHIROKI, K. and SHIMOJO, H. (1977). Transformation of a rat cell line by an adenovirus type 12 DNA fragment. *Virology*, 82, 214–20.

YEE, S.-P. and BRANTON, P. E. (1983). Co-purification of protein kinase activity with the 58000 dalton polypeptide coded for by the early E1B region of human adenovirus type 5. *Journal of General Virology*, **64**, 2305–9.

YEE, S.-P., ROWE, D. T., TREMBLAY, M. L., McDERMOTT, M. and BRANTON, P. E. (1983). Identification of human adenovirus early region 1 products by using antisera against synthetic peptides corresponding to the predicted carboxy termini. *Journal of Virology*, **46**, 1003–13.

YOHN, D. S., FUNK, C. A. and GRACE, J. T. (1967). Sex-related resistance in hamsters to adenovirus-12 oncogenesis. II. Influence of virus dose. *Journal of Virology*, **1**, 1186–92.

YOHN, D. S., FUNK, C. A. and GRACE, J. T. (1968). Sex-related resistance in hamsters to adenovirus 12 oncogenesis. III. Influence of immunological impairment by thymectomy and cortisone. *Journal of Immunology*, **100**, 771–80.

YOHN, D. S., FUNK, C. A., KALRUNS, V. I. and GRACE, J. T. (1965). Sex-related resistance in hamsters to adenovirus-12 oncogenesis. Influence of thymectomy at three weeks of age. *Journal of the National Cancer Institute*, **35**, 617–24.

YOSHIDA, K. and FUJINAGA, K. (1980). Unique species of mRNA from adenovirus type 7 early region 1 in cells transformed by adenovirus type 7 DNA fragment. *Journal of Virology*, **36**, 337–52.

ZINKERNAGEL, R. M. and DOHERTY, P. C. (1974). Restriction of *in vitro* T cell-mediated cytotoxicity in lymphocytic choriomeningitis within a syngeneic or semi-allogeneic system. *Nature, London*, **248**, 701–2.

ONCOGENESIS BY MOUSE MAMMARY TUMOUR VIRUS

CLIVE DICKSON and GORDON PETERS

Imperial Cancer Research Fund Laboratories, PO Box 123, Lincoln's Inn Fields, London WC2A 3PX, UK

Oncogenic retroviruses infect a wide range of vertebrate species and are associated with a variety of distinct malignancies, some arising in a matter of days or weeks after virus infection, while others are only detected after a latent period of several months (Weiss *et al.*, 1982). Despite this variability in tumourigenic potential, all members of the Retroviridae are unified by their genome organisation and mode of replication. One of the salient features of the life cycle is the reverse transcription of the viral RNA into a double-stranded DNA intermediate which becomes integrated as a provirus into the chromosomal DNA of the host cell. The introduction of such large (up to 10 kb) segments of DNA at essentially random sites within the host genome clearly has the potential to be mutagenic, and there are indeed a number of documented examples of retrovirus proviruses acting as insertional mutagens (Jenkins *et al.*, 1981; Varmus, Quintrell and Ortiz, 1981; Copeland, Jenkins and Lee, 1983; Harbers *et al.*, 1984). In these, and presumably the majority of cases, the provirus exerts a negative influence on gene expression simply as a result of physical disruption. In somatic cells, provirus integration is likely to occur on only one of any pair of chromosomes, so that such disruptive events would be expected to be phenotypically silent, or on rare occasions perhaps lethal to the cell. However, there are also clear indications that the transcriptional control elements carried within the long terminal repeat (LTR) segments which flank each provirus may also act as positive regulators of cellular gene expression (Varmus, 1982). Thus in the rare instances in which proviral 'activation' of an adjacent cellular gene may confer a selective advantage on a particular cell, then these mutagenic events might be phenotypically observable, possible examples being hyperplastic growth and tumour formation. Such concepts currently provide the most tenable explanation for oncogenesis in situations where there is a long delay between viral infection and overt neoplasia.

At the other end of the spectrum some retroviruses induce

malignant disease extremely rapidly and with very high efficiency, properties which cannot be accounted for by insertional mutagenesis. Here, tumorigenicity appears to be associated with the expression of specific genetic information, called the viral oncogene (v-*onc*), encoded within the viral genome (Bishop, 1983). Although these oncogenes were first identified as components of retroviral genomes, their origins can be traced to homologous sequences, called proto-oncogenes or cellular oncogenes (c-*oncs*), pre-existing in the DNA of normal somatic cells. The acutely oncogenic retroviruses can there-fore be viewed as transducing vectors that have acquired, and now express, genetic information of cellular origin. Since the integrated provirus normally serves as an extremely active transcription unit, controlled by elements within the viral LTRs, transduction by a retrovirus can lead to expression of these sequences of cellular origin at levels far in excess of those encountered in normal cells. Whether this alone is sufficient to account for neoplasia or whether structural alterations incurred during transduction are equally, or more rele-vant, are currently the subject of intense investigation and debate.

To date, about twenty distinct v-*oncs* have been identified, each with its c-*onc* counterpart, and in several instances cognate sequen-ces have been independently acquired by different viruses (Bishop, 1983). In the vast majority of these viruses, the acquisition of cellular sequences has occurred at the expense of some of the viral replicative functions, rendering the viruses defective for replication. Such agents can only be propagated in the continued presence of a non-defective helper virus. Thus, despite their enormous impact on our under-standing of tumourigenesis and the identification of oncogenes, the acutely oncogenic retroviruses are relatively rare in natural settings, with some important exceptions (see Neil, this volume), and do not contribute significantly to the epidemiology of malignant disease. Most naturally occurring retroviruses do not contain specific onco-genes and tumourigenesis by such agents cannot be attributed to the simple expression of genetic information carried within the viral RNA. However, since all retroviruses are capable of acting as inser-tional mutagens, as mentioned above, it is possible to reconcile these contrasting situations by postulating that cellular oncogenes are not only subject to transduction, but also represent targets for muta-genesis.

The first indications that this was indeed the case arose from studies on B-cell lymphomagenesis by avian leukosis virus (ALV). In the majority of such tumours, the ALV provirus was found to be integrated adjacent to a cellular oncogene, c-*myc*, previously identified as a component of an acutely oncogenic avian retrovirus (Fung *et al.*, 1981; Hayward, Neel and Astrin, 1981; Neel *et al.*, 1981; Payne *et al.*, 1981). Moreover, insertion of the provirus was accompanied by elevated levels of c-*myc* expression, suggesting that this process may at least contribute to the development of neoplasia. Significantly, the same virus is capable of inducing an erythroleukaemia when injected into a specific line of inbred chickens and in this case provirus integration has occurred in the vicinity of a different c-*onc* gene, *erb*-B (Fung *et al.*, 1983). Such observations not only provide a unifying theme by implicating cellular oncogenes in both the acute and non-acute malignancies associated with retrovirus infections, but also suggest a direct link between the type of tumour and the oncogene involved. More particularly, these considerations have provided a strong impetus for investigating other non-acutely oncogenic retrovirus systems with a view to identifying which, if any, cellular genes might be implicated in the disease. By far the commonest neoplasms associated with these viruses are leukaemias and lymphomas of various types, one notable exception being the induction of breast carcinomas by mouse mammary tumour virus (MMTV). In view of this unusual tissue tropism and other unique features of MMTV (Dickson and Peters, 1983), we have been interested in examining its role as an insertional mutagen and in particular whether a correspondingly unique cellular oncogene may be activated during carcinogenesis.

MOUSE MAMMARY CARCINOGENESIS

MMTV is a B-type retrovirus which induces carcinomas of the breast in several inbred strains of mouse. The virus is usually transmitted as a congenital infection acquired via the milk during nursing, but in some mouse strains it can be expressed from endogenous loci inherited via the germ line (Hilgers and Bentvelzen, 1978). Manifestation of the disease is strongly influenced by several factors, including genetic predisposition, hormonal status, parity, and environmental factors, such as diet and stress (Nandi and McGrath, 1973; Moore *et al.*, 1979). The virus also demonstrates a particularly striking tissue tropism, being almost exclusively expressed in the mammary glands of female mice. Following congenital infection, more than 50% of the mammary epithelium may overtly express viral antigens, yet the only phenotypic changes that have been reported are slight increases in lobuloalveolar differentiation and occasional hyperplasia (Cardiff and Young, 1980; Squartini, Basolo and Bistocchi, 1983). Full expression of the viral replicative functions therefore appears to have little effect on the behaviour of productively infected cells, a situation typical of non-acutely oncogenic retroviruses.

MMTV differs from most other retroviruses in having the potential to encode an additional gene, distinct from the normal replicative genes and located within the LTR of the provirus (Dickson and Peters, 1981; Donehower, Huang and Hager, 1981; Fasel *et al.*, 1982; Kennedy *et al.*, 1982). As yet, the function of these sequences, given the acronym *orf* (for open reading frame), remains totally obscure, but a discrete, spliced mRNA species appropriate for the expression of *orf* has been detected in some normal and hyperplastic mammary tissues (van Ooyen, Michalides and Nusse, 1983; Wheeler *et al.*, 1983). While it is tempting to speculate that the product of *orf* expression may contribute to some aspects of tumourigenesis, possibly in hyperplastic or preneoplastic situations, it is important to stress that the candidate *orf* mRNA is not normally detected in mammary tumour cells (Robertson and Varmus, 1979; Sen *et al.*, 1979). The *orf* sequences therefore do not constitute a viral oncogene, as presently defined, and there are no cellular counterparts other than as components of endogenous MMTV proviruses.

Although female mice exposed to milk-borne MMTV suffer a mammary tumour incidence in excess of 90%, there is typically a delay, of around 4–9 months, between exposure to virus and the overt appearance of a tumour (Moore *et al.*, 1979). When tumours eventually arise they appear to be clonal proliferations derived from a single, or at most a few, infected cell(s) (Cohen *et al.*, 1979; Cohen and Varmus, 1980; Fanning, Puma and Cardiff, 1980; Groner *et al.*, 1980; Morris *et al.*, 1980). Thus, in terms of the total number of virus–cell interactions in the animal, neoplasia is a very infrequent consequence and is more likely to reflect some stochastic event rather than the expression of a viral gene product. The clonality of the tumours can be judged from Southern blot analysis of the MMTV proviruses in the tissue DNA and, as a general rule, all virally induced mammary tumours contain at least one, and commonly several, new proviruses acquired by infection (Cohen *et al.*, 1979; Cohen and Varmus, 1980; Fanning *et al.*, 1980; Groner *et al.*, 1980; Morris *et al.*, 1980). The fact that such acquired proviruses can be detected against the background of endogenous MMTV sequences not only is evidence of clonality but also suggests that, in each tumour, one of the exogenous proviruses might have been responsible for tumour-igenesis, presumably by acting as an insertional mutagen. Since the number of target genes which would confer a selective growth advantage on the cell when perturbed or activated in this way is likely to be very small, a high proportion of the tumours might be expected to have an acquired provirus integrated into a common region or regions of the mouse chromosome. By applying molecular cloning techniques to characterise the sites of provirus integration in MMTV-induced tumours, two such common integration regions have recently been identified designated *int*-1 (Nusse and Varmus, 1982; Nusse *et al.*, 1984) and *int*-2 (Peters *et al.*, 1983; Dickson *et al.*, 1984). The remainder of this presentation will therefore review some of the experimental results which have led to these conclusions and discuss their implications in terms of mechanisms of carcinogenesis.

IDENTIFICATION OF COMMON INTEGRATION REGIONS IN MMTV-INDUCED MAMMARY TUMOURS

Figure 1 shows a typical analysis of MMTV proviruses in mammary tumour DNAs, using the standard techniques of restriction enzyme digestion, agarose gel electrophoresis and blot hybridization. The enzyme *Eco*RI has a single recognition site close to the mid-point of the MMTV genome, so that each provirus will be cleaved into two fragments, the sizes of which will depend on the location of *Eco*RI sites in the neighbouring cellular DNA. For simplicity, only the 3' virus–cell junctions are depicted, recognised by using a molecularly cloned probe specific for the 3' region of the MMTV provirus (Peters *et al.*, 1983). From these and other data, it is clear that most MMTV-induced tumours contain multiple acquired proviruses, super-imposed on the endogenous pattern, any one of which could be implicated in tumourigenesis. Ideally, a tumour with only a single new provirus would present the simplest situation in which to investigate the possibility of insertional mutagenesis, but while this pertained in the identification of the *int*-1 region (Nusse and Varmus, 1982) such tumours turn out to be relatively rare. The experiments leading to the identification of *int*-2 were performed on two independent tumours, both of which contained two newly acquired proviruses (Peters *et al.*, 1983; Dickson *et al.*, 1984).

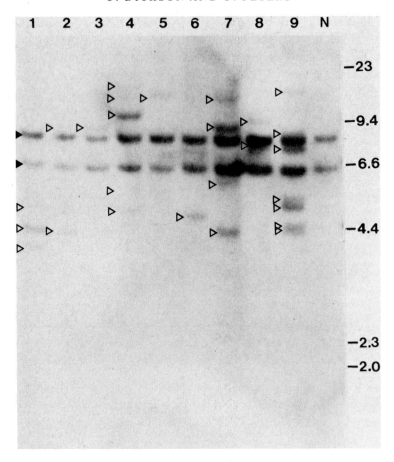

Fig. 1. Southern blotting analysis of MMTV proviruses in mouse mammary tumours. High molecular weight DNA from a series of virally induced mammary tumours (1–9) and normal tissue (N) from the BR6 strain of mouse was digested with the enzyme *Eco*RI and the resultant fragments were fractionated by electrophoresis in an agarose gel. After transfer to nitrocellulose (Southern, 1975) the virus–cell junctions were detected by hybridisation to a radioactive probe specific for the 3′ portion of the MMTV provirus (Peters *et al.*, 1983). Bands on the autoradiograph identified by the symbol ▶ indicate the 3′-specific junction fragments derived from the endogenous MMTV sequences common to all tissues in the BR6 mouse; ▷ indicates proviral junctions unique to each tumour, acquired by infection by milk-borne MMTV. Some tumour DNAs in this figure appear to contain a single exogenous provirus; however, the use of other restriction endonucleases usually shows this not to be the case and demonstrates how the endogenous fragments may effectively mask the presence of newly acquired sequences. The sizes of the various virus–cell junctions detected in this way were estimated relative to fragments of bacteriophage λ DNA generated by digestion with *Hin*dIII (shown in kilobases on the right of the figure).

Fig. 2. The strategy used to analyse sites of proviral integration in MMTV-induced mammary tumours. As depicted schematically in panel (*a*), an infecting provirus is likely to integrate into only one of any pair of chromosomes, introducing new restriction enzyme sites, indicated as ↓ and ↓. Using an enzyme, such as *Eco*RI, which cuts the viral sequences once (↓), each integrated provirus yields two virus–cell junctions, the sizes of which will depend on the

(a)

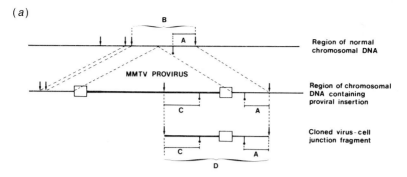

(b)

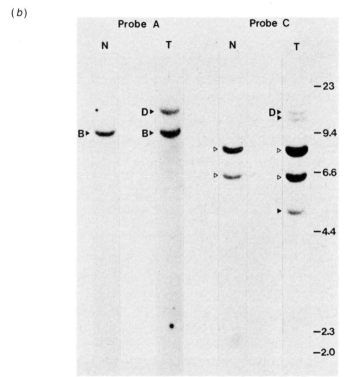

Fig. 2. (contd)
position of restriction sites in the neighbouring cellular DNA. Recombinant DNA clones of the 3'-specific junctions (such as fragment D) can therefore be isolated by hybridisation to a probe (C) specific for the 3' region of the provirus. Such clones subsequently permit the preparation of probe (A) specific for the adjacent cellular DNA.

When used in blot hybridisation analysis of normal cellular DNA (N) (cut at sites ↓), probe (A) will recognise a unique restriction fragment (B) present on both chromosomal copies. An example of this is shown in panel (b). In tumour DNA (T) on the other hand, the probe will recognise the normal unoccupied site (B) and a novel fragment (D) generated by provirus insertion. The same novel fragment will therefore be detectable using a virus-specific probe (C) as exemplified in panel (b). This latter analysis also illustrates the presence of endogenous MMTV sequences (▷) in the normal tissues of laboratory mice, and of multiple acquired proviruses (▶) in the virally induced mammary tumours.

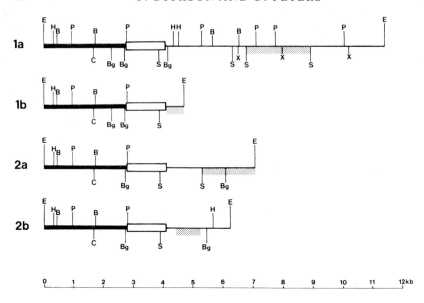

Fig. 3. Restriction enzyme analysis of cloned virus–cell junction fragments from two MMTV-induced tumours. Recombinant phage libraries were constructed from *Eco*RI digested DNA from two MMTV-induced mammary tumours in BALB/c mice (Peters *et al.*, 1983; Dickson *et al.*, 1984). Both tumours (1 and 2) contained two acquired proviruses (a and b), the 3'-specific junctions were isolated as clones and characterised by restriction enzyme digestion and hybridisation to virus-specific probes. Viral sequences are represented by bold lines, the viral LTR by the open boxes, and cellular DNA by the thinner lines. The recognition sites for the enzymes *Bam*HI (B), *Bgl*II (Bg), *Cla*I (C), *Eco*RI (E), *Hin*dIII (H), *Pst*I (P) and *Xba*I (X) are indicated. Fragments of unique sequence cellular DNA, depicted by the stippled boxes, were selected and sub-cloned for use as probes for the respective integration sites, as summarised in Table 1.

The approach adopted by the different laboratories was essentially identical, namely the isolation and characterisation of molecular clones containing the 3' virus–cell junction fragments from each provirus (see Fig. 2). From each of these clones, it was then possible to obtain fragments of unique-sequence cellular DNA flanking the provirus for use as hybridisation probes. Such probes are specific for the region of cellular DNA into which the respective provirus had integrated (operationally termed the unoccupied site) and, in normal tissues, they typically recognise discrete restriction fragments present on both chromosomal copies. In contrast, in tumours where a provirus has been inserted into this region, novel restriction fragments will be generated from the interrupted chromosome, together with the normal pattern derived from the unoccupied site on the other chromosome. Examples of this type of analysis are shown in Fig. 2.

For insertional mutagenesis to be a candidate for a mechanism for carcinogenesis, the same region or regions of chromosomal DNA

Table 1. *MMTV provirus integration sites in two virally induced mammary tumours in BALB/c mice*

Tumour	Designation of clone	Size in* kb	Size of unoccupied integration site	Proportion of BR6 tumours where MMTV insertion detected
1	1a	11.4	9.7	1/40
	1b	4.7	5.4	0/40
2	2a	7.0	10.0	11/40
	2b	6.2	6.2	0/40

*Only 3'-specific virus–cell junctions were obtained by molecular cloning, and all DNA fragments presented in this Table were generated by digestion with *Eco*RI.

might be expected to have sustained a provirus insertion in multiple independent tumours. However, as shown in Fig. 3, the four virus–cell junction fragments cloned from two independent mammary tumours (Dickson *et al.*, 1984) showed no obvious similarities in the restriction enzyme maps of flanking cellular DNA. The probes specific for the cellular DNAs (Fig. 3) each recognised unique unoccupied sites in normal DNA, as summarised in Table 1. Significantly, when these probes were used to screen a larger series of virally induced mammary tumours, by restriction enzyme digestion and blot hybridisation of tumour DNAs, one of the two proviruses in each tumour (1a and 2a), was found to reside in a region of chromosomal DNA which also harboured proviruses in other tumours. In the case of tumour 1, only one other example was detected, but with the probe from tumour 2, the results were more dramatic, indicating that the same 10kb segment of cellular DNA was interrupted by an MMTV provirus in 11 other tumours (Table 1). In contrast the two other proviruses, 1b and 2b, appeared to be in sites which were not interrupted in any of the other mammary tumours examined.

CHARACTERISATION OF PROVIRUS INTEGRATION REGIONS

The data discussed above were entirely consistent with the concept of insertional mutagenesis but, as such, suggested that more than one region of cellular DNA might be involved. To further delineate these common integration regions, molecular clones of DNA surrounding the insertion sites in the two tumours were isolated. Characterisation of the recombinant phage clones, obtained in conjunction with DNA blot analysis of mouse DNA, showed that the two regions were parts of a single domain, which we have termed *int-2*. In addition to obtaining a restriction enzyme map of the normal mouse cellular DNA in the common integration region, judicious use of viral probes from either the 3' or 5' regions of the MMTV provirus allowed the relative position, and transcriptional orientation of each provirus located within *int-2* to be determined. In a few cases, proviruses that had inserted close to but outside the region of the cloned DNA could only be approximately placed and orientated. The results of this analysis are illustrated schematically in Fig. 4 and show that while integration has occurred at multiple sites within *int-2*, the distribution of proviruses is clearly non-random, falling into two distinct but oppositely orientated groups, separated by 6–7 kb of cellular DNA. Of the mammary tumours entered into the analysis, approximately half showed integrations in the *int-2* region, each represented by a provirus integration event in Fig. 4.

An assumption implicit in these studies is that insertional mutagenesis will directly influence the expression or behaviour of a functional cellular gene, and therefore that the *int-2* region must contain such a gene. To examine this possibility, polyadenylated RNA was prepared from a series of virally induced mammary tumours and analysed by gel electrophoresis and blotting procedures. In view of the distribution of proviruses within the *int-2* region, hybridisation was carried out using a probe specific for the region of DNA lying between the two clusters of integration sites. Out of the seven tumours from which RNA was obtained, four were found to contain a 3.2 kb RNA species, plus some minor additional forms, which were specific for the *int-2* region (Dickson *et al.*, 1984). Moreover, the presence of *int-2* transcripts in these tumours correlated with the detection of an integrated provirus, the three negative examples showing no evidence for interruption in approximately 30 kb of cellular DNA around the *int-2* locus. These findings suggest that insertion of a provirus within the *int-2* region activates the expression of an otherwise silent cellular gene, a view supported by the failure to detect the 3.2 kb RNA in other adult mouse tissues, including mid-pregnancy mammary gland. However, a more extensive survey of normal and foetal tissues will be required before any

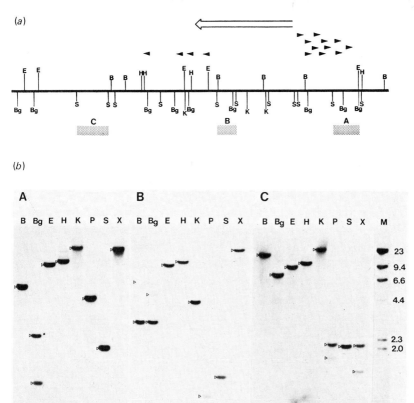

Fig. 4. Topography of the common integration region *int-2*. Using the cellular DNA probes 1a and 2a (Fig. 3 and Table 1), a series of overlapping recombinant DNA clones were isolated which spanned approximately 30 kb around the mouse *int-2* locus. These were characterised by restriction enzyme digestion and the data used to construct the physical map shown in panel (*a*). The cleavage sites for the enzymes *Bam*HI (B), *Bgl*II (Bg), *Eco*RI (E), *Hin*dIII (H), *Kpn*I (K) and *Sac*I (S) are indicated. Three unique DNA fragments, depicted by the stippled boxes A, B and C, were used as hybridisation probes in blot analyses of normal cellular DNA from BR6 mice (panel (*b*)). Complete agreement was noted between the observed and predicted restriction fragments detected with these probes. Note probe A is equivalent to 2a in Fig. 3, and probe C to 1a. ▷ indicates hybridising fragments.

Southern blot analyses of this type, and the use of probes for the 5' and 3' ends of the MMTV genome, enabled the site and orientation of proviruses within *int-2* to be determined. These are depicted as arrowheads (▶) above the restriction map in panel (*a*), each symbol denoting the location and transcriptional orientation of a provirus in a single, independent tumour. The approximate limits and direction of the *int-2* transcription unit (Dickson *et al.*, 1984) are indicated by the open arrow.

definitive statements can be made regarding the normal site or temporal control of the expression of this gene.

HORMONE-DEPENDENT TUMOURS

The majority of tumours entered into the analysis of the *int*-2 region were derived from the BR6 strain mouse. This strain, developed by Foulds in 1949, is unusual in that it gives rise to mammary tumours whose growth may be either dependent or independent of the hormones associated with pregnancy. The former arise and grow rapidly during pregnancy but appear to regress completely during lactation and subsequent gland involution. At the next pregnancy the tumours reappear and eventually may proceed through several cycles of growth and regression before ultimately becoming hormone independent. Consequently, it was of interest to determine how provirus integration in general, and more specifically integration into the *int*-2 domain, correlated with the appearance and progression of these hormone-dependent tumours. To answer this question, tumour biopsies were taken from several pregnancy-dependent nodules during each cycle of growth, and high molecular weight DNA was prepared for Southern blotting analysis (Peters *et al.*, 1984*b*). As anticipated, newly acquired proviral elements were present in all tumour series examined, including biopsies from the first detectable cycle of growth. More significantly, the same virus–cell junction fragments were present in all subsequent biopsies from any one site (see Fig. 5), suggesting that each cycle of growth represents the repeated expansion of a single, clonal population of cells. Transition to hormone independence also occurred within the same clonal population, without any overt changes in the patterns of integrated proviruses, although in several instances additional MMTV proviruses were acquired in later cycles, presumably as a result of superinfection.

Since approximately 50% of all BR6 mammary tumours, whether hormone dependent or hormone independent, showed evidence of MMTV integration in *int*-2, it was important to establish whether this event occurred early in the development of the disease, or at a later stage during the progression to autonomous growth. Several of the cycling tumour series under investigation were found to have an insertion in *int*-2 and, significantly, the acquired provirus was detectable at the earliest appearance of the tumour (Fig. 5*B*) and remained apparently unaltered during subsequent cycles. It would therefore

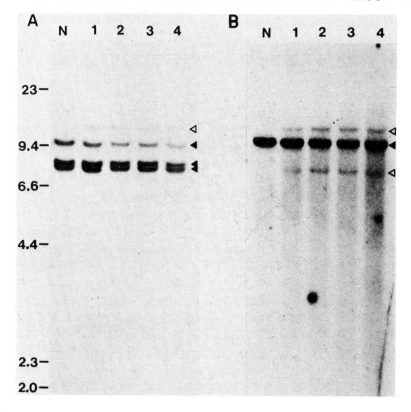

Fig. 5. MMTV proviruses in recurring pregnancy-dependent mammary tumours. Biopsies amounting to approximately 50% of the tumour mass were taken at each of four cycles of pregnancy-dependent growth (1–4) of a tumour in a BR6 mouse. DNA prepared from these biopsies, and from normal tissue of the same mouse (N) was subjected to restriction enzyme digestion and agarose gel electrophoresis as described elsewhere (Peters *et al.*, 1984*a*). Nitrocellulose filters carrying these DNAs were hybridised either with a 3′ specific viral DNA probe (panel A) to detect endogenous (▶) and newly acquired (▷) virus–cell junction fragments or with an *int*-2 specific probe (panel B). The latter detected the normal unoccupied site, as defined by a 10 kb *Eco*RI fragment (▶) and two additional fragments (▷) resulting from provirus insertion within *int*-2 in this tumour. The position of fragments of *Hin*dIII digested λ DNA and their sizes in kilobases are indicated on the left of the figure.

In a truly clonal tumour, the intensities of the novel fragments detected with either probe would be expected to be approximately half that of the diploid, endogenous sequences present in normal tissue. In practice, this quantification can be distorted by the inclusion of normal tissue in the tumour biopsy. However, the salient feature of these analyses is that each recurrence of the tumour represents proliferation of the same clonal population of MMTV-infected cells and that integration within *int*-2 is an early event in the development of the disease.

appear that activation of the *int*-2 gene may represent a crucial, early event in mammary carcinogenesis, but may not in itself be sufficient to induce frank malignancy, since the proliferation of the affected cells is still markedly influenced by the hormonal status of the animal.

TWO ANALOGOUS BUT DISTINCT INTEGRATION
REGIONS EXIST IN MOUSE DNA

A comparison of the above findings for the *int*-2 locus with those described for *int*-1 by Nusse *et al.* (1984) reveals many analogies. Perhaps the most striking is the relative positioning and orientation of proviruses located in these regions in different tumours. In both cases, the distribution is non-random, falling into two distinct clusters separated by a central stretch of DNA in which no integrations have been detected. While proviruses in each group share the same transcriptional orientation, the two groups are transcribed in opposite directions, pointing outward from the central segment. Moreover, in both *int*-1 and *int*-2, provirus integration on either side of the central region is accompanied by expression of polyadenylated RNA transcripts specific for this section of chromosomal DNA (Dickson *et al.*, 1984; Nusse *et al.*, 1984).

Despite these similarities, *int*-1 and *int*-2 show no obvious structural homology and do not cross-hybridise with one another or with other known viral oncogenes. They are also located on different mouse chromosomes, *int*-1 being on chromosome 15 (Nusse *et al.*, 1984) while *int*-2 is on 7 (Peters, Kozak and Dickson, 1984a). Of the tumours used to establish the existence of *int*-1, approximately 75% showed insertions into this region, while of the 40 tumours used in the *int*-2 analysis only 25% mapped to *int*-1 compared to 50% in *int*-2. The reasons for this variation in distribution between the two sites are obscure but may reflect differences in both viral and murine strains used. In both sets of data, there were examples of tumours which did not contain a provirus that mapped to either locus. Three possible explanations for this discrepancy can be envisaged. The first is that the maximum limits of the *int*-1 and *int*-2 domains may be greater than as presently defined and that extension of these analyses may eventually account for all virally induced tumours. Secondly, it is feasible that additional common integration regions exist, which may be influenced by provirus integration in these other tumours. Alternatively, the type of insertional mutagenesis described for *int*-1 and *int*-2 may be only one of several quite different mechanisms by which virally induced mammary tumours might arise. In this context, it is worth recalling that only about half of feline T-cell lymphomas associated with feline leukaemia virus show transduction or insertional mutagenesis of the *myc* gene (Neil, this volume), while those induced by bovine leukaemia virus (Burny *et al.*, this volume) and human T-cell leukaemia virus (Rattner *et al.*, this volume), neither of which appear to encode an oncogene, have so far revealed no evidence for common integration regions.

CONCLUSIONS AND SPECULATIONS

The molecular analysis of provirus integration sites in MMTV-induced mammary tumours has indicated that, in a majority of these tumours, an acquired provirus resides in either of two limited regions of chromosomal DNA, designated *int*-1 and *int*-2. Such findings are entirely consistent with the notion that proviral activation or insertional mutagenesis of particular cellular genes may be at least one of the causative factors in the development of the disease. As presently defined, both *int*-1 and *int*-2 span approximately 30 kb of cellular DNA but they are unique loci, located on different mouse chromosomes, and share no apparent homology with known viral and cellular oncogenes.

The central tenet of insertional mutagenesis is that, while proviral integration occurs essentially at random in the host cell DNA and can perturb virtually any cellular gene, on rare occasions such events may lead to detectable alterations in the growth properties of individual cells. One of the major points of discussion is therefore whether the frequent detection of proviruses in *int*-1 or *int*-2 reflects selection following chance integration or whether these regions of DNA constitute particularly favourable sites for provirus insertion. Although there are known examples of targeted integration, such as in baboon endogenous virus-infected human cell lines (Cohen and Murphey-Corb, 1983), retroviruses in general show no obvious preferences or specificity with respect to the DNA sequences surrounding the sites of integration. MMTV is no exception, and it is clear that, even within the defined integration regions *int*-1 and *int*-2, provirus insertion can occur at many different sites. Moreover, most of the tumours analysed in these studies contained multiple additional proviruses located elsewhere in the genome (Fig. 1). In the two rigorously studied examples reported here, these additional proviruses, represented by probes 1b and 2b, did not appear to reside in common integration regions (see Table 1).

Perhaps the most compelling evidence that *int*-1 and *int*-2 may be implicated in carcinogenesis, and not simply preferred targets for MMTV integration, is the finding that elements of each region are expressed as polyadenylated RNA, apparently as a consequence of provirus integration. Since neither RNA has yet been detected in tumours where there has been no proviral interruption of the respective integration locus, nor in normal, mid-pregnancy mammary glands, it seems likely that the influence of the MMTV provirus is to activate the expression of genes that are otherwise silent in these tissues. Within the limits of the analyses, activation has occurred irrespective of the location of the provirus within the defined region. However, despite the ostensibly random nature of the integration process, the distribution of proviruses detected in either *int*-1 or *int*-2 is clearly non-random. This apparent paradox can be explained if the ability of the provirus to exert a positive influence on gene expression is constrained by its position and orientation relative to the gene in question. The current evidence indicates that, in both *int*-1 and *int*-2, the major portion of the transcription unit lies between the two oppositely orientated clusters of provirus integration sites.

The fact that insertions within the exons of a gene would have primarily negative effects provides a plausible explanation for the absence of provirus integration within the central sections of *int*-1 and *int*-2, but the orientation of the proviruses outwards from this region of DNA is harder to rationalise. One obvious conclusion is that they cannot act by simple promoter insertion as is the case for the ALV provirus and c-*myc* in the majority of chicken B-cell lymphomas (Fung *et al.*, 1981; Hayward *et al.*, 1981; Neel *et al.*, 1981; Payne *et al.*, 1981). There the positioning of the provirus upstream and in the same orientation as the cellular gene allows the expression of hybrid mRNAs initiated at a viral promoter and reading into cellular sequences (Fig. 6). Curiously, this process seems to require concomitant deletions in the provirus, presumably to circumvent transcriptional interference between the two promoters in the viral LTRs (Cullen, Lomedico and Ju, 1984). In contrast, almost all the MMTV proviruses detected within *int*-1 and *int*-2 are ostensibly intact, and their positioning is more analogous to the minority (approx. 10%) of B-cell lymphomas in which promoter insertion does not operate (Payne, Bishop and Varmus, 1982).

At the present time, the most tenable explanation for the transcriptional activation of *int*-1 and *int*-2 is that of enhancement. So-called enhancer elements, capable of exerting a positive, *cis*-acting influence on the expression of adjacent genes, have recently been identified within the genomes of several DNA viruses, in some retroviral LTRs, and as components of particular eukaryotic genes (Khoury and Gruss, 1983). Available data indicate that the effect of such elements can be manifest over distances of several kilobases,

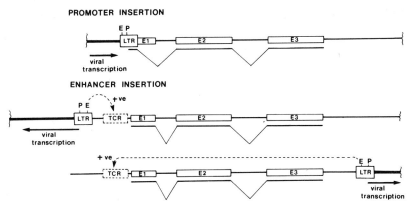

Fig. 6. Possible mechanisms for proviral activation of cellular genes. The figure shows a schematic portrayal of a cellular gene comprising three exons (E1, E2 and E3) whose expression is controlled by a transcriptional control element (TCR) at the 5′ end of the gene. In promoter insertion, a retroviral LTR disrupts the normal structure and presumably replaces the TCR such that expression of the gene comes under the control of the viral promoter (P) and enhancer (E) in the LTR. As a result, elevated levels of the cellular RNA are synthesised but as a hybrid transcript initiated within the viral LTR. In contrast, enhancer insertion envisages a less direct mechanism in which the presence of the viral enhancer (E) either upstream or downstream of the cellular gene exerts a positive influence on the synthesis of the normal RNA transcript. Since enhancer effects operate primarily on proximal promoters, the direction of viral transcription from the provirus must be such that the viral promoter (P) does not intervene between the enhancer (E) and the cellular transcriptional control unit (TCR).

irrespective of orientation, and there is increasing evidence that they may display inherent species or tissue specificity (Banerji, Olson and Schaffner, 1983; Gillies *et al.*, 1983; Queen and Baltimore, 1983; Spandidos and Wilkie, 1983). It is therefore tempting to speculate that the MMTV LTR may encode an enhancer-like activity which might function most effectively in mammary epithelial tissues. Moreover, the observation that enhancer effects can be blocked by the interposition of strong transcriptional promoters (de Villiers *et al.*, 1982; Wasylyk *et al.*, 1983) offers a rationalisation for the orientation of proviruses in *int*-1 and *int*-2, such that the viral promoter does not intervene between the enhancer and the cellular gene (Fig. 6). We hope that the current studies on the architecture of these genes will help to resolve many of these issues.

Whatever the details of the mechanism, the studies reviewed here suggest that the activation of either of two cellular genes may independently contribute to the development of mammary carcinomas. It will be interesting to determine whether these genes have comparable functions, operate at alternative positions of a single pathway, or act at different steps in tumour progression. However, at least in the case of *int*-2, it would appear that while activation may be a critical event, it is not in itself sufficient to give rise to a frank malignancy, since the resultant tumours may remain dependent on hormones for growth. Some other factors or genetic alterations may be involved in the subsequent transition to autonomous growth. One intriguing candidate would be the dominant transforming gene detected in mouse mammary tumour DNA by DNA transfection experiments (Lane, Sainten and Cooper, 1981), strengthening the analogy with ALV-induced lymphomagenesis where provirus activation of c-*myc* is associated with expression of a different dominant transforming gene (Cooper and Neiman, 1981). At present, such notions remain entirely speculative but, together with the potential of MMTV directly to influence the mammary epithelium, they represent a scheme of events which is compatible with the multifactorial nature of carcinogenesis, and which may soon be explicable at the molecular level.

REFERENCES

BANERJI, J., OLSON, L. and SCHAFFNER, W. (1983). A lymphocyte-specific cellular enhancer is located downstream of the joining region in immunoglobulin heavy chain genes. *Cell*, **33,** 729–40.

BISHOP, J. M. (1983). Cellular oncogenes and retroviruses. *Annual Reviews of Biochemistry*, **52,** 301–54.

CARDIFF, R. D. and YOUNG, L. J. T. (1980). Mouse mammary tumor biology: A new synthesis. In *Viruses in Naturally Occurring Cancers*, ed. M. Essex, G. Todaro and H. Zur Hausen, pp. 1105–14. New York: Cold Spring Harbor Laboratories.

COHEN, J. C. and MURPHEY-CORB, M. (1983). Targeted integration of baboon endogenous virus in the *BEVI* locus on human chromosome 6. *Nature*, **301,** 129–32.

COHEN, J. C., SHANK, P. R., MORRIS, V. L., CARDIFF, R. and VARMUS, H. E. (1979). Integration of the DNA of mouse mammary tumor virus in virus-infected normal and neoplastic tissue of the mouse. *Cell*, **16,** 333–45.

COHEN, J. C. and VARMUS, H. E. (1980). Proviruses of mouse mammary tumor virus in normal and neoplastic tissues from GR and C3Hf mouse strains. *Journal of Virology*, **35,** 298–305.

COOPER, G. M. and NEIMAN, P. E. (1981). Two distinct candidate transforming genes of lymphoid leukosis virus-induced neoplasms. *Nature*, **292,** 857–8.

COPELAND, N. G., JENKINS, N. A. and LEE, B. K. (1983). Association of the lethal yellow (A^y) coat color mutation with an ecotropic murine leukemia virus genome. *Proceedings of the National Academy of Sciences, USA*, **80,** 247–9.

CULLEN, B. R., LOMEDICO, P. T. and JU, G. (1984). Transcriptional interference in avian retroviruses – implications for the promoter insertion model of leukaemogenesis. *Nature*, **307,** 241–5.

DE VILLIERS, J., OLSON, L., BANERJI, J. and SCHAFFNER, W. (1982). Analysis of the transcriptional enhancer effect. *Cold Spring Harbor Symposium on Quantitative Biology*, **46,** 11–19.

DICKSON, C. and PETERS, G. (1981). Protein-coding potential of mouse mammary tumor virus genome RNA as examined by *in vitro* translation. *Journal of Virology*, **37,** 36–47.

DICKSON, C. and PETERS, G. (1983). Proteins encoded by mouse mammary tumour virus. *Current Topics in Microbiology and Immunology*, **106**, 1–34.

DICKSON, C., SMITH, R., BROOKES, S. and PETERS, G. (1984). Tumorigenesis by mouse mammary tumor virus: Proviral activation of a cellular gene in the common integration region *int*-2. *Cell*, **37**, 529–36.

DONEHOWER, L. A., HUANG, A. L. and HAGER, G. L. (1981). Regulatory and coding potential of the mouse mammary tumor virus long terminal redundancy. *Journal of Virology*, **37**, 226–38.

FANNING, T. G., PUMA, J. P. and CARDIFF, R. D. (1980). Selective amplification of mouse mammary tumor virus in mammary tumors of GR mice. *Journal of Virology*, **36**, 109–14.

FASEL, N., PEARSON, K., BUETTI, E. and DIGGELMANN, H. (1982). The region of mouse mammary tumor virus DNA containing the long terminal repeat includes a long coding sequence and signals for hormonally regulated transcription. *EMBO Journal*, **1**, 3–7.

FOULDS, L. (1949). Mammary tumours in hybrid mice: The presence and transmission of the mammary tumor agent. *British Journal of Cancer*, **3**, 230–9.

FUNG, Y.-K. T., FADLY, A. M., CRITTENDEN, L. B. and KUNG, H.-J. (1981). On the mechanism of retrovirus-induced avian lymphoid leukosis: Deletion and integration of the proviruses. *Proceedings of the National Academy of Sciences, USA*, **78**, 3418–22.

FUNG, Y.-K. T., LEWIS, W. G., CRITTENDEN, L. B. and KUNG, H.-J. (1983). Activation of the cellular oncogene c-*erb*B by LTR insertion: Molecular basis for induction of erythroblastosis by avian leukosis virus. *Cell*, **33**, 357–68.

GILLIES, S. D., MORRISON, S. L., OI, V. T. and TONEGAWA, S. (1983). A tissue-specific transcription enhancer element is located in the major intron of a rearranged immunoglobulin heavy chain gene. *Cell*, **33**, 717–28.

GRONER, B., BUETTI, E., DIGGELMANN, H. and HYNES, N. E. (1980). Characterization of endogenous and exogenous mouse mammary tumor virus proviral DNA with site-specific molecular clones. *Journal of Virology*, **36**, 734–45.

HARBERS, K., KUEHN, M., DELIUS, H. and JAENISCH, E. (1984). Insertion of retrovirus into the first intron of $\alpha 1\,(1)$ collagen gene leads to embryonic lethal mutation in mice. *Proceedings of the National Academy of Sciences, USA*, **81**, 1504–8.

HAYWARD, W. S., NEEL, B. G. and ASTRIN, S. M. (1981). Activation of a cellular oncogene by promoter insertion in ALV induced lymphoid leukosis. *Nature*, **290**, 475–80.

HILGERS, J. and BENTVELZEN, P. (1978). Interaction between viral and genetic factors in murine mammary cancer. *Advances in Cancer Research*, **26**, 143–95.

JENKINS, N. A., COPELAND, N. G., TAYLOR, B. A. and LEE, B. K. (1981). Dilute(d) coat colour mutation of DBA/2J mice is associated with the site of integration of an ecotropic MuLV genome. *Nature*, **293**, 370–4.

KENNEDY, N., KNEDLITSCHEK, G., GRONER, B., HYNES, N. E., HERRLICH, P., MICHALIDES, R. and VAN OOYEN, A. J. J. (1982). Long terminal repeats of endogenous mouse mammary tumor virus contain a long open reading frame which extends into adjacent sequences. *Nature*, **295**, 622–4.

KHOURY, G. and GRUSS, P. (1983). Enhancer elements. *Cell*, **33**, 313–14.

LANE, M.-A., SAINTEN, A. and COOPER, G. M. (1981). Activation of related transforming genes in mouse and human mammary carcinomas. *Proceedings of the National Academy of Sciences, USA*, **78**, 5185–9.

MOORE, D. H., LONG, C. A., VAIDYA, A. B., SHEFFIELD, J. B., DION, A. S. and LASFARGUES, E. Y. (1979). Mammary tumor viruses. *Advances in Cancer Research*, **29**, 347–418.

MORRIS, V. L., VLASSCHAERT, J. E., BEARD, C. L., MILAZZO, M. F. and BRADBURY,

W. C. (1980). Mammary tumors from BALB/c mice with a reported high mammary tumor incidence have acquired new mammary tumor virus DNA sequences. *Virology*, **100**, 101–9.

NANDI, S. and McGRATH, C. M. (1973). Mammary neoplasia in mice. *Advances in Cancer Research*, **23**, 353–414.

NEEL, B. G., HAYWARD, W. S., ROBINSON, H. L., FANG, J. and ASTRIN, S. M. (1981). Avian leukosis virus-induced tumors have common proviral integration sites and synthesize discrete new RNAs: Oncogenesis by promoter insertion. *Cell*, **23**, 323–34.

NUSSE, R., VAN OOYEN, A., COX, D., FUNG, Y.-K. T. and VARMUS, H. (1984). Mode of proviral activation of a putative mammary oncogene (*int*-1) on mouse chromosome 15. *Nature*, **307**, 131–6.

NUSSE, R. and VARMUS, H. E. (1982). Many tumors induced by the mouse mammary tumor virus contain a provirus integrated in the same region of the host genome. *Cell*, **31**, 99–109.

VAN OOYEN, A., MICHALIDES, R. J. A. M. and NUSSE, R. (1983). Structural analysis of a 1.7-kilobase mouse mammary tumor virus-specific RNA. *Journal of Virology*, **46**, 362–70.

PAYNE, G. S., BISHOP, J. M. and VARMUS, H. E. (1982). Multiple arrangements of viral DNA and an activated host oncogene in bursal lymphomas. *Nature*, **295**, 209–14.

PAYNE, G. S., COURTNEIDGE, S. A., CRITTENDEN, L. B., FADLY, A. M., BISHOP, J. M. and VARMUS, H. E. (1981). Analysis of avian leukosis virus DNA and RNA in bursal tumors: Viral gene expression is not required for maintenance of the tumor state. *Cell*, **23**, 311–22.

PETERS, G., BROOKES, S., SMITH, R. and DICKSON, C. (1983). Tumorigenesis by mouse mammary tumor virus: Evidence for a common region for provirus integration in mammary tumors. *Cell*, **33**, 369–77.

PETERS, G., KOZAK, C. and DICKSON, C. (1984*a*). Mouse mammary tumor virus integration regions *int*-1 and *int*-2 map on different mouse chromosomes. *Molecular and Cellular Biology*, **4**, 375–8.

PETERS, G., LEE, A. E. and DICKSON, C. (1984*b*). Activation of cellular gene by mouse mammary tumour virus may occur early in mammary tumour development. *Nature*, **309**, 273–5.

QUEEN, C. and BALTIMORE, D. (1983). Immunoglobulin gene transcription is activated by downstream sequence elements. *Cell*, **33**, 741–8.

ROBERTSON, D. L. and VARMUS, H. E. (1979). Structural analysis of the intracellular RNAs of murine mammary tumor virus. *Journal of Virology*, **30**, 576–89.

SEN, G. C., SMITH, S. W., MARCUS, S. L. and SARKAR, N. H. (1979). Identification of the messenger RNAs coding for the *gag* and *env* gene products of the murine mammary tumor virus. *Proceedings of the National Academy of Sciences, USA*, **76**, 1736–40.

SOUTHERN, E. M. (1975). Detection of specific sequences among DNA fragments separated by gel electrophoresis. *Journal of Molecular Biology*, **98**, 503–17.

SPANDIDOS, D. A. and WILKIE, N. M. (1983). Host-specificities of papillomavirus, Moloney murine sarcoma virus and Simian virus 40 enhancer sequences. *EMBO Journal*, **2**, 1193–9.

SQUARTINI, F., BASOLO, F. and BISTOCCHI, M. (1983). Lobuloalveolar differentiation and tumorigenesis. Two separate activities of mouse mammary tumor virus. *Cancer Research*, **43**, 5879–82.

VARMUS, H. E. (1982). Recent evidence for oncogenesis by insertion mutagenesis and gene activation. *Cancer Surveys*, **1**, 309–19.

VARMUS, H. E., QUINTRELL, N. and ORTIZ, S. (1981). Retroviruses as mutagens:

Insertion and excision of a nontransforming provirus alter expression of a resident transforming provirus. *Cell*, **25**, 23–36.

WASYLYK, B., WASYLYK, C., AUGEREAU, P. and CHAMBON, P. (1983). The SV40 72 bp repeat preferentially potentiates transcription starting from proximal natural or substitute promoter elements. *Cell*, **32**, 503–14.

WEISS, R., TEICH, N., VARMUS, H. and COFFIN, J. (eds.) (1982). *Molecular Biology of Tumor Viruses: RNA Tumour Viruses*. New York, Cold Spring Habor Laboratories.

WHEELER, D. A., BUTEL, J. S., MEDINA, D. M., CARDIFF, R. D. and HAGER, G. L. (1983). Transcription of mouse mammary tumor virus: Identification of a candidate mRNA for the LTR gene product. *Journal of Virology*, **46**, 42–9.

BOVINE LEUKAEMIA VIRUS: A TANTALIZING STORY

A. BURNY*†, C. BRUCK†, Y. CLEUTER†,
D. COUEZ†, J. DESCHAMPS†, J. GHYSDAEL†,
D. GREGOIRE†‡, R. KETTMANN*†,
M. MAMMERICKX†‡, G. MARBAIX† and
D. PORTETELLE*†

*Faculty of Agronomy, Gembloux, Belgium
†Department of Molecular Biology, University of Brussels, Belgium
‡National Institute for Veterinary Research, Uccle-Brussels, Belgium

Enzootic bovine leukaemia (EBL) has been recognized as a neoplasm of infectious origin for half a century. The agent was identified in 1969 by Miller et al. (1969) as a virus inducing a humoral immune response in the recipient host (Ferrer, Avila and Stock, 1972; Miller and Olson, 1972). That bovine leukaemia virus (BLV) was indeed a retrovirus was suggested by Kaaden, Dietzschold and Straub (1972) and demonstrated by Kettmann et al. (1975, 1976). It was also soon established, as suggested by repeated epidemiological observations, that BLV was an exogenous agent, establishing EBL as a typical infectious disease (Callahan et al., 1976; Kettmann et al., 1976). A transmissible cancer and its inducing agent were thus described in domestic cattle and sheep.

The scientific impact of the bovine leukaemia system was recently broadened by the discovery of the human T-cell lymphotropic viruses HTLV-I, HTLV-II and HTLV-III (see Yoshida et al. and Ratner et al. this volume). BLV, HTLV-I and HTLV-II share a number of biological and biochemical characteristics which suggests that the three viruses belong to a new family of retroviruses. At the time of writing the relatedness of BLV to HTLV-III is not known.

In this review we shall focus on the characteristics of BLV proviral DNA sequences and BLV proteins. We shall also illustrate the potency of an enzyme-linked immunosorbent assay (ELISA) detection method using a monoclonal antibody to gp51 epitope E, in a very large survey of BLV infection in Belgian cattle. Reviews on EBL and BLV were published previously (Burny et al., 1978, 1980, 1984; Mussgay and Kaaden, 1978; Ferrer, 1980; Ghysdael et al., 1984). We refer the reader to them for less recent information.

THE VIRUS

BLV has not been satisfactorily classified into any of the recognized retroviral groups but shows morphological resemblance to the type C leukaemia viruses of other animal species (Ferrer, Stock & Lin, 1971; Weiland and Überschär, 1976; Dekegel *et al.*, 1977; Calafat, Hageman and Ressang, 1974). Like other retroviruses, BLV contains a 60–70 S RNA, an RNA-dependent DNA polymerase and several unglycosylated and glycosylated major structural proteins with molecular weights from 10 000 to 51 000. However, BLV differs immunologically from practically all known retroviruses with the exceptions of ovine leukaemia virus (OLV, an entity probably identical to BLV) (Rohde *et al.*, 1978) and the more recently discovered human T-cell leukaemia virus (HTLV-I and perhaps HTLV-II) (Oroszlan *et al.*, 1979, 1982; Copeland *et al.*, 1983*a,b*; Schultz *et al.*, 1984). From these data, it was concluded that BLV is the prototype of a distinct family of retroviruses, in which the extent of genetic drift was estimated to be low by comparing p12 sequences of very distant virus 'strains', one from America the other one from Belgium (Burny *et al.*, 1984).

The structural organization of BLV has been examined by electron microscopy and by crosslinking reagents which can chemically link closely associated amino groups (Uckert *et al.*, 1982). Protein––protein interactions were analysed by the combined techniques of chemical cross linking together with two-dimensional diagonal polyacrylamide gel electrophoresis. The nearest neighbourhoods between lipid and protein and between RNA and protein were also investigated by means of chemical crosslinking (Uckert *et al.*, 1984). The results of these studies are reported in Table 1 (Uckert *et al.*, 1984).

From these data, a structural model was proposed for BLV (Uckert *et al.*, 1984). Its salient features are:

(i) the outer membrane consists mainly of gp30–gp51 dimers – complexes formed of two molecules of both gp51 and gp30 also exist;

(ii) internal structural proteins exist as homologous and heterologous dimers;

(iii) p15 exhibits two molecular forms (Prachar and Hlubinova, 1980; Uckert *et al.*, 1982, 1984) and is involved in complexes with itself, with p10, with lipids and with RNA;

(iv) p12, the nucleic acid binding protein (Morgan *et al.*, 1983) is found complexed to RNA.

Table 1. *Nearest neighbourhoods identified in BLV*

Reagent	Crosslinked complexes
Protein–protein crosslinking	
DTBPI,	(gp30–gp51)
MBI	(gp30–gp51)$_2$
	(gp30)$_2$
	(p24)$_2$
	(pp15)$_2$
	(p12)$_2$
	(p10)$_2$
	(p10–pp15)
	(p10–p24)
Lipid–protein crosslinking	
DMS	PE–pp15
RNA–protein crosslinking	
DEB	RNA–pp15
	RNA–p12

DEB: diepoxybutane; DMS: dimethylsuberimidate; DTBPI: dimethyldithiobispropionimidate; MBI: methylmercapto-butyrimidate; PE: phosphatidylethanolamine.

BLV PROVIRUS

BLV provirus structure

BLV replicates in a manner similar to that of all other known retroviruses: virus RNA is copied to double-stranded DNA by the process of reverse transcription and proviral DNA copies are inserted into the chromosome of the host cell (for a review see Weiss *et al.*, 1982). As is the rule for all retroviruses, BLV proviruses contain the viral genes flanked by long terminal repeat (LTR) sequences. The latter are covalently linked to cell DNA when the provirus is integrated into the host cell genome. The total length of BLV proviral DNA amounts to about 9 kb. Restriction maps have been published for America, (FLK–BLV), a Belgian and Japanese cloned proviruses. As expected, the maps derived for the FLK–BLV and the Japanese proviruses are very similar because the Japanese clone was from an Holstein cow. The Belgian isolate showed some variations, especially in the *gag* and *pol* gene regions (Kettmann, Couez and Burny, 1981; Deschamps, Kettmann and Burny, 1981; Kashmiri, Mehdi and Ferrer, 1984; Sagata *et al.*, 1983; E. Gelmann, R. Kettmann and A.

Burny, unpublished). Such observations do not necessarily mean that protein sequences vary significantly from one BLV variant to another. Indeed, comparison of p12 amino-acid sequences of FLK–BLV with those derived from the sequence of the Belgian BLV clone confirms that variations among BLV strains are minimal and thus re-emphasizes the significance of the observed kinship between the p12s of BLV and HTLV (Copeland *et al.*, 1983*a*; Burny *et al.*, 1984).

DNA sequence analysis

Large portions of the BLV proviruses from the FLK and the Belgian isolates have been sequenced in the hope that a detailed characterization of the genome and the comparison between different isolates will yield clues to the mechanism of leukaemogenesis.

Sequence of LTR from the Belgian tumour-derived BLV provirus clone: Fig. 1 (Couez *et al.*, 1984)

Sequencing of LTRs was a high priority as it has been suspected (Czernilofsky *et al.*, 1980; Robinson *et al.*, 1982; Tsichlis and Coffin, 1980; Lenz *et al.*, 1984; Chen *et al.*, 1984) that the U_3 region of the LTR plays a key role in leukaemogenesis by a number of viruses.

The sequence starts at the *Eco*RI site located in the X region (see below) and extends up to a second *Eco*RI site in cell DNA. The 3' end of the LTR has been identified by comparing sequences of clones derived from the 5' and 3' LTR structures. However, uncertainties persist. If we consider that only one base separates the primer binding site (PBS^-) from the 3' end of the 5' LTR, then the LTR extends over 536 nucleotides from position -215 to position 321. If, on the contrary, we leave two bases between PBS^- and the 3' end of the 5' LTR, the inverted repeat becomes GCAAACA and its best correspondence is TGGTTGC, which would place the putative beginning of the LTR at position -305 and extend the LTR to 625 nucleotides. A number of direct and inverted repeats are located in U_3 together with the CAT and TATA boxes and a putative Z-DNA region. From the data of Tsimanis *et al.* (1983), the polyadenylation site can be located at position 234. It follows that the R region is 234 base pairs long, a

```
(a)  -471                    G   AATTCGAGCT   GCCCCTTATC   CAAACGCCCG   GCCTGTCTTG   -431

     -430  GTCTGTCCCC   GCGATCGACC   TATTCCTAAC   CGGTCCCCCT   TCCCCATAGG   -381

     -380  ACCGGTTACA   CGTGTGGTCC   AGTCCTAAGG   CCTTACAACG   CTTCCTCCAT   -331
                 Z DNA

     -330  GACCCTACGC   TCACCTGGTC   AGAATTGGTT   CGTAGCGGGA   AACTAAGACT   -281
                                                    1

     -280  TGATTCACCC   TTAAAATTAC   AGCTGTTAGA   AAATGAATGG   CTCTCCCGCC   -231
                               5'ITR1

     -230  TTTTTTGAGG   GGGAGTCATT   TGTATGAAAG   ATCATGCAGG   CCTAGCGCCG   -181
               PSB+          IR              Z-DNA

     -180  CCACCGCCCC   GTAAACCAGA   CAGACACGTC   AGCTGCCAGA   GAAGCTGCTG   -131
                3            2           4                4

     -130  ACGGCAGCTG   GTGGTCAGAA   TCCCCGTACC   TCGCCAACTT   CCCCTTTCCC   -81
                5                        5

     -80   GAAAAATCCA   CACCCTGAGC   TGCTGACCTC   ACCTGCTGAT   AAATTAATAA   -31
                                         U3   5'CAP

     -30   AATGCCGGCC   CTGTCGAGTT   AGCGGCACCA   GAAGCGTTCT   CCTCCTGAGA    20

      21   CCCTACTGCT   CAGCTCTCGG   TCCTGAGCTC   TCTTGCTCCC   GAGACCTTCT    70

      71   GGTCGGCTAG   CCGGCAGCGG   TCAGGTAAGG   CAAACCACGG   TTTGGAGGGT   120

     121   GGTTCTCGGC   TGAGACCGCC   GCGAGCTCTA   TCTCCGGTCC   TCTGACCGTC   170

     171   TCCACGTGGA   CTCTCTCTCT   TGCCTCCTGA   CCCCGCGCTC   CAAGCGCGTC   220

     221   TGCCTTGCAC   CCGCGTTTGT   TTCCTGTCTT   ACTTTCTGTT   TCTCGCGGCC   270
                                                              3'LTR

     271   CGCGCTCTCT   CCCTCGGCGC   CCTCTAGCGG   CCAGGAGAGA   CCCGCCAAACA  320
            CELL         L'H                                    IR
     321   ACAAAGCACT   GGACATGACT   TAGAGACTGA   ACAACAACCA   TCTTGACGTG   370

     374   GCAGGTGGGC   AGGTTTATAA   GAAGCACATT   GGTTTGAATT   C            411
```

```
                          45
(b)       TGCAGGGGGGGGGGGGGAGCTCTCTTGCTCCCGAGACCTT......
                       5'LTR 321
     ......GCGGCCAGGAGAGACCCGCAAACAATTGGGGGCTCGTCCGGATT
                        IR              PBS-
```

Fig. 1. (*a*) Nucleotide sequence of the integrated 3' LTR with adjacent viral and cellular DNA in T15–14 fragment. (*b*) Partial nucleotide sequence of the 5' LTR in a 1.2 kb *Pst*I fragment.

The sequence of the plus strand (same sense as viral RNA) is presented. Position 1 in the sequence corresponds to the 5' end of the RNA genome, the capping site (↓). Important features of the BLV LTR sequence are: IR, inverted repeat; Z-DNA, zone of Z-DNA; E, enhancer core sequence; U_3, unique sequence derived from the 3' end of genomic RNA; PSB^+, plus strand binding site; PSB^-, tRNA^{Pro} binding site; LH, L'H, locally homologous regions to U_3 regions of the simian sarcoma virus LTR. The possible promoter sequences, Hogness and CAT boxes are shown in boxes. Arrows labelled 1 to 5 indicate direct or inverted repeats identified by Tsimanis et al. (1983). The signs suggest a second possibility for LTR boundaries and inverted repeats. Its advantages are (1) LTR begins with TG and ends with CA, features that are quite common; (2) two bases (rather than one) separate PSB^- from IR (see part (*b*)). (W. Haseltine, personal communication.)

value very similar to that found for the HTLV-I LTR (R = 228 base pairs). A computer-assisted search was used to look for homologous sequences in the BLV-LTR and the LTRs of Moloney leukaemia virus, simian sarcoma virus, Rauscher leukaemia virus, gibbon ape leukaemia virus, baboon endogenous virus and HTLV-I. No significant homology was revealed between these LTRs.

Sequence of the gene for envelope glycoprotein gp51 (Rice *et al.*, 1984)

The sequence of the *env* gene has been located by comparison of the translated DNA sequence with amino-acid sequence data on purified gp51 and gp30 (Schultz, Copeland and Oroszlan, 1984). It has been predicted that gp51 contains 268 amino-acids. It shows distant but nonetheless statistically significant homology to HTLV-I gp51 (Schultz *et al.*, 1984).

Sequence of the gene for transmembrane protein gp30 (Rice *et al.*, 1984)

The sequence coding for gp30 is at the 3' end of the open reading frame that contains the information for gp51 at its 5' end. The DNA regions coding for the respective envelope proteins have been delineated, using the N-terminal amino-acid sequence data of purified proteins (Schultz *et al.*, 1984). By this approach, it has been predicted that gp30 contains 214 amino-acids and will show structural features typical of type C viral transmembrane proteins. A two-dimensional model of protein folding for HTLV-I envelope proteins has been presented by W. Haseltine (personal communication). From the data obtained in our study (Rice *et al.*, 1984), it can be inferred that BLV and HTLV-I envelope proteins have very similar three-dimensional configurations.

Sequence of the X region (Rice *et al.*, 1984)

BLV provirus contains a region of 1817 base pairs between the presumptive terminator of the *env* gene (3' side of gp30) and the 5' end of U_3. (When the limits of LTR are taken to be those proposed by Couez *et al.* (1984).) This sequence contains two open reading frames: one frame where the potential protein is 128 amino-acids long and a second frame where the potential protein is 154 amino-acids. RNA processing involving additional exons could increase the size of the potential proteins considerably. Some remarkable similarities between the BLV and HTLV sequences in this region have been noted (Rice *et al.*, 1984). An intensive search is going on to identify the putative protein product. Is it at all related to the 18 000 M_r protein seen by Ghysdael *et al.* (1978, 1979) in translation experiments of viral RNAs *in vitro*? What role does it play as a candidate in leukaemogenesis by BLV?

BLV PROTEINS

Core proteins

The virus contains *gag*, *pol* and *env* genes and an *X* region. It has been proposed that four proteins are encoded in the *gag* gene region. They are designated p15, p10, p24 and p12 (Burny *et al.*, 1980; Morgan *et al.*, 1983; Schultz and Oroszlan, 1984; Uckert *et al.*, 1984).

p15 Two different p15 molecules seem to exist in BLV, as determined by gel electrophoresis data (Prachar and Hlubinova, 1980; Uckert *et al.*, 1982, 1984). p15 is a basic protein and has been identified as the major virus phosphoprotein (Uckert and Wunderlich, 1979). Chemical crosslinking experiments have located p15 in the vicinity of membrane lipids but also close to viral RNA (Uckert *et al.*, 1984). It has been observed that BLV p15 is myristilated and suggested that this myristilation of p15 might be relevant to the phenomenon of transformation (Schultz and Oroszlan, 1984). In the *gag* sequence, p15 is the first protein at the NH$_2$-terminus of the *gag* precursor.

p10 This is a highly basic protein with DNA binding capacity (Long, Henderson and Oroszlan, 1980). As shown in Table 1, it is found in homologous dimers (p10–p10) and heterologous complexes (p10–p15). A lower molecular weight has sometimes been attributed to this molecule. It is hoped that this puzzling situation will be soon clarified by protein and DNA sequencing data.

p24 It is the major core protein of BLV called in the early literature the ether-resistant antigen (Burny *et al.*, 1980). It has been characterized in detail and partially sequenced (Gilden *et al.*, 1975; Oroszlan *et al.*, 1979). Its relatedness to the HTLV-I p24 is striking. Monoclonal antibodies have been raised against two different epitopes (C. Bruck, M. Mammerickx, S. Mathot, D. Portetelle and A. Burny, unpublished). p24 exists as homologous dimers and complexes to p10.

p12 This is the RNA-binding protein. It has been sequenced and its primary sequence compared to the corresponding DNA. As expected, this is a proline-rich protein. The presence of two stop codons at the 3' end of the sequence indicates that p12 is on the carboxyl side of the *gag* protein. p12 exists as homologous dimers and is complexed to RNA in native viral particles.

Envelope proteins

Glycoproteins gp51 and gp30 correspond to the major envelope and transmembrane proteins, respectively. They are synthesized as a single glycosylated precursor of molecular weight 72 000, Pr 72^{env} (Ghysdael *et al.*, 1978; Mamoun *et al.*, 1983; Bruck *et al.*, 1984). It has been suggested that gp51 is a cleavage product of a larger molecule of molecular weight 60 000 (Schultz *et al.*, 1984). From amino-acid (Schultz *et al.*, 1984) and nucleotide sequencing data (Rice *et al.*, 1984), it was found that gp51 has 268 amino-acids and that gp30 has 214. The cleavage site between both proteins is an arginine, as observed in HTLV-I (Seiki *et al.*, 1983), in avian (Hunter *et al.*, 1983; Schwartz, Tizard and Gilbert, 1983), and in murine (Oroszlan *et al.*, 1980; Shinnick *et al.*, 1981) type C viruses.

HTLV-I and BLV gp51s display a number of similarities in amino-acid sequence. The two transmembrane proteins show 36% identity, in an alignment requiring 6 gaps. More distant relatedness is also seen between BLV gp30 and both murine leukaemia virus p15E and Rous sarcoma virus gp36.

Knowing the entire sequence of BLV gp51 will allow identification of the three biologically crucial epitopes (F, G, H) of the protein (Bruck *et al.*, 1982a,b). BLV variants have been identified which carry mutations affecting F, G or H. DNA sequence analysis of these variants and comparison with the sequence presented here should allow the presence of specific amino-acids to be correlated with regions of defined function and/or three-dimensional structure (Bruck *et al.*, 1984).

ORIGIN AND HOST RANGE OF BLV

The information pertinent to the origin of BLV derives from molecular hybridization experiments and stresses that the virus is exogenous to all animal species examined (Callahan *et al.*, 1976; Kettmann *et al.*, 1976). The discovery of human T-cell leukaemia viruses (HTLV-I and HTLV-II) (Poiesz *et al.*, 1980, 1981; Hinuma *et al.*, 1981; Kalyanaraman *et al.*, 1982), and their similarities with BLV, strongly suggest that the three viruses are members of the same family and derive from a common ancestor. They share biological and biochemical peculiarities (similar amino-acid sequences and tertiary structures in virus-coded proteins, induction of syncytia, presence of the X region between the *env* gene and the 3′ LTR, etc.) (Seiki *et al.*, 1983; Nagy *et al.*, 1983; Rice *et al.*, 1984; W. Haseltine, personal communication) and spread horizontally among target human or animal populations.

BLV behaves as a naturally infectious and neoplastic agent in cattle populations and induces tumours in sheep and goats after experimental infections. In ruminants, therefore, the virus exhibits oncogenic properties. Moreover, BLV can infect other animal hosts in nature, such as capybaras and water buffaloes (Marin *et al.*, 1982). Successful experimental transmission has been reported to occur with chimpanzees, macaques, pigs and rabbits (Burny *et al.*, 1984).

The possible infectivity of BLV for man is obviously an important question. All evidence accumulated so far indicates that natural infection of man by this virus has not been observed. Needless to say, this does not mean that recombination events between human viruses and BLV can not occur and lead to appearance of oncogenic entities detrimental to man.

THE DISEASE

Leukaemia and lymphosarcoma in cattle

BLV is an exogenous retrovirus which induces in cattle a natural disease called enzootic bovine leukosis (EBL), bovine leukaemia, bovine malignant lymphoma or bovine lymphosarcoma. It is a chronic disease evolving over extended periods of time (1–8 years) (Burny *et al.*, 1980).

BLV-infected cattle

Infection occurs through contact with BLV-infected animals. Spread of infection within a herd can most of the time be traced back to a newly introduced BLV carrier. The more intensive the husbandry practices, the faster the spread of infection. In temperate climates where agriculture is intensive, transmission of BLV is iatrogenic. In tropical climates, biting insects appear to be an additional mode of transmission (Burny *et al.*, 1984). In most cases if not all, BLV persists indefinitely in the host, inducing a strong humoral immune response with high plasma levels of antibodies specific for BLV. In the first 6 months to 1 year post-infection, the presence of virus can only be detected by serological techniques. Southern blot analysis

(Southern, 1975) of DNA from circulating white cells remains negative for the first 6 months to 1 year. Later on, BLV proviral DNA becomes detectable.

BLV-infected cattle with persistent lymphocytosis (PL)

Persistent lymphocytosis, a condition linked to the presence of BLV, is apparently genetically determined. Some 30 to 70% of infected animals will sooner or later develop the condition. Only a fraction of circulating lymphocytes harbour copies of the provirus DNA at the level of one to four copies per cell. The virus-infected B-cell population is polyclonal and exhibits no gross chromosomal abnormalities (Hare, Yang and Mc Feely, 1967). The integrated provirus may be maintained in a repressed state by the action of a repressor protein (Gupta and Ferrer, 1982; Gupta, Kashmiri and Ferrer, 1984).

BLV-infected cattle in the tumour phase

Development of tumours in the BLV-infected host is a rare event. The proportion of virus carriers that develop tumours during their normal life span is estimated to be less than 1%. This increases, of course, with age; it also depends on genetic make-up and probably undefined environmental factors.

Several important characteristics of BLV-induced tumours are as follows:

(1) Most tumours are a rather homogeneous lymphoid cell population, harbouring one to four copies of BLV proviral DNA.

(2) According to cytological and functional criteria, the vast majority of EBL tumours are of the B-cell type (Parodi *et al.*, 1982). That this is indeed the case was confirmed by monoclonal antibodies specific for the bovine B-cell (S. Black, personal communication).

(3) Whether unique or multiple, the proviral copy(ies) may be complete or deleted. Deletions affecting the 3′ side of the proviral DNA have not been encountered so far.

(4) With respect to provirus integration site, bovine tumours are monoclonal populations of cells. The integration sites, however, are not conserved from one tumour to the next and different tumours harbour the provirus in different chromosomes (Grégoire *et al.*, 1984).

(5) BLV proviruses present in tumours are in a repressed state. Neither viral RNA nor viral proteins can be detected, irrespective of the method used (Kettmann *et al.*, 1980, 1982).

Leukaemia and lymphosarcoma in sheep

BLV-infected animals

Natural infection of sheep flocks by BLV has been reported (Burny *et al.*, 1980), but seems to be very rare. Experimental transmission trials have indeed amply demonstrated that the virus does not spread horizontally within sheep flocks: BLV-injected sheep can be housed with control sheep, with no evidence of seroconversion in the latter.

The most striking features encountered in BLV-infected sheep are: (1) The provirus can be found as unintegrated linear molecules in the circulating leukocytes (Fig. 2). (2) Antibodies to biologically important epitopes of BLV gp51 appear first. Later on, the humoral immune response widens; antibodies to all viral proteins are produced, sometimes at very high titres. We believe that high antibody titres together with the presence of unintegrated provirus reflect intensive virus production in the haematopoietic organs. It should be remembered that the best BLV production system is based on the high capacity of sheep cells to replicate the virus (Van der Maaten and Miller, 1976). The high efficiency of this host–virus system is probably correlated (Chen *et al.*, 1984; Lenz *et al.*, 1984) with the high sensitivity of sheep to BLV-induced tumours. (3) Persistent lymphocytosis is not encountered in sheep. Once the number of circulating leukocytes rises, it invariably indicates the proliferation of the tumour cell clone.

The tumour phase

The five characteristics mentioned for cattle tumours hold true for sheep tumours, the major discrepancy between the animal species being the notably high sensitivity of sheep to the oncogenic potential of BLV. The sheep tumour cell is of the B type (A. Parodi, personal communication), as routinely observed from monoclonal antibody assays. However, transformation of a cell target in the early steps of lymphoid cell differentiation may lead to tumours of the T-cell type (Horvath, 1982).

Leukaemia and lymphosarcoma in the goat

BLV-infected goats

Natural infection by BLV has not been observed in goats. This animal species can be experimentally infected by the virus, but shows an extreme resistance to the onset of tumours. The experimental goat

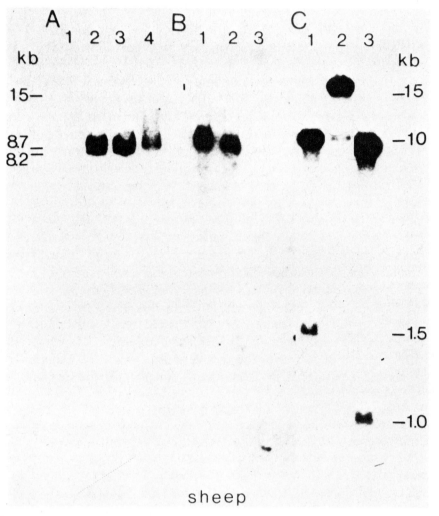

Fig. 2. Hybridization patterns of ^{32}P-labelled cloned BLV DNA on sheep DNAs. Leukocyte DNAs (10 μg each) of normal sheep (A, lane 1); sheep 72 (A, lane 2), sheep 73 (A, lane 3) and sheep 75 (A, lane 4) (Ab[+] animals) and tumour DNAs (10 μg each) from sheep 319 (C, lane 1), sheep 375 (C, lane 2) and sheep 20 (C, lane 3) were exhaustively digested by EcoRI, submitted to electrophoresis in a 0.8% agarose gel and analysed by hybridization. Autoradiographs are shown. Panel B represents the analysis of 10 μg each of undigested leukocyte DNAs from sheep 72 (B, lane 1), sheep 73 (B, lane 2) and sheep 75 (B, lane 3).

seroconverts shortly after infection and builds up a very strong humoral reaction to BLV structural antigens. Within a few months, BLV provirus can easily be demonstrated in Southern blots of peripheral lymphocyte DNA. Persistent lymphocytosis does not exist.

The tumour phase

Two tumour cases have been described so far (Olson *et al.*, 1981; Kettman *et al.*, 1984). They developed almost 10 years post-infection, a very long latency period indeed if compared to the normal life span of a goat. All the features of BLV tumours observed in cattle can also be seen in BLV-infected goats.

Cell target for BLV

It is now well established that the preferred target for BLV replication is a cell belonging to the B-lymphocyte lineage. Practically no information is available, however, about the host cell and the reasons why it acts as a target. The major obstacle to progress in this direction is the lack of techniques to grow specific subsets of B-cells. Our observation that only circulating lymphocytes of infected sheep carry unintegrated BLV proviral DNA is an indication that infection can occur very late in the process of cell differentiation. We recently succeeded in growing tumour cells from the peripheral blood of a case of bovine leukosis. The cells contained four proviruses, apparently complete. After three months in culture, the growing cells were a pure population of BLV-carrying cells showing a BLV restriction pattern identical with that found in the unique tumour of the animal of origin. No expression of any part of the BLV genome could be demonstrated even after three months in culture. These data again emphasize that BLV expression is not required for maintenance of the tumour stage and stress the point that lack of BLV expression in the tumour cell lies at the provirus level, at least in some cases.

The recent discovery (Palacios *et al.*, 1984) that interleukin-3 supports growth of mouse pre-B-cell clones *in vitro* might open the road to major progress in the understanding of the initial steps of host–virus interplay.

Mechanism of virus-induced disease

The mechanism of leukaemogenesis by BLV is not understood. It now seems unlikely that tumour induction is related to the chromosomal site at which provirus DNA is integrated in fully developed tumours, unless activation of many different alternative oncogenes can lead to the neoplastic state. This being a remote possibility, we are inclined to the view that a viral gene product may be required at some stage during tumourigenesis, perhaps from the *X* region, by analogy with HTLV-I (see Yoshida *et al.* and Rattner *et al.*, this volume). Since the expression of provirus DNA is repressed in fully developed tumour cells, the gene product would not be required for maintenance of the transformed state. Alternatively, or additionally, some other mechanism, akin to the B-cell receptor model proposed by McGrath and Weissman (1979), may play a role.

EPIDEMIOLOGICAL STUDIES

As discussed in a recent review (Burny *et al.*, 1984), very sensitive methods of detection of BLV antigens or antibodies are available. The most popular serological tests are immunodiffusion and ELISA. The sensitivity of the latter compares easily with that of radio-immunoassay (RIA) and is faster and cheaper. It is also specific, with very low background, provided that a well-chosen monoclonal antibody is first adsorbed onto the plastic wells (Portetelle *et al.*, 1983). This technique has been used this year by one of us (M.M.) to screen for BLV reactors among 750 000 cattle. Pools of up to 85 sera were routinely examined. In several cases, a single positive serum among 80 negative ones has been detected (M. Mammerickx, D. Portetelle and A. Burny, unpublished).

CONCLUSIONS

The period covered by this review has been very fruitful for BLV research. Considered for a number of years as an enigma, the mode of action of BLV is now being illuminated. It is felt that the X region of the provirus encodes a protein the action of which is critical for cell transformation. As stated before, we are confronted with a new type of transforming agent. The virus carries a function directly involved in transformation but still induces a chronic leukaemia. The choice is between the following possibilities: (1) Cell transformation is not a rare event but transformed cells are eliminated by the immune system of the host. (2) Cell transformation is a rare event. The function fulfilled by X is necessary but not sufficient for cell transformation. Uncovering the mechanisms involved will be difficult and will require a better understanding of the intricacies of cell differentiation.

The data obtained from studies of gp51 have important implications for the design of a BLV vaccine. The repeated observation that passive antibodies to BLV structural proteins efficiently protect the calf or the lamb from infection by the administered virus has proved that vaccination should be feasible. Analysis of bovine and ovine immune sera has shown that protective sera contain immuno-globulins reactive against epitopes, F, G and H of BLV gp51, the same three epitopes that apparently play a role in the biological activities of the virus (infectivity and syncytia induction). The design of an efficient long-lasting BLV vaccine is not an easy task. It requires the large-scale production of defined and adequately folded epitopes of gp51. It has been suggested (P. Fischinger, personal communi-cation) that some epitopes of gp30 could be *ad hoc* determinants (interspecies determinants) to be used in vaccine trials, especially considering their highly conserved structure. This has obviously still to be tested.

A third and immediate outcome of BLV research has to do with epidemiology. The design of a very specific, sensitive, fast and cheap ELISA test has allowed large-scale surveys in the cattle population. Pools of as many as 80 sera can be analysed as a single assay. The presence of only one positive serum can be detected. Accurate and easy testing is a major part in any sanitation program in veterinary medicine. The data obtained in this part of the work show that BLV infection can be detected and controlled, a major step toward elimin-ation of enzootic bovine leukaemia.

Finally, our previous observation that BLV can be infectious to rabbits and induce lymphopaenia should be reconsidered in view of the involvement of members of the BLV family in induction of the acquired immuno-deficiency syndrome.

The accumulating results derived from BLV and HTLV research confirm that both viruses are related and probably derive from a common ancestor. Any progress made in one system may be used in the other; they cross-feed continuously. BLV research is not only rewarding for itself and profitable to animal health; it also provides valuable observations to be used in human medicine in relation to HTLV-induced diseases.

REFERENCES

BRUCK, C., MATHOT, S., PORTETELLE, D., FRANSSEN, J. D., HERION, P. and BURNY, A. (1982*a*). Monoclonal antibodies define eight independent antigenic sites on the bovine leukaemia virus envelope glycoprotein gp51. *Virology*, **122**, 342–52.

BRUCK, C., PORTETELLE, D., BURNY, A. and ZAVADA, J. (1982*b*). Topographical analysis by monoclonal antibodies of BLV-gp51 epitopes involved in viral functions. *Virology*, **122**, 353–62.

BRUCK, C., PORTETELLE, D., MAMMERICKX, M., MATHOT, S. and BURNY, A. (1984). Epitopes of bovine leukaemia virus glycoprotein gp51 recognized by sera of infected cattle and sheep. *Leukaemia Research*, **8**, 315–23.

BRUCK, C., RENSONNET, N., PORTETELLE, D., CLEUTER, Y., MAMMERICKX, M., BURNY, A., MAMOUN, R., GUILLEMAIN, B., VAN DER MAATEN, M. J. and GHYSDAEL, J. (1984). Biologically active epitopes of bovine leukaemia virus glycoprotein gp51: their dependence on protein glycosylation and genetic variability. *Virology*, **136**, 20–31.

BURNY, A., BEX, F., CHANTRENNE, H., CLEUTER, Y., DEKEGEL, D., GHYSDAEL, J., KETTMANN, R., LECLERCQ, M., LEUNEN, J., MAMMERICKX, M. and PORTETELLE, D. (1978). Bovine leukaemia virus involvement in enzootic bovine leukosis. *Advances in Cancer Research*, **28**, 251–311.

BURNY, A., BRUCK, C., CHANTRENNE, H., CLEUTER, Y., DEKEGEL, D., GHYSDAEL, J., KETTMANN, R., LECLERCQ, M., LEUNEN, J., MAMMERICKX, M. and PORTETELLE, D. (1980). Bovine leukaemia virus: molecular biology and epidemiology. In *Viral Oncology*, ed. G. Klein, pp. 231–89. New York, Raven Press.

BURNY, A., BRUCK, C., CLEUTER, Y., COUEZ, D., DEKEGEL, D., DESCHAMPS, J., GHYSDAEL, J., GILDEN, R. V., KETTMANN, R., MARBAIX, G., MAMMERICKX, M. and PORTETELLE, D. (1984). Leukaemogenesis by bovine leukaemia virus. In *Mechanisms of viral leukaemogenesis*, vol. 1, ed. J. Goldmann and O. Jarrett, pp. 229–60. Edinburgh, Churchill Livingstone.

CALAFAT, J., HAGEMAN, P. C. and RESSANG, A. A. (1974). Structure of C-type virus particles in lymphocyte cultures of bovine origin. *Journal of the National Cancer Institute*, **52**, 1251–7.

CALLAHAN, R., LIEBER, M. M., TODARO, G. J., GRAVES, D. C. and FERRER, J. F. (1976). Bovine leukaemia virus genes in the DNA of leukaemic cattle. *Science*, **192**, 1005–7.

CHEN, I. S. Y., MCLAUGHLIN, J. and GOLDE, D. W. (1984). Long terminal repeats of human T-cell leukaemia virus II genome determine target cell specificity. *Nature*, **309**, 276–9.

COPELAND, T. D., MORGAN, M. A. and OROSZLAN, S. (1983*a*). Complete amino acid sequence of the nucleic acid binding protein of bovine leukaemia virus. *FEBS Letters*, **156**, 37–40.

COPELAND, T. D., OROSZLAN, S., KALYANARAMAN, V. S., SARNFADHARAN, M. G. and GALLO, R. C. (1983*b*). Complete amino acid sequence of human T-cell leukaemia virus structural protein p15. *FEBS Letters*, **162**, 390–5.

COUEZ, D., DESCHAMPS, J., KETTMANN, R., STEPHENS, R., GILDEN, R. and BURNY, A. (1984). Nucleotide sequence analysis of the long terminal repeat of integrated bovine leukaemia provirus DNA and of adjacent viral and host sequences. *Journal of Virology*, **49**, 615–20.

CZERNILOFSKY, A. P., DE LORBE, W., SWANSTROM, R., VARMUS, H. E., BISHOP, J. M., FISCHER, E. and GOODMAN, H. M. (1980). The nucleotide sequence of an untranslated but conserved domain at the 3' end of the avian sarcoma virus genome. *Nucleic Acids Research*, **8**, 2967–84.

DEKEGEL, D., MAMMERICKX, M., BURNY, A., PORTETELLE, D., CLEUTER, Y., GHYSDAEL, J. and KETTMANN, R. (1977). Morphogenesis of bovine leukaemia virus (BLV). In *Bovine Leukosis: Various Methods of Molecular Virology*, ed. A. Burny, pp. 31–42. Luxembourg, Commission of the European Communities.

DESCHAMPS, J., KETTMANN, R. and BURNY, A. (1981). Experiments with cloned complete tumour-derived bovine leukaemia virus information prove that the virus is totally exogenous to its target animal species. *Journal of Virology*, **40**, 605–9.

FERRER, J. F. (1980). Bovine lymphosarcoma. *Advances in Veterinary Science and Comparative Medicine*, **24**, 1–68.

FERRER, J. F., AVILA, L. and STOCK, N. D. (1972). Serological detection of type-C viruses found in bovine cultures. *Cancer Research*, **32**, 1864–70.

FERRER, J. F., STOCK, N. D. and LIN, P. S. (1971). Detection of replicating C-type viruses in continuous cell cultures established from cows with leukaemia: Effect of the culture medium. *Journal of the National Cancer Institute*, **47**, 613–21.

GHYSDAEL, J., BRUCK, C., KETTMANN, R. and BURNY, A. (1984). Bovine leukaemia virus. *Current Topics in Microbiology and Immunology*, in press.

GHYSDAEL, J., KETTMANN, R. and BURNY, A. (1978). Translation of bovine leukemia virus genome information in heterologous protein synthesizing systems programmed with viral RNA and cell-lines persistently infected by BLV. *Annals de Recherches Veterinaire*, **9**, 627–34.

GHYSDAEL, J., KETTMANN, R. and BURNY, A. (1979). Translation of bovine leukaemia virus virion RNAs in heterologous protein-synthesizing systems. *Journal of Virology*, **29**, 1087–98.

GILDEN, R. V., LONG, C. W., HANSON, M., TONI, R., CHARMAN, H., OROSZLAN, S., MILLER, J. M. and VAN DER MAATEN, M. J. (1975). Characterization of the major internal protein and RNA-dependent DNA polymerase of bovine leukaemia virus. *Journal of General Virology*, **29**, 305–14.

GREGOIRE, D., COUEZ, D., DESCHAMPS, J., HEUERSTZ, S., HORS-CAYLA, M. C., SZPIRER, J., SZPIRER, C., BURNY, A., HUEZ, G. and KETTMANN, R. (1984). Different bovine leukaemia virus-induced tumors harbor the provirus in different chromosomes. *Journal of Virology*, **50**, 275–9.

GUPTA, P. and FERRER, J. F. (1982). Expression of bovine leukaemia virus genome is blocked by a non-immunoglobulin protein in plasma from infected cattle. *Science*, **215**, 405–7.

GUPTA, P., KASHMIRI, S. V. S. and FERRER, J. F. (1984). Transcriptional control of the bovine leukaemia virus genome: role and characterization of a non-immunoglobulin plasma protein from bovine leukaemia virus-infected cattle. *Journal of Virology*, **50**, 267–70.

HARE, W. C. D., YANG, T. J. and MC FEELY, R. A. (1967). A survey of chromosome bandings in 47 cases of bovine lymphosarcoma (leukaemia). *Journal of the National Cancer Institute*, **38**, 383–92.

HINUMA, Y., NAGATA, K., HANAOKA, M., NAKAI, N., MATSUMOTO, T., KINOSHITA, K., SHIRAWA, S. and MIYOSHI, I. (1981). Adult T-cell leukemia antigen in an ATL cell-line and detection of antibodies to the antigen in human sera. *Proceedings of the National Academy of Sciences, USA*, **78**, 6476–80.

HORVATH, Z. (1982). Experimental transmission of bovine leukosis to sheep (clinicopathological consideration). *Current Topics in Veterinary Medicine and Animal Science*, **15**, 269–81.

HUNTER, E., HILL, E., HARDWICK, M., BHOWN, A., SCHWARTZ, D. E. and TIZARD, R. (1983). Complete sequence of the Rous sarcoma virus *env* gene: Identification of structural and functional regions of its product. *Journal of Virology*, **48**, 920–36.

KAADEN, O., DIETZSCHOLD, B. and STRAUB, O. C. (1972). Beitrag zur Isolierung einer RNA-abhängigen DNA-polymerase aus Lymphozyten eines an lymphatischer leukose erkrankten Rindes. *Zentralblatt für Bakteriologie Originale A*, **220**, 101–5.

KALYANARAMAN, V. S., SARNGADHARAN, M. G., ROBERT-GUROFF, M., MIYOSHI, I., BLAYNEY, D., GOLDE, D. and GALLO, R. C. (1982). A new subtype of human

T-cell leukaemia virus (HTLV-II) associated with a T-cell variant of hairy cell leukaemia. *Science*, **218**, 571–3.

KASHMIRI, S. V. S., MEHDI, R. and FERRER, J. F. (1984). Molecular cloning of covalently closed circular DNA of bovine leukaemia virus. *Journal of Virology*, **49**, 583–7.

KETTMANN, R., COUEZ, D. and BURNY, A. (1981). Restriction endonuclease mapping of linear unintegrated proviral DNA of bovine leukaemia virus. *Journal of Virology*, **38**, 27–33.

KETTMANN, R., DESCHAMPS, J., CLEUTER, Y., COUEZ, D., BURNY, A. and MARBAIX, G. (1982). Leukaemogenesis by bovine leukaemia virus: Proviral DNA integration and lack of RNA expression of viral long terminal repeat and 3' proximate cellular sequences. *Proceedings of the National Academy of Sciences, USA*, **79**, 2465–9.

KETTMANN, R., DESCHAMPS, J., COUEZ, D., CLAUSTRIAUX, J. J., PALM, R. and BURNY, A. (1983). Chromosome integration domain for bovine leukaemia provirus in tumours. *Journal of Virology*, **47**, 146–50.

KETTMANN, R., MAMMERICKX, M., DEKEGEL, D., GHYSDAEL, J., PORTETELLE, D. and BURNY, A. (1975). Biochemical approach to bovine leukaemia. *Acta Haematologica*, **54**, 201–9.

KETTMANN, R., MAMMERICKX, M., PORTETELLE, D., GREGOIRE, D. and BURNY, A. (1984). Experimental infection of sheep and goat with bovine leukaemia virus: Localization of proviral information in the target cells. *Leukaemia Research*, in press.

KETTMANN, R., MARBAIX, G., CLEUTER, Y., PORTETELLE, D., MAMMERICKX, M. and BURNY, A. (1980). Genomic integration of bovine leukaemia provirus and lack of viral RNA expression in the target cells of cattle with different responses to BLV infection. *Leukaemia Research*, **4**, 509–19.

KETTMANN, R., PORTETELLE, D., MAMMERICKX, M., CLEUTER, Y., DEKEGEL, D., GALOUX, M., GHYSDAEL, J., BURNY, A. and CHANTRENNE, H. (1976). Bovine leukaemia virus: An exogenous RNA oncogenic virus. *Proceedings of the National Academy of Sciences, USA*, **73**, 1014–18.

LENZ, J., CELANDER, D., CROWTHER, R. L., PATARCA, R., PERKINS, D. W. and HASELTINE, W. A. (1984). Determination of the leukaemogenicity of a murine retrovirus by sequences within the long terminal repeat. *Nature*, **308**, 467–70.

LONG, C. W., HENDERSON, L. E. and OROSZLAN, S. (1980). Isolation and characterization of low molecular-weight DNA-binding proteins from retroviruses. *Virology*, **104**, 491–6.

McGRATH, M. S. and WEISSMAN, I. L. (1979). AICR leukaemogenesis: Identification and biological significance of thymic lymphoma receptors for AKR retrovirus. *Cell*, **17**, 65–75.

MAMOUN, R. Z., ANSTIER, T., GUILLEMAIN, B. and DUPLAN, J. F. (1983). Bovine lymphosarcoma: Expression of BLV-related proteins in cultured cells. *Journal of General Virology*, **64**, 1895–905.

MARIN, C., DE LOPEZ, N., DE ALVAREZ, L., CASTANOS, H., ESPANA, W., LEON, A. and BELLO, A. (1982). Humoral spontaneous response to bovine leukaemia virus infection in zebra, sheep, buffalo and capybara. *Current Topics in Veterinary Medicine and Animal Science*, **15**, 310–20.

MILLER, J. M., MILLER, L. D., OLSON, C. and GILLETTE, K. G. (1969). Virus-like particles in phytohaemagglutinin-stimulated lymphocyte cultures with reference to bovine lymphosarcoma. *Journal of the National Cancer Institute*, **43**, 1297–305.

MILLER, J. M. and OLSON, C. (1972). Precipitating antibody to an internal antigen of the C-type virus associated with bovine lymphosarcoma. *Journal of the National Cancer Institute*, **49**, 1459–62.

MORGAN, M. A., COPELAND, T. D. and OROSZLAN, S. (1983). Structural and antigenic analysis of the nucleic acid binding proteins of bovine and feline leukaemia viruses. *Journal of Virology*, **46**, 177–86.

MUSSGAY, M. and KAADEN, O. R. (1978). Progress in studies on the aetiology and serologic diagnosis of enzootic bovine leukosis. *Current Topics in Microbiology and Immunology*, **79**, 43–72.

NAGY, K., CLAPHAM, P., CHEINSONG-POPOV, R. and WEISS, R. (1983). Human T-cell leukaemia virus type I: Induction of syncytia and inhibition by patients' sera. *International Journal of Cancer*, **32**, 321–8.

OLSON, C., KETTMANN, R., BURNY, A. and KAJA, R. (1981). Goat lymphosarcoma from bovine leukaemia virus. *Journal of the National Cancer Institute*, **67**, 671–5.

OROSZLAN, S., COPELAND, T. D., HENDERSON, L. E., STEPHENSON, J. R. and GILDEN, R. V. (1979). Amino-terminal sequence of bovine leukaemia virus major internal protein: Homology with mammalian type C virus p30s. *Proceedings of the National Academy of Sciences, USA*, **76**, 2996–3000.

OROSZLAN, S., HENDERSON, L. E., COPELAND, T. D., SCHULTZ, A. M. and RABIN, E. M. (1980). Processing and structure of murine leukaemia virus *gag* and *env* gene encoded polyproteins. In *Biosynthesis, Modification and Processing of Cellular and Viral Polyproteins*, ed. G. Koch and D. Richter, pp. 219–33. New York, Academic Press.

OROSZLAN, S., SARNGADHARAN, M. G., COPELAND, T. D., KALYANARAMAN, V. S., GILDEN, R. V. and GALLO, R. C. (1982). Primary structure analysis of the major internal protein p24 of human type C T-cell leukaemia virus. *Proceedings of the National Academy of Sciences, USA*, **79**, 1291–4.

PALACIOS, R., HENSON, G., STEINMETZ, M. and McKEARN, J. P. (1984). Interleukin-3 supports growth of mouse pre-B-cell clones *in vitro*. *Nature*, **309**, 126–31.

PARODI, A. L., MIALOT, M., CRESPEAU, F., LEVY, D., SALMON, H., NOGUES, G. and GERARD-MARCHAND, R. (1982). Attempt for a new cytological and cyto-immunological classification of bovine malignant lymphoma (BML) (lymphosarcoma). *Current Topics in Veterinary Medicine and Animal Science*, **15**, 561–72.

POIESZ, B. J., RUSCETTI, F. W., GAZDAR, A. F., BUNN, P. A., MINNA, J. D. and GALLO, R. C. (1980). Detection and isolation of type-C retrovirus particles from fresh and cultured lymphocytes of a patient with cutaneous T-cell lymphoma. *Proceedings of the National Academy of Sciences, USA*, **77**, 7415–19.

POIESZ, B. J., RUSCETTI, F. W., REITZ, M. S., KALYANARAMAN, V. S. and GALLO, R. C. (1981). Isolation of a new type-C retrovirus (HTLV) in primary uncultured cells of a patient with Sezary T-cell leukemia. *Nature*, **294**, 268–71.

PORTETELLE, D., BRUCK, C., MAMMERICKX, M. and BURNY, A. (1983). Use of monoclonal antibody in an ELISA test for the detection of antibodies to bovine leukaemia virus. *Journal of Virological Methods*, **6**, 19–29.

PRACHAR, J. and HLUBINOVA, K. (1980). Glycoprotein and protein composition of BLV. *Neoplasma*, **27**, 669–74.

RICE, N. R., STEPHENS, R. M., COUEZ, D., DESCHAMPS, J., KETTMANN, R., BURNY, A. and GILDEN, R. V. (1984). The nucleotide sequence of the *env* gene and post-*env* region of bovine leukaemia virus. *Virology*, in press.

ROBINSON, H. L., BLAIS, B. M., TSICHLIS, P. L. and COFFIN, J. M. (1982). At least two regions of the viral genome determine the oncogenic potential of avian leukosis viruses. *Proceedings of the National Academy of Sciences, USA*, **79**, 1225–9.

ROHDE, W., PAULI, G., PAULSEN, J., HARMS, E. and BAUER, H. (1978). Bovine and ovine leukaemia viruses. I. Characterization of viral antigens. *Journal of Virology*, **26**, 159–64.

SAGATA, N., OGAWA, Y., KAWAMURA, J., ONUMA, M., IZAWA, H. and IKAWA, Y.

(1983). Molecular cloning of bovine leukaemia virus DNA integrated into the bovine tumor cell genome. *Gene*, **26**, 1–10.

SCHULTZ, A. M., COPELAND, T. D. and OROSZLAN, S. (1984). The envelope proteins of bovine leukaemia virus: Purification and sequence analysis. *Virology*, **135**, in press.

SCHULTZ, A. M. and OROSZLAN, S. (1984). Myristylation of gag–onc fusion proteins in mammalian transforming retroviruses. *Virology*, **133**, 431–7.

SCHWARTZ, D. E., TIZARD, R. and GILBERT, W. (1983). Nucleotide sequence of Rous sarcoma virus. *Cell*, **32**, 853–69.

SEIKI, M., HATTORI, S., HIRAYAMA, Y. and YOSHIDA, M. (1983). Human adult T-cell leukaemia virus: Complete nucleotide sequence of the provirus genome integrated in leukaemia cell DNA. *Proceedings of the National Academy of Sciences, USA*, **80**, 3618–22.

SHINNICK, T. M., LERNER, R. A. and SUTCLIFFE, J. G. (1981). Nucleotide sequence of Moloney leukaemia virus. *Nature*, **293**, 543–8.

SOUTHERN, E. M. (1975). Detection of specific sequences among DNA fragments separated by gel electrophoresis. *Journal of Molecular Biology*, **98**, 503–18.

TSICHLIS, P. N. and COFFIN, J. M. (1980). Recombinants between endogenous and exogenous avian tumor viruses: Role of the C region and other portions of the genome in the control of replication and transformation. *Journal of Virology*, **33**, 238–49.

TSIMANIS, A., BICHKO, V., DREILINA, D., MELDRAIS, J., LOZHA, V., KUKAINE, R. and GREN, E. (1983). The structure of cloned 3′-terminal RNA region of bovine leukaemia virus (BLV). *Nucleic Acids Research*, **11**, 6079–87.

UCKERT, W., WESTERMANN, P. and WUNDERLICH, V. (1982). Nearest neighbour relationships of major structural proteins within bovine leukaemia virus particles. *Virology*, **121**, 240–50.

UCKERT, W. and WUNDERLICH, V. (1979). Proteins of bovine leukaemia virus: p15 is the major phosphorylated protein. *Acta Biologica et Medica Germanica*, **38**, k42.

UCKERT, W., WUNDERLICH, V., GHYSDAEL, J., PORTETELLE, D. and BURNY, A. (1984). Bovine leukaemia virus (BLV) – A structural model based on chemical crosslinking studies. *Virology*, **133**, 386–92.

VAN DER MAATEN, M. J. and MILLER, J. M. (1976). Replication of bovine leukaemia virus in monolayer cell cultures. *Bibliotheca Haematologica*, **43**, 360–362.

WEILAND, F. and ÜBERSCHÄR, S. (1976). Ultrastructural comparison of bovine leukaemia virus (BLV) with C-type particles of other species. *Archives de Virologie*, **52**, 187–90.

WEISS, R., TEICH, N., VARMUS, H. and COFFIN, J. (eds.) (1982). *The Molecular Biology of RNA Tumor Viruses*, second edn. New York, Cold Spring Harbor Laboratories.

MOLECULAR ASPECTS OF FELINE LEUKAEMIA VIRUSES AND THEIR ASSOCIATED DISEASES

JAMES C. NEIL

Beatson Institute for Cancer Research, Wolfson Laboratory for Molecular Pathology, Bearsden, Glasgow G61 1BD, UK

Each retrovirus–host system that has been studied in detail has had its own particular contribution to our knowledge of these viruses and their associated diseases. The induction and isolation of mutants of avian retroviruses have yielded most to our understanding of the viral gene functions involved in replication and cell transformation, while the murine retroviruses have had most to offer towards an appreciation of the complex host factors in susceptibility or resistance to disease. The major contribution of the feline retrovirus system is as a natural disease model. Unlike most of the avian and murine virus studies, feline retrovirus research can be done with naturally occurring tumours from a randomly outbred species. In this respect, the cat may represent one of the best models for the human population. The development of molecular analysis of FeLV and its associated diseases has been relatively recent, but the new perspective offered by this work should guarantee continued interest.

Feline leukaemia viruses induce a wide spectrum of diseases in their host, the domestic cat. The interaction of these viruses and their host has been studied in some detail and it has become clear that the outcome of infection depends on a variety of factors including virus strain, age, dose, and route of exposure. In this paper, I will consider the molecular mechanisms in FeLV-induced disease, with particular emphasis on our recent studies on the role of FeLV in T-cell leukaemias. Most FeLV isolates carry only the replicative genes *gag*, *pol* and *env* and induce neoplastic disease at low frequency and after a long latent period. However, some strains show markedly increased pathogenicity, and FeLV appears to be the most prolific of the known retroviruses in recombination with proto-oncogenes. The study of feline leukaemia and sarcoma viruses has identified eight different oncogenes and will probably yield still more. Capture of the *myc* oncogene by FeLV is unexpectedly common and is sufficiently frequent to account for a significant proportion of FeLV-positive

T-cell leukaemias. In addition, some cases of this disease show integration of FeLV next to the c-*myc* gene, which may not only activate the *myc* gene but represent an intermediate step in the generation of the FeLV/*myc* recombinant viruses. Still more virus-positive leukaemias are unaccounted for at the molecular level and may involve other, possibly uncharacterised, oncogenes that can substitute for *myc* in the oncogenic process.

Another area of study likely to yield valuable insight is the analysis of the sequence and structure of the envelope (*env*) gene products of different FeLV isolates. Interference subgroup, a property determined by the viral envelope, correlates closely with viral pathogenicity in several important cases. Construction of recombinant viruses *in vitro* should delineate the structural features associated with these unusual disease properties. Molecular analysis may also shed light on the origins of envelope gene variants of FeLV that show greatly increased or novel pathogenicity.

BIOLOGY OF FELINE LEUKAEMIA VIRUSES

Diseases associated with FeLV infection

A list of diseases associated with FeLV infection is given in Table 1. This information has been summarised from a detailed review of the pathogenesis of FeLV-related diseases (Jarrett, 1984). As can be seen from the table, neoplastic disease of each haematopoetic lineage is mirrored by a degenerative condition affecting the same cell series. In fact, non-neoplastic diseases such as immunosuppression or anaemia failure cause more frequent problems in nature than the leukaemogenic effects of FeLV. This observation may have implications for the early events in the progression to neoplastic disease. Depletion of mature haemopoietic cell populations as a result of viral infection, and increased cell proliferation in response to this, may lead to increased rates of mutation arising from viral integration or other factors.

The most common neoplastic disease associated with FeLV is T-cell lymphosarcoma. Most tumours of this type are virus-positive. Two common forms are seen; thymic lymphosarcomas, where the tumour mass is predominantly in a single site, and multicentric lymphosarcomas, in which multiple lymph nodes are involved. Alimentary lymphosarcomas, generally of B-cell origin, are at least as

Table 1. *Diseases of the cat associated with FeLV infection*

Neoplastic	Degenerative
Thymic lymphosarcoma	Thymic atrophy
Multicentric fibrosarcoma	Immunosuppression
Alimentary lymphosarcoma	Agranulocytosis
Myeloid leukaemia	Pancytopaenia
Erythroleukaemia	Erythroid aplasia

common as the T-cell tumours, but are mostly FeLV-negative. The relationship of FeLV to this type of tumour is therefore rather unclear and may involve some indirect mechanism if it is causally related to the tumour at all. Other types of leukaemia are less commonly seen, but may be associated with the presence of FeLV variants. One such example is the rapid myeloid leukaemia induced by a field isolate of FeLV, FeLV-GM1 (Onions *et al.*, 1985). Multicentric fibrosarcomas are quite rare, particularly in younger cats, but these appear to be associated very frequently with highly oncogenic feline sarcoma viruses that have each presumably arisen *de novo* in an FeLV-infected cat.

Epidemiology of FeLV infection

The epidemiology of FeLV infection in the cat population reveals two distinct patterns (Hardy *et al.*, 1973). In closed multiple-cat households into which FeLV is introduced, the likelihood of exposure to virus is high, as is the incidence of persistent infection. It is in this situation that most FeLV-related disease problems are manifested. In contrast, free-range cats show lower rates of viral exposure and a much lower incidence of persistent infection. These differences may be explained by the higher doses of virus to which cats are exposed in closed households and the age at which exposure occurs. Very young cats are most susceptible to infection with a cut-off at around four months of age when a proportion of cats become refractory to infection in the absence of predisposing immunosuppression (Hoover *et al.*, 1976; Schaller *et al.*, 1978). A feature of the infectious cycle of FeLV that has come to light only recently is the ability of the virus to persist in latent form even in the presence of a virus-neutralising antibody response. Upon treatment with immunosuppressive agents, frank viraemia and shedding of infectious virus can reappear (Post and Warren, 1980; Rojko *et al.*, 1982). Latently infected cats are unlikely to represent an important reservoir of infectious virus in the population. Until recently, however, diagnostic tests for virus infection would have failed to detect these cats.

Immune responses of cats to FeLV infection

Antibody responses to FeLV infection have been measured by a variety of methods, including virus neutralisation, membrane immunofluorescence, solid- or liquid-phase radioimmunoassay or complement-dependent cytolytic assays. The conclusions of these studies have been varied and occasionally controversial. Antibody response appears to be a good indicator of exposure to FeLV, but the protective value of the response depends on additional factors, since an antibody response can often be detected in cats with persistent viraemia (Russell and Jarrett, 1978; Hardy *et al.*, 1976). It would seem, however, that a virus-neutralising antibody response is protective in uninfected animals. It has been proposed that antibodies detected by a membrane immunofluorescence test on FeLV-infected cells have protective value against neoplastic disease, even in viraemic cats (Essex *et al.*, 1975). Cell-mediated immunity to FeLV infection has not been studied in detail although this is expected to play an important role in resistance to infection (Jarrett, 1984).

Feline leukaemia virus subgroups

FeLV isolates can be classified A, B or C, on the basis of interference subgroup (Sarma and Log, 1973). Thus, cells infected with a virus of subgroup A are resistant to infection with any other subgroup A virus. Interference subgroup, determined by the viral envelope (*env*) gene, reflects the cell surface receptors utilised by FeLV to gain access to the host cell and correlates with other important biological properties. Table 2 lists the host range and frequency of isolation of FeLV subgroups A, B and C. Although viruses of a single subgroup can be separated and grown independently in cultured feline embryonic fibroblast cells, subgroups B and C show an association with FeLV-A *in vivo* that may be based on, respectively, the dependence of FeLV-B on FeLV-A for transmission and the generation of FeLV-C isolates by recombination between FeLV-A and cat geno-

Table 2. *FeLV subgroups – biological properties*

Subgroup	Host range *in vitro*	Occurrence in natural isolates (%)
A	Ecotropic – cat	100
B	Amphotropic – cat, dog, human	50
C	Amphotropic – cat, human, guinea pig	1

mic DNA (Sarma and Log, 1973; Jarrett and Russell, 1978; Russell and Jarrett, 1978). The importance of the FeLV subgroups in disease is discussed further later.

INTERACTION OF FELV WITH ONCOGENES

Oncogene transduction

Transduction of oncogenes by retroviruses has been the factor that has allowed us to recognise the existence of this set of cellular genes. Until the elucidation of the nature and origin of the *src* gene, which was apparently acquired from chicken DNA by an avian retrovirus (Stehelin *et al.*, 1976), the idea of genes of this sort was still hypothetical (Huebner and Todaro, 1969). Nevertheless, the process of oncogene transduction by retroviruses has generally been thought of as rather rare and of no numerical significance in naturally occurring tumours. The feline retrovirus system provides evidence against this idea, as can be seen from the frequency of isolation of oncogene-containing viruses from spontaneous, FeLV-associated tumours of the cat.

A list of the different oncogenes that have been acquired by FeLV through recombination with the host cell genome is shown in Table 3. Until recently, all of the oncogene-containing FeLV isolates had been found in cases of multicentric fibrosarcoma. A range of oncogenes can confer on FeLV the ability to induce sarcomas (reviewed by Besmer, 1984). Most belong to the family of protein kinases and related genes, of which the prototype is the Rous sarcoma virus *src* gene. Also represented among the feline sarcoma viruses are the Kirsten *ras* gene (Ki-*ras*) and the growth factor-related *sis* gene (Besmer *et al.*, 1983). One isolate, GR-FeSV, contains two distinct

Table 3. *Transduction of oncogenes by FeLV*

Oncogene	Number of isolates	Associated disease
fes	3	Sarcoma
fms	1	Sarcoma
abl	1	Sarcoma
sis	1	Sarcoma
Ki-ras	1	Sarcoma
fgr	1	Sarcoma
kit	1	Sarcoma
myc	>10	T-cell leukaemia

host–derived sequences, a tyrosine-specific protein kinase domain (*fgr*) and a domain related to the actin gene family (Naharro, Robbins and Reddy, 1984). These sequences are expressed as a fusion protein with the configuration gag-actin-fgr. The significance of the actin sequence for the transformation potential of GR-FeSV remains to be established but suggests the intriguing possibility that there might be a link between cytoskeletal elements and growth control. In this respect, it is notable that a *ras*-related gene of yeast is flanked in the yeast genome by actin and tubulin genes (Gallwitz, Donath and Sander, 1983). GR-FeSV is the only example of a feline virus that has captured two sequences that are apparently unlinked in normal host DNA. The avian tumour viruses provide several examples of such tripartite structures. Thus, avian erythroblastosis virus contains *erb*-A and *erb*-B (Vennstrom and Bishop, 1982), MH2 avian carcinoma virus contains *myc* and *mil* (Coll *et al.*, 1983) and E26 avian myeloblastosis virus contains *myb* and *ets* (Leprince *et al.*, 1983; Nunn *et al.*, 1983). The relative contribution of each cellular gene to the pathogenic spectrum of the virus has been investigated in detail only for AEV, where *erb*-B alone is sufficient for oncogenic activity. The *erb*-A sequence produces no tumours without *erb*-B but it does appear to enhance the effects of *erb*-B and to alter the phenotype of the erythroblastic tumour cells that are produced. (Frykberg *et al.*, 1983). It is also likely that the *mil* and *ets* sequences are responsible for the relatively wide disease spectra of MH2 and E26 viruses.

Examination of the tumours most commonly associated with infectious FeLV, T-cell leukaemias, has shown a remarkably high incidence of transduction of the *myc* gene. Out of a total of 16 virus-positive leukaemias, we found 5 cases in which FeLV/*myc* recombinant viruses were evident in the tumour DNA. Independent examples of transduction of *myc* by FeLV have been provided by the tumour series examined by J. Casey (Louisiana State University) and J. Mullins (Harvard School of Public Health) (Levy, Gardner and Casey, 1984; Mullins *et al.*, 1984*b*; Neil *et al.*, 1984). The relative frequency of *myc* transduction by FeLV leads to the speculation that this process may be driven by some homology between FeLV and the c-*myc* locus. We are examining the recombination junction points in several isolates to test this possibility. A currently popular model for the generation of oncogene-containing viruses involves integration upstream of the cellular proto-oncogene in the same transcriptional orientation, followed by a deletion that fuses viral and oncogene coding sequences (Swanstrom *et al.*, 1983). Homology between the c-*onc* locus and the helper viral genome around the 5′ recombination site could conceivably assist or direct this process. Examples from several retrovirus systems lend support to this idea (Besmer, 1984). An equal number of examples fail to show such a relationship, but it must be pointed out that a deletion of viral genomic sequences subsequent to the original oncogene capture could obscure underlying common features of this sort. We have determined the 5′ FeLV/*myc* junction in one isolate and will compare this to the normal c-*myc* sequence when this is available. The case for which most information is available is shown in Fig. 1. The diagram compares the restriction maps of a typical 'helper' FeLV and a corresponding FeLV/*myc* recombinant virus, both of which were cloned from tumour DNA from a spontaneous thymic lymphosarcoma. The virus represented by clone CT4 contains the entire *myc* coding sequence derived from, it is assumed, exons 2 and 3 of feline c-*myc*. The acquired *myc* sequence replaces the 3′ end of the *pol* gene and possibly all of the *env* gene, although the 3′ *myc*/FeLV junction has not yet been located. Sequences immediately upstream of the CT4 v-*myc* coding sequences appear not to be derived from the intron sequences upstream of c-*myc* exon 2 but may be derived from exon 1, although this remains to be established. If this is correct, it would suggest that the original recombination between FeLV and c-*myc* occurred in exon 1 and that the parental FeLV provirus had integrated either in or upstream of exon 1. The splice acceptor site for c-*myc* exon 2 has been lost in this process, but the expected *env* splice acceptor site remains intact in FeLV CT4, as it is 5′ to the recombination site, suggesting that CT4 v-*myc* could be expressed from a spliced mRNA, in a manner similar

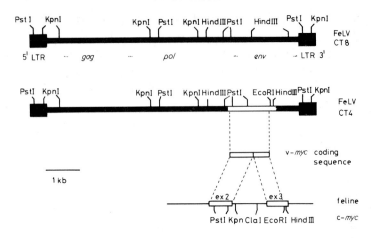

Fig. 1. Comparative restriction enzyme and genetic maps of a helper-type FeLV provirus (CT8) and a recombinant provirus carrying the *myc* gene (CT4). The recombinant virus is similar in overall length to the helper virus but has a deletion of most, if not all, of the *env* gene sequences. The 5' junction of FeLV and *myc* sequences is in the 3' end of *pol* in FeLV and immediately upstream of the expected translational initiation site in the *myc* sequence (M. Stewart, A. Wheeler and J.C.N., unpublished results). The exact location of the 3' FeLV/*myc* junction is unknown, as our clone extends only to the *Eco*RI site. However, we know that the virus contains the entire *myc* coding sequence, since the translational termination occurs around 30 bp 5' to the *Eco*RI site. In this configuration, it may be expected that the *myc* gene product will be expressed from a subgenomic mRNA derived by splicing. The genetic structure of this virus is analogous to that of OK10 avian carcinoma virus (Bister *et al.*, 1980). The map of feline c-*myc* shown below is derived by hybridisation analysis of the cloned feline c-*myc* locus using probes specific for exons 2 and 3 of human c-*myc*. Exon 1 is thought not to contain coding sequences but to be important in regulating the level of c-*myc* product at the translational level. The loss of exon 1 sequences in v-*myc* may therefore render it refractory to this control. It is not yet known whether FeLV v-*myc* contains any primary sequence changes in the coding sequence as compared to c-*myc*. The structure of CT4 is quite different from the FeLV/*myc* recombinant provirus cloned by Casey and coworkers (Levy *et al.*, 1984) and another case analysed in some detail by us (T3). In these cases it appears that the FeLV/*myc* 5' junction is in the FeLV *gag* sequences, and it would appear from comparison with our restriction maps that the virus characterised by Levy *et al.* has sustained some loss of *myc* coding sequences. These findings do not rule out the possibility that the FeLV/*myc* viruses have been generated by homologous recombination between specific regions of FeLV and *myc*. Internal deletions of the viral genome subsequent to oncogene capture may obscure an underlying common feature of this sort.

to the expression of the *env* gene of the helper virus (M. Stewart, A. Wheeler, and J.C.N., unpublished). With these structural features our clone closely resembles avian carcinoma virus OK10 (Bister *et al.*, 1980). In contrast, another case from our series, T3, contains a virus in which it appears that the residual portion of the *gag* gene has been fused to the *myc* sequence. In this context *myc* may be expressed as a *gag–myc* fusion protein, as in avian myelocytomatosis virus MC29. As already discussed, this variety of genome structures does not rule out the possibility of a common 5' recombination point, since this could be masked by subsequent deletions within the viral genome.

Very little information is available on the biological properties of the FeLV/*myc* recombinant viruses. Cell lines releasing these viruses (T3, F422) can transmit the *myc*-containing viral genomes to feline embryo cells in culture, but these cells do not become morphologically transformed. Experiments are in progress in which FeLV/*myc* viruses have been inoculated into young cats. Preliminary results show that the T3 and F422 viruses can induce thymic lymphosarcomas after only a short latent period (D. Onions, G. Lees and J. C. N., unpublished results). It is even possible that FeLV/*myc* viruses could be horizontally transmitted among the cat population. Oncogene-containing viruses have generally been thought to arise *de novo* and to fail to be transmitted because they lead to the rapid death of their host. Only where tumour virologists intervene would these agents be preserved for further analysis. However, less pathogenic viruses might allow the host to survive long enough for infectious spread to occur. Whatever the explanation, the strong association of the *myc* gene with feline T-cell leukaemia is worthy of further study.

Insertional mutagenesis by FeLV

Although examples of oncogene capture by FeLV abound, considerably less is known about FeLV as an insertional mutagen. Our analysis of a series of naturally occurring and experimentally induced tumours revealed two thymic lymphosarcomas in which FeLV had integrated close to the c-*myc* gene (Neil *et al.*, 1984). Figure 2 shows a comparison of the human and feline c-*myc* loci. The feline c-*myc* map has been established by analysis with probes specific for human c-*myc* exons 1, 2 and 3 and some limited sequence data. However, it is already clear that the overall structure of the locus is quite similar to that of human c-*myc*. Available sequence data indicate a very high degree of homology between the coding sequences of the human and feline *myc* genes (M. Stewart, A. Wheeler and J.C.N., unpublished results). The question we wish to address is how FeLV activates the c-*myc* gene in these cases. The level of c-*myc* expression detected in total tumour RNA is not strikingly high in these two cases (T7, T8), but it is noticeably higher than in several cases of FeLV-negative thymic lymphosarcoma. We believe that measurements of the level of *myc* RNA in the tumours are rather unreliable, since in a solid

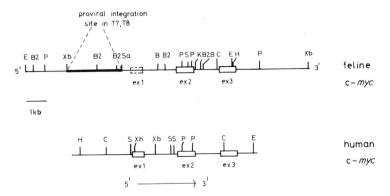

Fig. 2. Comparative maps of feline and human c-*myc* loci. Most of the *Hind*III-*Eco*RI fragment containing the human c-*myc* locus has been sequenced (Colby *et al.*, 1983; Watt *et al.*, 1983) and the positions of exons 1, 2 and 3 determined directly. The feline c-*myc* map has been derived by cross-hybridisation analysis using subfragments of human c-*myc* and avian myelocytomatosis virus MC29 v-*myc*. The location of the putative exon 1 of FeLV c-*myc* is as yet uncertain, but is predicted to be situated between the *Bam*HI and *Sal*I restriction enzyme sites by low stringency hybridisation with human c-*myc* exon 1. This is of obvious interest as we have identified two cases of thymic lymphosarcoma in which FeLV integrates somewhere between the *Xba*I and *Sal*I sites upstream of exon 2. If these viruses integrate within exon 1 they would remove the normal c-*myc* promoter from the body of the gene. Alternatively, insertion of viral enhancer sequences or disruption of some regulatory sequences upstream of exon 1 may be the important event. In both the cases analysed (T7, T8) there is no apparent change in c-*myc* transcript size and no hybrid transcript detected in which LTR and c-*myc* sequences are fused. However, the level of c-*myc* RNA is in these cases noticeably higher than in FeLV-negative thymic lymphosarcomas. (G. Lees, D. Onions and J.C.N., unpublished results). The location and orientation of these proviruses are also of relevance when considering the mechanism by which the FeLV/*myc* recombinant viruses are generated.

tumour mass many quiescent or dying cells may be present. Where it has been possible to culture cells from the tumours, much higher levels of *myc* transcription have been detected, notably in cases where FeLV/*myc* recombinant viruses have been present. The c-*myc* transcripts from T7 and T8 show no obvious size abnormality and no linkage to viral LTR sequences (G. Lees, D. Onions and J.C.N., unpublished results). This suggests that T7 and T8 differ from the majority of cases of insertional mutagenesis of *myc* by avian leukosis viruses, where integration in exon 1 or intron 1 removes the normal c-*myc* promoter and generates a hybrid U5-c-*myc* transcript (Hayward, Neel and Astrin, 1981; Payne, Bishop and Varmus, 1982). There are some examples, however, where ALV has integrated in the opposite transcriptional orientation or downstream of c-*myc* and where it is assumed that enhancer elements in the U3 region of the viral LTR effect the activation of *myc*. The structure of a FeLV LTR is shown in Fig. 3. Enhancer elements analogous to those of other retroviruses are assumed to be located in the U3 region.

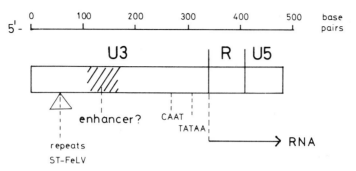

Fig. 3. Structure of an FeLV long terminal repeat. The LTR of the Gardner–Arnstein strain of FeLV is shown (GA-FeLV), as published recently (Elder and Mullins, 1983). The salient features for the role of FeLV in leukaemogenesis are the transcriptional control sequences. A typical eukaryotic promoter consensus sequence is located 5' to the R region which begins at the RNA cap site. In addition, sequences homologous to the enhancer elements of murine leukaemia virus LTRs are located in the U3 region. These are assumed to play a similar role in FeLV, stimulating viral expression at the expense of normal cellular transcription. Since enhancer elements appear to act independently of orientation and at some distance, viral insertion in the vicinity of a proto-oncogene may be sufficient for activation. It is worth noting that there are marked differences between the FeLV LTRs that have been sequenced, although all show some common features. Large-scale insertions, deletions or reiteration of U3 sequences have been demonstrated. The site of additional repeated sequences in the LTR of the Snyder-Theilen strain of FeLV is shown (ST-FeLV). The biological significance of these differences for FeLV is not yet known. It has been shown that small differences in the LTRs of murine leukaemia viruses are responsible for marked differences in pathogenicity, presumably mediated by their effects on viral expression in the target cells for transformation (Lenz *et al.*, 1984).

We have found a possibly important difference between experimental and natural tumours in that the experimental tumours show evidence of insertional mutagenesis of c-*myc* by FeLV, while the natural tumours show a predominance of transduction. These results have been confirmed and extended by recent studies in another laboratory (J. Mullins, personal communication), where four out of seven experimental tumours showed insertion of FeLV close to c-*myc*. One experimental case that showed a FeLV/*myc* recombinant virus came from an early experiment performed by Rickard, where there is the possibility that the inoculum already contained a recombinant virus. This is suggested because another of the experimental cases, derived using the same virus source, gave rise to the F422 line which has subsequently been shown to contain a FeLV/*myc* virus (Mullins *et al.*, 1984a; Neil *et al.*, 1984). In contrast to the experimental cases, Mullins' further examination of a large series of natural tumours showed a high incidence of transduction of *myc* by FeLV at a level similar to that seen in our initial study. Several possible explanations can be considered. The cat population may harbour horizontally transmitted FeLV/*myc* recombinant viruses that emerge as the agents of a significant number of naturally occurring tumours. On the other hand, if these viruses arise *de novo*, another explanation must be sought. It may be that the experimental FeLV tumours, like their counterparts in the avian and murine systems, have been induced by viruses that have been selected for rapid and pronounced pathogenicity. These variants may have, for example, transcriptional control sequences that function particularly well in the target cell for transformation (cf. Lenz *et al.*, 1984). These viruses may enhance expression of adjacent genes with high efficiency. Viral strains commonly encountered in the field may be less effective in this function and thus integration upstream of c-*myc* may generally be insufficient to promote tumour formation without some further genetic change. The generation of a recombinant virus may serve this purpose in some cases at least. The transcriptional activation of *myc* may be more efficient where *myc* is part of the viral genome, since expression involves viral enhancer and promoter sequences. The capture of an oncogene as a component of the viral genome allows the additional possibilities of amplification of the oncogene and insertion at novel chromosomal sites. Whatever the explanation, the discrepancy between the observations on natural and experimental tumours indicates the value of the FeLV system, where both can be compared.

There is no reason to believe that the same molecular changes will affect an oncogene in each independent case of a disease, even where the clinical and pathological features of the disease are highly characteristic. A useful parallel can be drawn with studies on the activation of c-*myc* by chromosome translocation in Burkitt's lymphoma or murine plasmacytoma. In these cases, it appears that *myc* activation can occur by any of a variety of mechanisms (reviewed by Perry, 1983). In some cases, translocation results in truncation of c-*myc* upstream of exon 2, removing exon 1 and revealing the existence of cryptic promoters in intron sequences that can be used for the expression of c-*myc*. In some of these cases, it seems that a tissue-specific enhancer from an immunoglobulin gene cluster has been brought together with c-*myc* in the course of translocation (Saito *et al.*, 1983). Disruption of normal c-*myc* transcriptional control may well be the basis of oncogenic activation, although the details must differ in individual cases. Mutations within the c-*myc* locus were invoked as a possible cause of activation (Rabbitts, Hamlyn and Baer, 1983), although this idea is presently in disfavour, at least with regard to mutations within the coding sequence. Some potentially tumour-specific mutations in c-*myc* have been noted after proviral insertion in ALV-induced bursal lymphomas (Westaway, Payne and Varmus, 1984). In the light of these examples we are analysing T7 and T8 to determine the crucial events in *myc* activation by FeLV.

Thymic lymphosarcomas are also a common result of infection of mice or rats with retroviruses and both these systems are being investigated actively to see whether insertional mutagenesis plays a role in the pathology of these diseases and, if so, which cellular genes function as targets. There is some evidence to implicate the *myc* gene in a number of cases and at least two genetic loci, distinct from c-*myc*, have been identified as common viral integration sites in a series of tumours (Tsichilis, Gunter-Strauss and Hu, 1983; Cuypers *et al.*, 1984). It remains to be shown whether these viral integrations result in oncogene activation, but it suggests, as we might expect, that a number of alternative genes might play a role in the same type of tumour. The majority of cases of feline leukaemia have no obvious disruption of the *myc* locus and it seems highly probable that other oncogenes, known or as yet uncharacterised, play a role in this complex of diseases.

POSSIBLE ROLES OF REPLICATIVE GENES IN FELV-RELATED DISEASES

Alternative mechanisms of oncogenesis

Although numerous oncogene-containing FeLVs have been identified in naturally occurring tumours, the possibility remains that FeLV can promote the development of tumours by quite different mechanisms. Apart from the examples of insertional mutagenesis by FeLV discussed in the last section there is evidence to suggest that the *env* gene may be involved in the pathogenesis of some FeLV-related diseases. The role of the *env* gene product in the replicative cycle of FeLV is to permit entry to the host cell, but the precedent of the murine retrovirus, Friend Spleen Focus-Forming Virus (Friend SFFV), that induces cell proliferation by virtue of a recombinant *env* gene suggests that *env* products may be capable of playing a more direct role in oncogenesis. Since *env* controls entry to the host cell, it might be expected that this gene could determine the high degree of tissue-specificity of the transforming activity of many oncogenic retroviruses. However, restriction of viral gene expression and replication can occur at a number of different levels and experiments designed to locate the determinants of tissue specificity in oncogenesis by murine leukaemia viruses have implicated the proviral LTRs rather than the *env* gene (Chatis *et al.*, 1983; Des Grosseillers *et al.*, 1983). It has been shown quite recently that enhancer elements can function in a tissue-specific manner, and that this is also the case for LTRs of leukaemogenic viruses (see Fig. 2*b*). Since viral replication is likely to be important for spread of virus to the appropriate target tissue, all the replicative genes are essential and any defect in a replicative gene could score as a determinant of reduced pathogenicity if isogenic pairs of viruses are compared. This should be kept in mind when experiments involving hybrid viruses, constructed *in vitro*, are being performed to locate sequences necessary for pathogenicity. Whether any particular gene product, such as the *env* gene product, has the ability to stimulate haematopoietic cell growth directly is not addressed by experiments of this sort.

Parallel between FeLV-B and murine mink cell focus-forming (MCF) viruses

A recent analysis of the DNA sequence of the FeLV-B envelope gene revealed a high degree of homology to the *env* gene of murine MCF viruses (Elder and Mullins, 1983). Homology is found in the 5′ portion of the *env* gene in a stretch of sequence that, in the MCF virus, represents a substitution relative to the common murine ecotropic viruses. Since an ecotropic virus is assumed to be the parent virus of MCF, the *env* sequence substitution is thought to be derived from the mouse cell genome by recombination. A parallel situation may exist in the cat since the prototype subgroup A virus, FeLV-A/Glasgow-1, which is defined as ecotropic by its host range properties, appears to lack the MCF-specific sequences. This is illustrated in Fig. 4, where a heteroduplex comparison of proviral clones of FeLV-A and FeLV-B isolates is shown. A single-stranded region is seen which corresponds, according to measurement and comparison to the published sequence of FeLV-B (Elder and Mullins, 1983; Nunberg *et al.*, 1984), to the 5′ end of the *env* gene. The parallel between murine and feline viruses can also be telling when one compares the biological properties of MCF and FeLV-B isolates. To consider the feline viruses first, although FeLV-B isolates on their own are not spread by infection among cats, these viruses induce tumours, albeit inefficiently, after direct inoculation in the absence of FeLV-A.

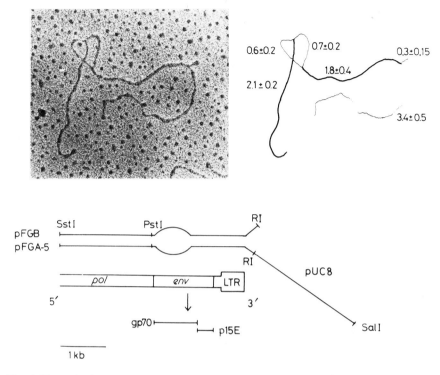

Fig. 4. Heteroduplex comparison of cloned proviral DNA of FeLV-A/Glasgow-1 (pFGA-5) and FeLV-B/Gardner–Arnstein (pFGB). To simplify this analysis, fragments corresponding only to the *pol* and *env* genes and the 3' LTR have been compared. Measurements on the corresponding line drawing are in kilobases, relative to standard size markers. Double-stranded regions are indicated by heavy lines and single-stranded by fainter lines. The single-stranded loop corresponds to the 5' half of the FeLV *env* gene. Additional regions of single-stranded DNA are seen at the 3' end of each clone, corresponding to unrelated cellular flanking sequences and, in the case of pFGA-5, plasmid vector sequences. An explanatory diagram is given underneath. The expected locations of the coding regions for the mature *env* gene products, gp70 and p15E, are shown underneath. This analysis suggests that the subgroup-specific determinants will be located on the amino-terminal domain of gp70. I am indebted to Carolyn Normand for performing the heteroduplex analysis.

Moreover, some of the most highly oncogenic isolates of FeLV, disregarding those which contain oncogenes, are FeLV-A/B mixtures. Similarly, murine MCF viruses, although originally characterised as highly pathogenic variants, do not always display this property but are united only by their classification in a single interference group like FeLV-B (Rein, 1982). If this comparison is carried further, it could be inferred that the mouse and the cat carry related elements in their genomic DNA that can participate in recombination with the common ecotropic viruses to generate variants with altered host range or pathogenicity. These genomic elements may be endogenous retroviral sequences that are common between mouse and cat, at least in the crucial region of the *env* gene. Alternatively, there is the intriguing possibility that the common elements reside in some normal cellular gene encoding a cell-surface protein. Cloning of genomic sequences related to the MCF-specific region will allow these ideas to be tested directly.

FeLV-GM1: an env gene recombinant virus?

A novel field isolate of FeLV which induces myeloid leukaemia after only a short latent period has been described recently (Onions *et al.*, 1985). The original case was a myeloid leukaemia with some erythroid involvement. This may indicate that the virus transforms a relatively early progenitor cell giving rise to cells of both pathways. Alternatively there may be multiple potential targets for FeLV-GM1. The rapid onset of disease after inoculation of FeLV-GM1 is reminiscent of the avian acute leukaemia viruses, which carry host-derived oncogenes, or the Friend SFFV, which achieves its transforming effects with a recombinant *env* gene (Linemeyer *et al.*, 1982). Observations on cats inoculated with FeLV-GM1, which is a complex of FeLV-A and FeLV-B, shows that clinically apparent disease correlates with the titre of the B envelope component in plasma (D. Onions, personal communication). A subset of inoculated cats remains healthy for an extended period after infection, during which time only FeLV-A is detected in plasma. Onset of disease in these cases is concomitant with the appearance of the subgroup B virus. Unlike other subgroup B isolates, the B component of FeLV-GM1 has never been seen to replicate to high titre, either *in vitro* or *in vivo*. Attempts to separate the B from the A component by passage through human cells have also been unsuccessful. These properties suggest that the B component of FeLV-GM1 may be defective for replication. Its specific association with the onset of disease implicates the subgroup B virus as the highly pathogenic agent in the virus complex. Because of the parallel between FeLV-B and MCF viruses, outlined above, it seems highly plausible that the virus acts via a recombinant *env* gene although it is also possible that the B genome carries an oncogene in addition to the functional *env* product. A third possibility is that the role of the B component is to allow access of another component of the virus complex to the crucial target cell. We have constructed probes specific for FeLV-B from restriction fragments corresponding to the

5' end of the *env* gene and will use these to clone the B component of FeLV-GM1 to test these ideas directly. The clearest demonstration would be provided if the disease can be reproduced using molecularly cloned viruses.

FeLV-C isolates and their association with aplastic anaemia

FeLV isolates of subgroup C are rare in nature and, like subgroup B viruses, always occur as phenotypic mixtures with FeLV-A (Sarma and Log, 1973). Again, the subgroup C viruses can be isolated and grown separately *in vitro* and will induce disease in newborn cats after a short latent period (Mackey *et al.*, 1975). FeLV-C isolates examined so far have produced a rapid and fatal red cell aplasia. This unusual pathogenicity has been shown to involve depletion of erythroid precursor cells at a fairly early stage of differentiation (BFU-E) (Onions *et al.*, 1982). Correlation of the disease with interference subgroup, a phenotype carried by the virion envelope, suggests that the effect may be mediated by an *env* gene product, gp70 or p15E. Since recombinant *env* products can induce the proliferation of a specific cell type, illustrated by the response of erythroid cells to Friend SFFV, it can be considered that an alternative effect is also possible. The rapid onset of the FeLV-C disease, which can be monitored in haematopoietic cell colony assays, suggests that the effect is not mediated by the immune system, at least at the level of serum antibody. The pronounced and rapid disease induced in neonatal cats suggests that the FeLV-C-related disease will be relatively easily ascribed to a particular viral gene or function and, with this aim in mind, we are preparing recombinants between FeLV-A and FeLV-C.

Endogenous FeLV-related sequences

Feline DNA contains some 15–20 copies of sequences closely related to FeLV. The structure of most of these elements appears to be like that of an intact FeLV provirus – some 8 to 9 kb in length, and flanked by long terminal repeats (Mullins *et al.*, 1984*b*), although some copies appear to have large sequence deletions relative to infectious FeLV (Soe *et al.*, 1983). The principal differences between endogenous and exogenous FeLV genomes reside in the U3 region of the LTR. Probes constructed from this region are specific for either exogenous or endogenous viruses and this allows a straightforward strategy for cloning exogenous proviruses from cat DNA (Casey *et al.*, 1981). Use of the LTR as a probe also reveals the presence of multiple copies of endogenous LTRs, possibly 150–200 copies, that are presumably unlinked to other viral genomic sequences. These sequences appear to be mobile in the cat genome and can be seen to alter on passage of cat cells in culture (Mullins *et al.*, 1984*b*). Unlike the endogenous proviruses of the mouse, the expression of FeLV-related sequences is generally tightly repressed. I know of only one tumour in which endogenous FeLV-related sequences have been shown to be expressed, and this was not one normally associated with FeLV – a squamous cell carcinoma (J. Mullins, personal communication). Despite their transcriptional silence, the endogenous sequences might well function as reservoirs for recombination with FeLV-A to generate the envelope variants and other viruses of increased pathogenicity.

Summary

Feline leukaemia and sarcoma viruses have provided a rich source for the identification of host genes that can confer oncogenic potential on retroviruses. The prospects are that further new oncogenes will emerge although the repeated identification of the same gene(s) in independent isolates suggests that the latter will be the more likely outcome. The role of FeLV in the most common malignancy associated with viral infection, T-cell lymphosarcoma, has been brought into focus by the identification of recombinant FeLV proviruses carrying the *myc* oncogene in naturally occurring tumours. We are concentrating much of our effort on analysis of these novel viruses and on the other cases of T-cell and B-cell tumours that must involve distinct if not dissimilar mechanisms. The study of FeLV continues to reveal unexpected aspects of the interaction between virus and host. In addition to the frank viraemic state, FeLV can induce latent infections and can under some circumstances induce tumours in apparently virus-negative animals (D. Onions, G. Lees and J.C.N., in preparation). Molecular analysis is required in these cases to show the presence of exogenous viral infection. The frequent induction of degenerative rather than neoplastic disease has been known for some time to be a common feature of FeLV infection and a particular feature of certain FeLV isolates. The discovery of other retroviruses that function in a similar way suggests that FeLV will again prove to be a useful natural and experimental model. Analysis at the molecular level is now an essential component of this study.

I am most grateful for a fruitful collaboration, with D. Onions, G. Lees and O. Jarrett of the Department of Veterinary Pathology, University of Glasgow, which is the basis of much of the work presented here. I would like to thank those involved in the FeLV research at the Beatson Institute, D. Forrest, R. McFarlane, M. O'Hara, M. Stewart, M. Warnock and A. Wheeler, and N. Wilkie for his interest and support. I thank A. Balmain, D. Onions, M. Stewart and J. Paul for helpful comments on the manuscript. Our research is supported by the Cancer Research Campaign and the Leukaemia Research Fund.

REFERENCES

BESMER, P. (1984). Acute transforming feline retroviruses. *Current Topics in Microbiology and Immunology*, in press.

BESMER, P., SNYDER, H. W., JR, MURPHY, J. E., HARDY, W. D., JR and PARODI, A. (1983). The Parodi–Irgens feline sarcoma virus and simian sarcoma viruses have homologous oncogenes but in different contexts of the viral genomes. *Journal of Virology*, **46**, 606–13.

BISTER, K., RAMSAY, G., HAYMAN, M. J. and DUESBERG, P. H. (1980). OK10, an avian acute leukemia virus with a unique genetic structure. *Proceedings of the National Academy of Sciences, USA*, **77**, 7142–6.

CASEY, J. W., ROACH, A., MULLINS, J. I., BURCK, K. B., NICOLSON, M. O., GARDNER, M. B. and DAVIDSON, N. (1981). The U3 portion of feline leukaemia virus DNA identifies horizontally acquired proviruses in leukaemic cats. *Proceedings of the National Academy of Sciences, USA*, **78**, 3418–22.

CHATIS, P. A., HOLLAND, C. A., HARTLEY, J. W., ROWE, W. P. and HOPKINS, N. (1983). Role of the 3′ end of the genome in determining disease specificity of Friend and Moloney murine leukemia viruses. *Proceedings of the National Academy of Sciences, USA*, **80**, 4408–11.

COLBY, W. D., CHEN, E. Y., SMITH, D. H. and LEVINSON, A. D. (1983). Identification and nucleotide sequence of a human locus homologous to the v-*myc* oncogene of avian myelocytomatosis virus MC29. *Nature*, **301**, 722–5.

COLL, J., RIGHI, M., DE TAISNE, C., DISSOUS, C., GEGONNE, A. and STEHELIN, D. (1983). Molecular cloning of the avian acute transforming retrovirus MH2 reveals a novel cell-derived sequence (v-*mil*) in addition to the *myc* oncogene. *The EMBO Journal*, **2**, 2189–94.

CUYPERS, H. T., SELTEN, G., QUINT, W., ZIJLSTRA, M., MAANDAG, E. R., BOELENS, W., VANWEZENBEEK, P., MALIEF, C. and BERNS, A. (1984). Murine leukemia virus-induced lymphomagenesis: Integration of proviruses in a distinct chromosomal region. *Cell*, **37**, 141–50.

DESGROSSEILLERS, L., RASSART, E. and JOLICOEUR, P. (1983). Thymotropism of murine leukemia virus is conferred by its long terminal repeat. *Proceedings of the National Academy of Sciences, USA*, **80**, 4203–7.

ELDER, J. and MULLINS, J. I. (1983). Nucleotide sequence of the envelope gene of Gardner–Arnstein feline leukemia virus B reveals unique sequence homologies with a murine mink cell focus-forming virus. *Journal of Virology*, **46**, 871–80.

ESSEX, M., SLISKI, A., COTTER, S. M., JAKOWSKI, R. M. and HARDY, W. D., JR (1975). Immunosurveillance of naturally-occurring feline leukemia. *Science*, **190**, 790–2.

FRYKBERG, L., PALMIERI, S., BEUG, H., GRAF, T., HAYMAN, M. J. and VENNSTROM, B. (1983). Transforming capacities of avian erythroblastosis virus mutants deleted in the *erb*A or *erb*B oncogenes. *Cell*, **32**, 227–38.

GALLWITZ, D., DONATH, C. and SANDER, C. (1983). A yeast gene encoding a protein homologous to the human c-*has*/*bas* proto-oncogene product. *Nature*, **306**, 704–7.

HARDY, W. D., JR, HESS, P. W., MACEWAN, E. G., MCCLELLAND, A. J., ZUCKER-MAN, E. E., ESSEX, M., COTTER, S. M. and JARRETT, O. (1976). Biology of feline leukemia virus in the natural environment. *Cancer Research*, **36**, 582–8.

HARDY, W. D., JR, OLD, L. J., HESS, P. W., ESSEX, M. and COTTER, S. M. (1973). Horizontal transmission of feline leukaemia virus. *Nature*, **244**, 266–9.

HAYWARD, W. S., NEEL, B. G. and ASTRIN, S. M. (1981). Activation of a cellular *onc* gene by promoter insertion in ALV-induced lymphoid leukosis. *Nature*, **290**, 475–80.

HOOVER, E. A., OLSEN, R. G., HARDY, W. D., JR, SCHALLER, J. P. and MATHES, L. E. (1976). Feline leukemia virus infection: Age-related variation in response of cats to experimental infection. *Journal of the National Cancer Institute*, **57**, 365–9.

HUEBNER, R. J. and TODARO, G. J. (1969). Oncogenes of RNA tumour viruses as determinants of cancer. *Proceedings of the National Academy of Sciences, USA*, **64**, 1087–94.

JARRETT, O. (1984). Pathogenesis of feline leukaemia virus-related diseases. In *Mechanisms of Viral Leukaemogenesis*, ed. J. M. Goldman and O. Jarrett, pp. 135–54. Edinburgh, Churchill Livingstone.

JARRETT, O. and RUSSELL, P. H. (1978). Differential growth and transmission in cats of feline leukaemia viruses of subgroups A and B. *International Journal of Cancer*, **21**, 466–72.

LENZ, J., CELANDER, D., CROWTHER, R. L., PATARCA, R., PERKINS, D. W. and HASELTINE, W. (1984). Determination of the leukaemogenicity of a murine retrovirus by sequences within the long terminal repeat. *Nature*, **308**, 467–70.

LEPRINCE, D., GEGONNE, A., COLL, J., DETAISNE, C., SCHNEEBERGER, A., LAGROU, C. and STEHELIN, D. (1983). A putative second cell-derived oncogene of the avian leukaemia retrovirus E26. *Nature*, **306**, 395–7.

LEVY, L. S., GARDNER, M. B. and CASEY, J. W. (1984). Isolation of a feline leukaemia provirus containing the oncogene *myc* from a feline lymphosarcoma. *Nature*, **309**, 853–6.

LINEMEYER, D. L., MENKE, J. G., RUSCETTI, S. K., EVANS, L. H. and SCOLNICK, E. M. (1982). Envelope gene sequences which encode the gp52 protein of Friend spleen focus-forming virus are required for the induction of erythroid cell proliferation. *Journal of Virology*, **43**, 223–33.

MACKEY, L., JARRETT, W., JARRETT, O. and LAIRD, H. M. (1975). Anemia associated with feline leukemia virus infection in cats. *Journal of the National Cancer Institute*, **54**, 209–17.

MULLINS, J. I., BRODY, D. S., BINARI, R. C., JR and COTTER, S. M. (1984a). Viral transduction of c-*myc* gene in naturally occurring feline leukaemias. *Nature*, **309**, 856–8.

MULLINS, J. I., CASEY, J. W., SANTON, J. B., BURCK, K. B., NICOLSON, M. O. and DAVIDSON, N. (1984b). Cat endogenous FeLV-related sequences are arranged as proviruses and contain U3 regions in long terminal repeats distinct from those of FeLV. *Virology*, in press.

NAHARRO, G., ROBBINS, K. C. and REDDY, E. P. (1984). Gene product of v-*fgr onc*: Hybrid protein containing a portion of actin and a tyrosine-specific protein kinase. *Science*, **223**, 63–6.

NEIL, J. C., HUGHES, D., MCFARLANE, R., WILKIE, N., ONIONS, D. E., LEES, G. and JARRETT, O. (1984). Transduction and rearrangement of the *myc* gene by feline leukaemia virus in naturally-occurring T-cell leukaemias. *Nature*, **309**, 814–20.

NUNBERG, J. J., WILLIAMS, M. E. and INNIS, M. A. (1984). Nucleotide sequences of the envelope genes of two isolates of feline leukaemia virus subgroup B. *Journal of Virology*, **49**, 629–32.

NUNN, M. F., SEEBURG, P. H., MOSCOVICI, C. and DUESBERG, P. H. (1983). Tripartite structure of the avian erythroblastosis virus E26 transforming gene. *Nature*, **306**, 391–5.

ONIONS, D. E., JARRETT, O., TESTA, N., FRASSONI, F. and TOTH, S. (1982). Selective effects of feline leukaemia virus on early erythroid precursors. *Nature*, **296**, 156–8.

ONIONS, D. E., TESTA, N., JARRETT, O. and TOTH, S. (1985). Rapid induction of myeloid leukaemia by a novel feline leukaemia virus. *Nature*, in press.

PAYNE, G. S., BISHOP, J. M. and VARMUS, H. E. (1982). Multiple arrangements of viral DNA and an activated host oncogene (c-*myc*) in bursal lymphomas. *Nature*, **295**, 209–13.

PERRY, R. P. (1983). Consequences of *myc* invasion of immunoglobulin loci: facts and speculation. *Cell*, **33**, 647–9.

POST, J. E. and WARREN, L. (1980). Reactivation of latent feline leukemia virus. In *Feline Leukemia Virus*, ed. W. D. Hardy Jr, M. Essex and A. J. McClelland, pp. 151–5. Amsterdam, North-Holland Elsevier.

RABBITTS, T. H., HAMLYN, P. H. and BAER, R. (1983). Altered nucleotide sequences of a translocated c-*myc* gene in Burkitt's lymphoma. *Nature*, **306**, 760–5.

REIN, A. (1982). Interference grouping of murine leukemia viruses: A distinct recep-receptor for the MCF-recombinant viruses on mouse cells. *Virology*, **120**, 251–7.

ROJKO, J. L., HOOVER, E. A., QUACKENBUSH, S. L. and OLSEN, R. G. (1982). Reactivation of latent feline leukaemia virus infection. *Nature*, **298**, 385–8.

RUSSELL, P. H. and JARRETT, O. (1978). The occurrence of feline leukaemia virus neutralising antibodies in cats. *International Journal of Cancer*, **22**, 351–7.

SAITO, H., HAYDAY, A. C., WIMAN, K., HAYWARD, W. S. and TONEGAWA, S. (1983). Activation of the c-*myc* gene by translocation: A model for translational control. *Proceedings of the National Academy of Sciences, USA*, **80**, 7476–80.

SARMA, P. S. and LOG, T. (1973). Subgroup classification of feline leukemia and sarcoma viruses by viral interference and neutralization tests. *Virology*, **54**, 160–9.

SCHALLER, J. P., MATHES, J. E., HOOVER, E. A., KOESTNR, A. and OLSEN, R. G. (1978). Increased susceptibility to feline leukemia virus infection in cats exposed to methylnitrosurea. *Cancer Research*, **38**, 996–8.

SOE, L. H., DEVI, B. G., MULLINS, J. I. and ROY-BURMAN, P. (1983). Molecular cloning and characterisation of endogenous feline leukaemia virus sequences from a cat genomic library. *Journal of Virology*, **46**, 829–40.

STEHELIN, D., VARMUS, H. E., BISHOP, J. M. and VOGT, P. K. (1976). DNA related to the transforming gene(s) of avian sarcoma virus is present in normal avian DNA. *Nature*, **260**, 170–3.

SWANSTROM, R., PARKER, R. C., VARMUS, H. E. and BISHOP, J. M. (1983). Transduction of a cellular oncogene: the genesis of Rous sarcoma virus. *Proceedings of the National Academy of Sciences, USA*, **80**, 2519–23.

TSICHLIS, P., GUNTER-STRAUSS, P. and HU, L. F. (1983). A common region for proviral integration in MoMuLV-induced rat thymic lymphomas. *Nature*, **302**, 445–9.

VENNSTROM, B. and BISHOP, J. M. (1982). Isolation and characterization of chicken DNA homologous to the putative oncogenes of avian erythroblastosis virus. *Cell*, **28**, 135–43.

WESTAWAY, D., PAYNE, G. and VARMUS, H. E. (1984). Proviral deletions and on-cogene base substitutions in insertionally mutagenized c-*myc* alleles may contribute to the progression of avian bursal tumours. *Proceedings of the National Academy of Sciences, USA*, **81**, 843–7.

WATT, R., NISHIKURA, K., SORRENTINO, J., AR-RUSHDI, A., CROCE, C. M. and ROVERA, G. (1983). The structure and nucleotide sequence of the 5' end of the human c-*myc* oncogene. *Proceedings of the National Academy of Sciences, USA*, **80**, 6307–11.

THE MOLECULAR BIOLOGY OF HUMAN T-CELL LEUKEMIA VIRUS

M. YOSHIDA, S. HATTORI, T. KIYOKAWA, T. WATANABE and M. SEIKI

Department of Viral Oncology, Cancer Institute, Kami-Ikebukuro, Toshima-ku, Tokyo 170, Japan

Adult T-cell leukemia (ATL) is a clinical entity of T-cell malignancy proposed by Takatsuki and his colleagues (Uchiyama *et al.*, 1977) on the basis of its clinical and hematological characteristics. The disease is also called adult T-cell leukemia/lymphoma (ATLL) because of its leukemic lymphomatous nature. The most striking feature of this disease is the clustering of cases found in the south-western part of Japan (Uchiyama *et al.*, 1977; Hinuma *et al.*, 1981) and more recently in the Caribbean (Blattner *et al.*, 1982). This geographical localization suggests that some unique factors are involved in ATL.

A few years ago, retroviruses were independently isolated, by two groups, from human T-cell leukemias. One retrovirus is human T-cell leukemia virus (HTLV), isolated by Poiesz *et al.* (1980) from a patient with cutaneous T-cell lymphoma; the other is adult T-cell leukemia virus (ATLV), isolated from a Japanese case of ATL (Yoshida, Miyoshi and Hinuma, 1982). Subsequently, HTLV and ATLV were shown to be similar to each other in immunological reactivities (Robert-Guroff *et al.*, 1982) and nucleic acid hybridization (Popovic *et al.*, 1982). We have molecularly cloned the provirus genome (Seiki, Hattori and Yoshida, 1982) and determined the total nucleotide sequence of the ATLV provirus genome (Seiki *et al.*, 1983). We recently reported that ATLV and HTLV type I are identical with respect to the locations of gene-specific sequences and the sites of some restriction enzymes (Watanabe, Seiki and Yoshida, 1983, 1984), so hereafter we use the term HTLV for the virus previously reported as ATLV.

The retrovirus HTLV is exogenous for humans (Reitz *et al.*, 1981; Yoshida *et al.*, 1982) and distinct from known animal retroviruses in the structure of its provirus genome (Seiki *et al.*, 1982, 1983). Furthermore, HTLV was shown, by extensive surveys of antibodies against the viral proteins (Hinuma *et al.*, 1982; Kalyanaraman *et al.*, 1982), to be closely associated with a unique T-cell malignancy ATL. The association of the virus with ATL was also demonstrated by detection of the provirus genome in leukemic cells of ATL patients (Yoshida *et al.*, 1982, 1984*a*, *b*; Wong-Staal *et al.*, 1983). In this chapter, we will summarize our recent work on the structure of the viral genome, the causative roles of HTLV in ATL development, and the identification of the viral proteins.

GENOME STRUCTURE OF HTLV

Since the close association of HTLV with ATL was shown by sero-epidemiology, identification of the genetic structure and the gene products seemed to be important in understanding the origin of the virus and the mechanisms by which it causes leukemogenesis (Yoshida, 1983). For this purpose, we isolated a clone (λATK-1) of the provirus genome integrated into fresh leukemic cells of an ATL patient (Seiki *et al.*, 1982). A simple restriction cleavage map of the inserted fragment in λATK-1 was constructed and the viral sequence was subcloned into pBR322. The total nucleotide sequence of the provirus genome was determined using these subclones, and found to contain 9032 nucleotides (Seiki *et al.*, 1983).

The provirus DNA in λATK-1 contained two direct repeats of the long terminal repeat (LTR) (U3-R-U5) sequence, one at each end (Fig. 1), similar to those of animal retroviruses (Varmus, 1982). The LTR structure is thought to play essential roles in the integration of provirus DNA into the host chromosomal DNA and also in regulation of transcription of the provirus genome.

In general, replication-competent retroviruses have a common genomic organization, i.e. *gag*, *pol*, and *env*, in this order, from the 5′ end of the genome (Vogt, 1977). The DNA sequence of HTLV contained three larger open reading frames and an additional four smaller ones. The three large reading frames probably correspond to the *gag*, *pol*, and *env* genes because of their positions and characteristic amino-acid sequences (Seiki *et al.*, 1983). The *gag* gene was predicted to code for a precursor polypeptide of 48 kD, which would be processed into 14 kD, 24 kD, and 9 kD proteins in this order from the 5′ end of the *gag* gene. The *pol* gene has coding capacity for a 99 kD protein, but a probable protein can not be predicted since no information is available on possible splicing of mRNA or processing of a possible precursor. The *env* gene can code for a 54 kD polypeptide, and the 5′ end of the *env* gene is located within the putative *pol* gene,

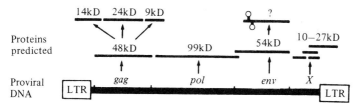

Fig. 1. Gene arrangement of HTLV genome, determined by nucleotide sequence, and proteins predicted from the DNA sequence. The genome contains 9032 bp with two LTRs; the molecular sizes of the predicted proteins are shown in kilodaltons (kD).

overlapping by 5 bp. The overlapping between *pol* and *env* was also found in the Rous sarcoma virus (Schwartz, Tizard and Gilbert, 1983) and Moloney leukemia virus genomes (Shinnick, Lerner and Sutcliff, 1981).

In addition to *gag*, *pol*, and *env*, the HTLV sequence has four extra open reading frames, as indicated in Fig. 1, which have capacities to code for proteins, pX-I–pX-IV, with molecular weights of 11 kD, 10 kD, 12 kD, and 27 kD, respectively. Although the presence of these proteins in infected or leukemic cells remains to be studied, some of them might have functions in the process of transformation of infected T-cells. If some of these sequences have features in common with the known *onc* genes in acute leukemia viruses, similar nucleotide sequences would be expected to be present in normal human DNA. However, the subcloned DNA fragment containing this region did not hybridize significantly with normal human DNA in blotting analysis. This preliminary result indicated that the *X* region is not an *onc* gene directly derived from a cellular DNA sequence. Similar experiments using other viral DNA fragments did not show significant homology with human cell DNA. Thus, it was concluded that HTLV has no typical *onc* gene derived from a cellular sequence (Seiki *et al.*, 1983).

HTLV PLAYS CAUSATIVE ROLES IN ATL DEVELOPMENT

Epidemiological studies of ATL showed the close association of HTLV with ATL; however, no typical *onc* gene was identified by sequence analysis of the total genome and by hybridization experiments (Seiki *et al.*, 1983).

To understand whether HTLV is directly involved in leukemogenesis of ATL and whether the virus is associated with any other

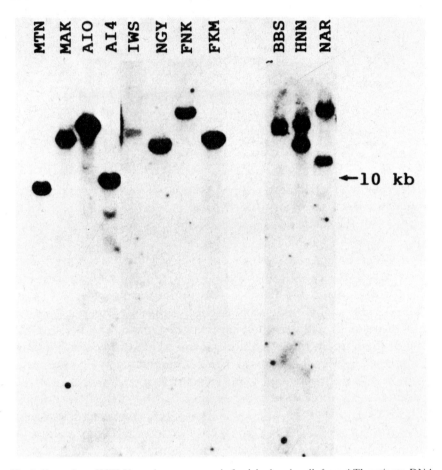

Fig. 2. Detection of HTLV provirus sequences in fresh leukemic cells from ATL patients. DNA extracted from fresh lymphocytes was digested with *Eco*RI, which does not cut the provirus genome, and analyzed by blotting using a representative HTLV probe. In most cases, leukemic cells contained a single copy of the integrated provirus and some others contained two or more copies.

types of lymphomas or leukemias, we surveyed the provirus genome integrated into fresh tumor cells of 210 cases of lymphomas and leukemias from areas in Japan in which the diseases were endemic. For these purposes, we used only fresh tumor cells for the following reasons: HTLV produced from only a few non-leukemic cells can be transmitted into T-cells in culture and the infected cells can reproduce as transformed T-cells (Miyoshi *et al.*, 1981; Markham *et al.*, 1983). Thus, cell lines or cells maintained in culture do not necessarily represent primary tumor cells. In fact, the available cell lines contained numerous copies of the integrated provirus (Watanabe *et*

Table 1. *Association of the HTLV genome with ATL and related diseases*

Group	Viral* genome	Anti† ATLA	Diagnosis	Cell type	Case No.	Sum
i	+	+	ATL	T	122	
			Non-Hodgkin	T	9	137
			CLL	T	5	
			MF	T	1	
ii	−	+	CLL	T	1	
			CLL	B	1	
			Hodgkin‡	(−)	2	9
			Non-Hodgkin	(−)	2	
			AML	(−)	3	
iii	−	−	CLL	T	6	
			CLL	B&N	6	
			Sezary	T	1	
			ALL	TBN	16	64
			non-Hodgkin	TBN	23	
			Hodgkin‡	(−)	1	
			Myeloma	(−)	4	
			CML	(−)	3	
			AML	(−)	4	
			Total			210

* Provirus genome in tumor cells; † Serum antibodies against ATL-associated antigens (ATLA); ‡ The amount of tumor cells in the specimen might not be sufficient to detect the provirus sequence. (−), not analyzed. Non-Hodgkin, non-Hodgkin's lymphoma; Hodgkin, Hodgkin's disease; CLL, chronic lymphocytic leukemia; MF, mycosis fungoides; AML, acute myeloblastic leukemia; Sezary, Sezary syndrome; CML, chronic myelogenous leukemia.

al., 1984), whereas fresh leukemic cells contained only one or two copies (Yoshida *et al.*, 1982, 1984*a*).

Cellular DNA isolated from fresh lymphocytes in peripheral blood was digested with *Eco*RI, or in a few cases with *Sst*I, both of which do not cut the proviral genome. By this assay, integrated provirus sequences are detected as discrete bands (Fig. 2).

Table 1 summarizes the results of surveying 210 cases of lymphoma and leukemia (Yoshida et al., 1984*b*). They were divided into three groups according to the presence or absence of the provirus genome in their leukemic cells and of antibodies to viral proteins: (i) patients with antibodies and the provirus genome in their tumor cells; (ii) patients with antibodies, but with no detectable provirus genome in their leukemic cell DNAs; and (iii) patients who had neither provirus nor antibodies. All 122 patients with typical ATL were included in group (i) of Table 1. Furthermore, these provirus sequences were detected as one or two discrete bands in digests with *Eco*RI (which does not cut typical provirus DNA), as shown in Fig. 2. These data clearly indicate that the leukemic cells were monoclonal, originating from a single cell infected with HTLV, because integration of the HTLV genome in non-leukemic cell DNA was random (Yoshida, 1983; Yoshida *et al.*, 1984). The monoclonal expansion of infected cells as tumor cells in all 122 cases strongly suggests that HTLV directly infects the target cell, which then becomes leukemic, and this suggests that HTLV has causative roles in ATL development (Yoshida *et al.*, 1984a, b), although the infection does not necessarily induce ATL. If viral involvement were indirect, e.g. mediated by a factor(s) released by infected cells, or just a coincidental infection in leukemic cells after transformation, some cases should have leukemic cells in which the provirus genome is absent or is integrated at multiple sites. Such indirect mechanisms are, therefore, unlikely to be of primary importance, although they could modulate the tumorigenic process.

In addition to ATL patients, group (i) in Table 1 included some cases of T-cell malignancies diagnosed as non-Hodgkin's lymphoma or chronic lymphocytic leukemia (CLL). However, these cases were all clinically and pathologically similar to ATL (Yamaguchi *et al.*, 1984), and can thus be included in ATL.

No cases in groups (ii) and (iii) were directly associated with HTLV, since the provirus genome was not detected in their primary tumor cells. Patients in group (ii) had antibodies against the viral proteins, but no provirus genome was detected in their tumor cells.

These findings are explained either by the patients developing leukemia as a result of some factors other than the virus, or by the patients being infected after developing leukemia. Since in the region of Japan where the disease is endemic, 10–20% of healthy adults have antibodies against the viral proteins (Hinuma *et al.*, 1982), the existence of such cases is to be expected. These data indicate that detection of the provirus in primary tumor cells is crucial in judging association of the virus with ATL-related diseases.

TEST FOR A POSSIBLE MECHANISM OF LEUKEMOGENESIS

As described above, HTLV plays causative roles in ATL development (Yoshida *et al.*, 1984a, b); however, the virus has no typical *onc* gene (Seiki *et al.*, 1983). On the basis of these findings, one of the most probable mechanisms of ATL leukemogenesis is insertional mutagenesis, in which the provirus genome is integrated into a specific locus on the chromosomal DNA and then activates an adjacent cellular *onc* gene. This mechanism has been demonstrated in avian lymphoma (Hayward, Neel & Astrin, 1981).

To test the possible common locus for provirus integration, cellular sequences of fresh leukemic cells flanking the integrated proviruses from one patient were used as probes to detect possible DNA rearrangements in other patients by independent provirus integration at the same locus. Two sets of probes were isolated from independent patients, each set of which could detect about 30 kb regions of cellular DNA. Using these probes, rearranged larger fragments were detected in the DNA of the control patient, from whom the original clones were isolated; no such rearranged fragments were detected in the DNA of the other 35 ATL patients (Seiki *et al.*, 1984), and some examples are shown in Fig. 3. From these observations, it can be concluded that it is unlikely that the HTLV provirus is integrated at a common locus in leukemic cells.

However, it is possible that the provirus integrates into multiple sites, but on a specific chromosome. To test this possibility, we analyzed the specificity of provirus integration at the chromosomal level by using the somatic cell hybrid strategy. Human × mouse somatic cell hybrids characterized for their chromosomal content were analyzed by hybridization to two probes isolated from flanking sequences adjacent to each proviral genome integrated in leukemic cells. The sequences which hybridized to the probes segregated independently in cell hybrids, demonstrating that integration is on different chromosomes (Seiki *et al.*, 1984). Analysis of the distribution of the probes in the cell hybrids demonstrated that one probe co-segregated with chromosome 7 and another co-segregated with chromosome 17.

The absence of chromosomal specificity of the provirus integration between two ATL patients is consistent with the absence of a common site shown by blotting analysis of the proviruses in leukemic cells, and strongly suggests that provirus integration in tumor cells occurs at random sites. This situation is very different from the common sites for provirus integration of avian leukosis viruses or mouse mammary tumor virus (MMTV), in which 80% of ALV-induced lymphomas have the provirus within a 14 kb region (Payne, Bishop and Varmus, 1982), and 40% or 70% of MMTV-induced

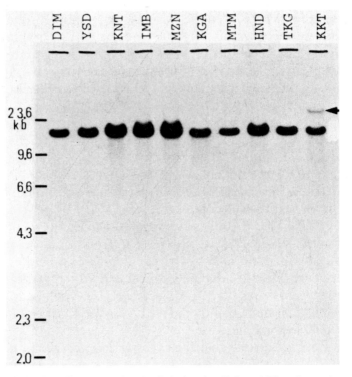

Fig. 3. Analysis of provirus integration sites in leukemic cells from ATL patients using probes of cellular flanking sequences. DNA samples from fresh leukemic cells of independent ATL patients were digested with *Sst*I and analyzed by blotting with ^{32}P-labeled flanking sequence isolated from the DNA of an ATL patient. Lane KKT is DNA from which the original flanking sequence was isolated, and other lanes are DNAs from other ATL patients.

tumors have it within 24 kb or 35 kb regions (Nusse and Varmus, 1982; Dickson and Peters, this volume). Since the direct activation of a cellular *onc* gene by a promoter or enhancer sequence in the provirus LTR requires a specific integration, results suggesting random integration are not consistent with the idea of simple activation of a particular cellular *onc* gene by the LTR sequence of the HTLV provirus. This conclusion may be consistent with the fact that fresh ATL cells do not express viral antigens or mRNA containing viral LTR sequences (M. Yoshida, unpublished).

IDENTIFICATION AND CHARACTERIZATION OF PROTEINS ENCODED BY THE VIRAL GENOME

On the basis of the absence of a common site for provirus integration in primary tumor cells (Seiki *et al.*, 1984), *cis*-acting function of the integrated provirus genome seems to be unlikely. Therefore, proteins coded by the HTLV genome are suspected to have some function in ATL development as *trans*-acting factor(s). The identification and characterization of the viral proteins are also important to establish a system for diagnosis and prevention of HTLV infection. For these two reasons, we tried to identify the *gag*, *env*, and *X* region products: a main strategy was to use synthetic peptides. Many peptides corresponding to parts of the putative proteins predicted from the nucleotide sequence of HTLV were synthesised and antisera against these synthetic peptides were prepared in rabbits. Antisera produced in this way were used to detect the viral proteins in HTLV-infected cell lines.

Gag gene products

Three decapeptides used in this work are shown in Fig. 4; antisera against these peptides, p-14-C, p-24-C and p-9-N, were found to detect p19, p24 and p15, respectively, in HTLV-producing cells (Hattori *et al.*, 1984). All these bands were detected by sera of ATL patients. In addition to these three proteins, some of the sera detected a 53 kD protein. This 53 kD band was demonstrated to be the precursor protein, pr53, for p19, p24 and p15, which were previously shown to be matured *gag* proteins (Kalyanaraman *et al.*, 1982; Schübach *et al.*, 1983) by pulse–chase experiments.

These results clearly demonstrated that the *gag* gene, located at the 5′ end of the HTLV genome, codes for pr53, which is processed into three *gag* proteins, p19, p24 and p15, in this order, from the 5′ end of the gene. These results are consistent with amino-acid sequencing analysis of the purified *gag* proteins (Oroszlan *et al.*, 1982).

Env gene products

Monospecific antiserum against the synthetic peptide which corresponds to the carboxy-terminus of the predicted *env* gene product was used to detect the products. This antipeptide serum revealed a protein of 62 kD in HTLV-producing cell lines (Hattori *et al.*, 1983, 1984). When the cell lines were pretreated with tunicamycin, the same antiserum revealed a 46 kD protein as the major band instead of the 62 kD protein (Fig. 5). From these observations, it was concluded that the *env* gene codes for a 46 kD protein, which is then glycosylated into gp62. The predicted molecular size of the *env*

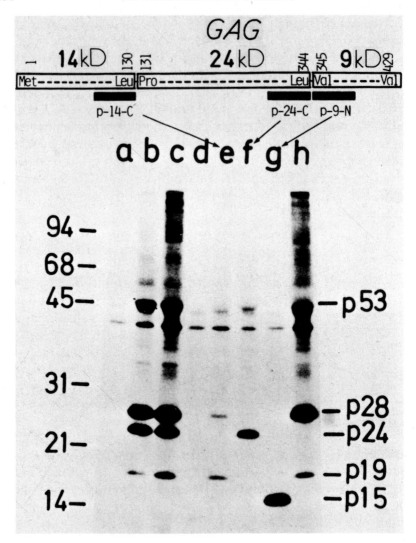

Fig. 4. Identification of *gag* proteins. Decapeptides corresponding to the parts indicated by thick bars were synthesized and rabbit antisera against these peptides were prepared. HTLV-producing cell line MT-2 was labeled with [^{35}S]cysteine and the immunocomplexes were analyzed by electrophoresis in a 12% polyacrylamide gel. The band of p28 is a *gag*-related polypeptide, specifically found in MT-2 cells, that contains multiple copies of variously deleted proviruses. a, normal human serum; b, c, sera from two ATL patients; d, normal rabbit serum; e, f, and g, antisera against p-14-C, p-24-C and p-9-N, respectively; h, monoclonal antibody to p19.

polypeptide is 54 kD, thus 46kD might be an already processed product. To determine whether gp62 was the mature product of the *env* gene or a precursor protein, the cells were labeled for a much

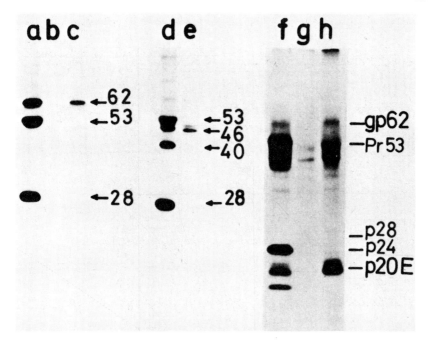

Fig. 5. Identification of *env* gene products. A decapeptide from the carboxy-terminus of the predicted *env* polypeptide was synthesized and antiserum against it was used to identify the *env* gene product. MT-2 cells labeled with [³⁵S]cysteine for 1 h (a–c), MT-2 cells pretreated with tunicamycin (d, e) and HUT102 cells labeled for 15 h (f, g, h) were used. Serum of an ATL patient (a, d, f), rabbit antiserum to *env* peptide (c, e, h) and normal rabbit serum (b, g) are included.

longer period (15 h). In extracts of these cells, antiserum against the peptide revealed a protein of 20 kD as well as a faint band of gp62. This is evidence that the carboxy-terminal portion of gp62 is processed into a 20 kD protein (p20E). The precursor–product relationship between gp62 and p20E was confirmed in a pulse–chase experiment (Hattori *et al.*, 1984).

For detection of another matured *env* gene product derived from the amino-terminal portion of gp62, we produced in *E. coli* a hybrid protein composed of the amino-terminal half of the *env* polypeptide and of β-galactosidase, as described in the following section. Using this *env*–β-galactosidase, antibodies recognizing the amino-terminal half of the *env* gene product were purified from the serum of an ATL patient. This purified antibody preparation detected a diffuse band of 46 kD, together with a faint band of gp62 (Hattori *et al.*, 1984). It was concluded that the 46 kD protein is encoded by the *env* gene. This protein was shown to be glycoprotein gp46 in a separate experiment.

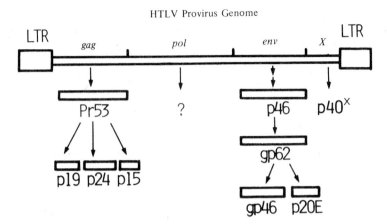

Fig. 6. Summary of HTLV *gag* and *env* gene products described in this chapter.

In summary, the *env* gene codes for a 46 kD precursor, which is glycosylated into gp62. The gp62 is further processed into gp46 and p20E. Recently, p40[x] was identified as the gene product of the *X* region in frame IV (Kiyokawa *et al.*, 1984). The proteins so far identified and the gene locations are summarized in Fig. 6.

PRODUCTION OF *ENV* POLYPEPTIDE IN *E. COLI* AND FUNCTION OF *ENV* GENE PRODUCT

In animals, it is thought that continued replication and spread of leukemia virus within individuals are requirements for subsequent development of viral leukemia (Vogt, 1977). On the basis of these findings, it seems important, for prevention of ATL, to understand the mechanism of the infection and replication of HTLV. The glyco-proteins encoded by the *env* gene of animal retroviruses are exposed on the surface of the viral particles (Kennel *et al.*, 1973) and are known to be essential in the early stages of viral infection for inter-action with receptors on the surface of target cells (Delarco and Todaro, 1976). To establish a system for the study of the *env* gene function, we constructed expression plasmids containing DNA seg-ments of the *env* gene, and the polypeptides derived from these sequences were produced efficiently in *E. coli* (Kiyokawa *et al.*, 1984).

The HTLV *env* gene was divided into two fragments, as illustrated in Fig. 7, and each fragment was inserted into the expression vectors pORF1 and pORF2, respectively (Weinstock *et al.*, 1983). The plasmids pEH9 and pEA1 thus constructed contained the genes for the following three sequences: a short peptide from the *E. coli* outer membrane protein, half of the *env* polypeptide, and β-galactosidase. The plasmids were used to transform *E. coli*, and new proteins with molecular weights of about 150 kD were produced in large amounts, as much as 10–20% of the total protein (Kiyokawa *et al.*, 1984). The sizes of these main proteins were the same as those expected from the DNA sequences. The hybrid proteins produced in this way, EH9 and EA1, both cross-reacted with serum from an ATL patient on Western blotting analysis, indicating that these hybrid proteins had antigenic sites similar to the native *env* gene products, although they were not glycosylated. Therefore these proteins can be used for diagnosis of HTLV infection by detecting *env* specific antibodies in human sera, since they are free from any possible contamination with human proteins.

SPECIFIC ANTISERA AGAINST *ENV* GENE PRODUCTS

Antiserum against the hybrid protein produced in *E. coli* was used to study the function of *env* gene products, and two properties were found: (i) *env* gene products induce cell fusion of certain types of cell line; and (ii) *env* gene products are exposed on the surface of infected cells and can be the target for cytotoxic antibodies.

Nagy *et al.* (1983) and Hoshino *et al.* (1983) reported that HTLV-producing cells can induce syncytia in certain types of cell line, especially cat S^+L^- cells. Addition of these antisera against amino- and carboxy-terminal halves of the *env* polypeptide to the medium during co-cultivation of cat S^+L^- cells with HTLV-infected cells markedly inhibited syncytia formation, whereas addition of normal rabbit serum only reduced syncytia weakly (Fig. 8*a*). These observations clearly demonstrate that HTLV *env* gene products are involved in induction of cell fusion of certain types of cell line, thus indicating that the HTLV *env* gene has a function similar to that of murine leukemkia virus (Rowe, Pugh and Hartley, 1970). As reported by other groups (Nagy *et al.*, 1983; Hoshino *et al.*, 1983), sera from ATL patients also inhibited syncytia formation. These findings are consistent with the presence of antibodies against the *env*-gene products of HTLV in sera of ATL patients (Hattori *et al.*, 1983, 1984).

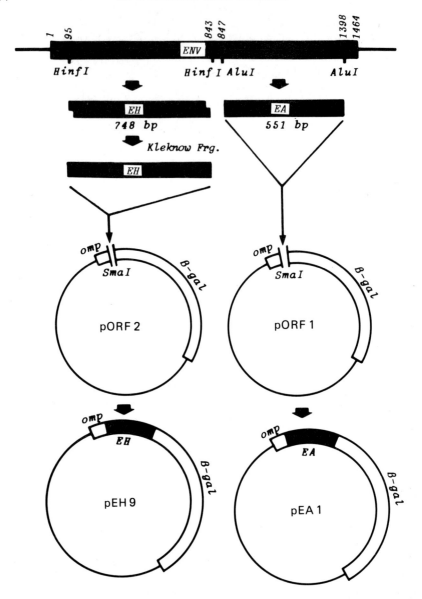

Fig. 7. Schematic illustration of plasmids for expression of the HTLV *env* gene in *E. coli*. Frg. = fragment.

Antisera against *env* polypeptides were found to be cytotoxic to cell lines producing HTLV in the presence of complement (Fig. 8*b*). Since these toxicities were specific to HTLV-infected cells and unrelated to the origin of the cells, it was concluded that the *env*-gene products of HTLV, gp62 or gp46, or both, were exposed on the cell

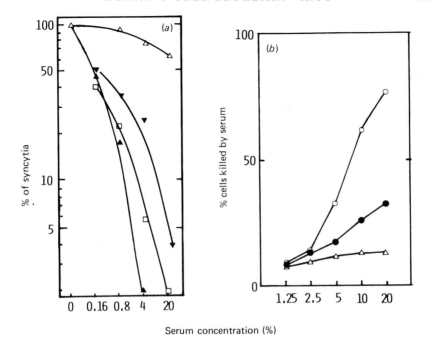

Serum concentration (%)

Fig. 8. (*a*) Inhibition by antiserum against *env* polypeptides of syncytia formation induced by HTLV. Rabbit antisera to amino-terminal (▼) and carboxy-terminal (▲) halves of the *env* polypeptide, serum from ATL patient (□), and normal rabbit serum (△) were included in the cultures to be tested.

(*b*) Complement-dependent cytotoxicities of antisera against amino-terminal (●) and carboxy-terminal (○) halves of the *env*-polypeptide. HTLV-producing cell line MT-2 was used as target cells and rabbit serum was used as source of the complement. Antiserum without complement was also included as control (△).

surface and were the targets in this system. Some sera from ATL patients were cytotoxic to HTLV-producing cells. The presence of cytotoxic antibodies in the sera of patients can explain the absence of viral antigens in primary ATL cells. In most cases of ATL, fresh tumor cells in peripheral blood are not expressing the HTLV antigens in amounts detectable by indirect immunofluorescence assay (Hinuma *et al.* 1981), but they can express the viral antigens after short-term (2–3 days) culture with calf serum. These phenomena can be reasonably explained by the occurrence of immunological masking with cytotoxic antibodies to HTLV-positive cells in the serum of patients. These considerations of cytotoxic antibodies suggest that gp62 or gp46, or both, can be targets for modulating the proliferation of the HTLV-infected or HTLV-transformed cells in response to serum antibodies.

Summary

Human T-cell leukemia virus (HTLV) was isolated from ATL patients and characterized at the molecular level. (i) HTLV contains *gag*, *pol*, and *env* genes. In addition to these, HTLV also contains an *X* region, which can code for one to four small proteins pX-I–pX-IV, but does not contain a typical *onc* gene. (ii) HTLV is closely associated with ATL and plays causative roles in ATL development in infection of the target cells. (iii) Direct activation of cellular *onc* genes by integrated HTLV proviruses seems to be unlikely, since the provirus integration has no locus common in the DNA of leukemic cells, but *X* product may *trans*-activate as yet unknown cellular genes. (iv) *Gag* gene codes for Pr53, which is processed into the matured *gag* proteins, p19, p24, and p15, in this order, from the 5′ end of the gene. (v) *Env* gene codes for a 46 kD protein, which is glycosylated into gp62. The gp62 is then processed into gp46 and p20E. (vi) *Env* gene products can induce cell fusion in certain cell types and they are exposed on the cell surface; thus they can modulate the proliferation of HTLV-infected cells in response to serum antibodies. (vii) *Env* gene products were produced in *E. coli* and can be useful for diagnosis or prevention of HTLV infection. (viii) The *X* region codes for a p40 protein in frame IV.

REFERENCES

BLATTNER, W. A., KALYANARAMAN, V. S., ROBERT-GUROFF, M., LISTER, T. A., GALTON, D. A. G., SARIN, P. S., CRAWFORD, M. H., CATOVSKY, D., GREAVES, M. and GALL, R. C. (1982). The human type-C retrovirus, HTLV, in blacks from the Caribbean region, and relationship to adult T-cell leukemia/lymphoma. *International Journal of Cancer*, **30**, 257–64.

CHEN, I. S. Y., McLAUGHLIN, J. and GOLDE, D. W. (1984). Long terminal repeats of human T-cell leukaemia virus II genome determine target cell specificity. *Nature*, **309**, 276–9.

DELARCO, J. and TODARO, G. J. (1976). Membrane receptors for murine leukemia viruses: Characterization using the purified viral envelope glycoprotein, gp71. *Cell*, **8**, 365–71.

HASELTINE, W. A., SODROSKI, J., PATARCA, R., BRIGGS, D., PERKINS D. and WONG-STAAL, F. (1984). Structure of the 3′ terminal region of HTLV-II human lymphotropic virus: Evidence for a new coding region. *Science*, **225**, 419–24.

HATTORI, S., IMAGAWA, K., SHIMIZU, F., HASHIMURA, E., SEIKI, M. and YOSHIDA, M. (1983). Identification of envelope glycoprotein encoded by *env* gene of human T-cell leukemia virus. *Gann (Japanese Journal of Cancer Research)*, **74**, 790–3.

HATTORI, S., KIYOKAWA, T., IMAGAWA, K., SHIMIZU, F., HASHIMURA, E., SEIKI, M. and YOSHIDA, M. (1984). Identification of gag and env gene products of human T-cell leukemia virus (HTLV). *Virology*, **136**, 338–47.

HAYWARD, W. S., NEEL, B. G. and ASTRIN, S. M. (1981). Activation of cellular *onc* gene by promoter insertion in ALV-induced lymphoid leukosis. *Nature*, **290**, 475–80.

HINUMA, Y., KOMODA, H., CHOSA, T., KONDO, T., KOHAKURA, M., TAKENAKA, T., KIKUCHI, M., ICHIMARU, M., YUNOKI, K., SATO, I., MATSUO, R., TAKIUCHI, Y., UCHINO, H. and HANAOKA, M. (1982). Antibodies to adult T-cell leukemia associated antigen (ATLA) in sera from patients with ATL and controls in Japan: A nation-wide sero-epidemiologic study. *International Journal of Cancer*, **29**, 631–5.

HINUMA, Y., NAGATA, K., HANAOKA, M., NAKAI, M., MATSUMOTO, T., KINOSHITA, K., SHIRAKAWA, S. and MIYOSHI, I. (1981). Adult T-cell leukemia: Antigen in an ATL cell line and detection of antibodies to the antigen in human sera. *Proceedings of the National Academy of Sciences, USA*, **78**, 6476–80.

HOSHINO, H., SHIMOYAMA, M., MIWA, M. and SUGIMURA, T. (1983). Detection of lymphocytes producing a human retrovirus associated with adult T-cell leukemia by syncytia induction assay. *Proceedings of the National Academy of Science, USA*, **80**, 7337–41.

KALYANARAMAN, V. S., SARNGADHARAN, M. G., NAKAO, Y., ITO, Y., AOKI, T. and GALLO, R. C. (1982). Natural antibodies to the structural core protein (p24) of the human T-cell leukemia (lymphoma) retrovirus found in sera of leukemia patients in Japan. *Proceedings of the National Academy of Sciences, USA*, **79**, 1653–7.

KENNEL, S. J., DEL VILLANO, B. C., LEVY, R. L. and LERNER, R. A. (1973). Properties of an oncornavirus glycoprotein: Evidence for its presence on the surface of virions and infected cells. *Virology*, **55**, 464–75.

KIYOKAWA, T., HATTORI, S., SEIKI, M. and YOSHIDA, M. (1984). Envelope proteins of human T cell leukemia virus: Expression in *E. coli* and its application to studies of *env* gene function. *Proceedings of the National Academy of Sciences, USA*, in press.

KIYOKAWA, T., SEIKI, M., IMAGAWA, K., SHIMUZU, F. and YOSHIDA, M. (1984). Identification of a protein (p40ˣ) encoded by a unique sequence pX of HTLV type I. *Gann (Japanese Journal of Cancer Research)*, **75**, 747–51.

MARKHAM, P. D., SALAHUDDIN, S. Z., KALYANARAMAN, V. S., POPOVIC, M., SARIN, P. and GALLO, R. C. (1983). *International Journal of Cancer*, **31**, 413–20.

MIYOSHI, I., KUBONISHI, I., YOSHIMOTO, S., AKAGI, T., OHTSUKI, Y., SHIRAISHI, Y., NAGATA, Y. and HINUMA, Y. (1981). Type C virus particles in a cord T cell line derived by cocultivating normal human cord leukocytes and human leukemic T cells. *Nature*, **294**, 770–1.

NAGY, N., CLAPHAM, P., CHEINGSONG-POPOV, R. and WEISS, R. A. (1983). Human T cell leukemia virus type I: Induction of syncytia and inhibition by patients' sera. *International Journal of Cancer*, **32**, 321–8.

NUSSE, R. and VARMUS, H. E. (1982). Many tumors induced by the mouse mammary tumor virus contain a provirus integrated in the same region of the host genome. *Cell*, **31**, 99–109.

OROSZLAN, S., SAINGADHARAN, M. G., COPELAND, T., KALYANARAMAN, V. S., GILDEN, R. V. and GALLO, R. C. (1982). Primary structure analysis of the major internal protein p24 of human type C T-cell leukemia virus. *Proceedings of the National Academy of Sciences, USA*, **79**, 1291–4.

PAYNE, G. S., BISHOP, J. M and VARMUS, H. E. (1982). Multiple arrangements of viral DNA and an activated host oncogene in bursal lymphomas. *Nature*, **295**, 209–13.

POIESZ, B. J., RUSCETTI, F. W., GAZDAR, A. F., BUNN, P. A., MINNA, J. D. and GALLO, R. C. (1980). Detection and isolation of type C retrovirus particles from fresh and cultured lymphocytes of a patient with cutaneous T-cell lymphoma. *Proceedings of the National Academy of Sciences, USA*, **77**, 7415–19.

POPOVIC, M., REITZ, M. S., SARNGADHARAN, M. G., ROBERT-GUROFF, M., KALYAN-ARAMAN, V. S., NAKAO, Y., MIYOSHI, I., MINOWADA, J., YOSHIDA, M., ITO, Y. and GALLO, R. C. (1982). The virus of Japanese adult T-cell leukemia is a member of the human T-cell leukemia virus group. *Nature*, **300**, 63–6.

REITZ, M. S., POIESZ, B. J., RUSCETTI, F. M. and GALLO, R. C. (1981). Characterization and distribution of nucleic acid sequences of a novel type C retrovirus isolated from neoplastic human T lymphocytes. *Proceedings of the National Academy of Sciences, USA*, **78**, 1887–91.

ROBERT-GUROFF, M., NAKAO, Y., NOTAKE, K., ITO, Y., SLISKI, A. and GALLO, R. C. (1982). Natural antibodies to human retrovirus HTLV in a cluster of Japanese patients with adult T-cell leukemia. *Science*, **215**, 975–8.

ROWE, W. P., PUGH, W. E. and HARTLEY, J. W. (1970). Plaque assay techniques for murine leukemia viruses. *Virology*, **42**, 1136–9.

SCHÜBACH, J., KALYANARAMAN, V. S., SARNGADHARAN, M. G., BLATTNER, W. A. and GALLO, R. C. (1983). Antibodies against three purified proteins of the human type C retrovirus, human T-cell leukemia-lymphoma virus, in adult T-cell leukemia-lymphoma patients and healthy blacks from the Caribbean. *Cancer Research*, **43**, 886–91.

SCHWARTZ, D. E., TIZARD, R. and GILBERT, W. (1983). Nucleotide sequence of Rous sarcoma virus. *Cell*, **32**, 853–69.

SEIKI, M., EDDY, R., SHOWS, R. B. and YOSHIDA, M. (1984). Nonspecific integration of the HTLV provirus genome into adult T-cell leukemia cells. *Nature*, **309**, 640–2.

SEIKI, M., HATTORI, S., HIRAYAMA, Y. and YOSHIDA, M. (1983). Human adult T-cell leukemia virus: Complete nucleotide sequence of the provirus genome integrated in leukemia cell DNA. *Proceedings of the National Academy of Sciences, USA*, **80**, 3618–22.

SEIKI, M., HATTORI, S. and YOSHIDA, M. (1982). Human adult T-cell leukemia virus: Molecular cloning of the provirus DNA and the unique terminal structure. *Proceedings of the National Academy of Sciences, USA*, **79**, 6899–902.

SHINNICK, T. M., LERNER, R. A. and SUTCLIFF, J. G. (1981). Nucleotide sequence of Moloney murine leukemia virus. *Nature*, 293, 543–8.

SODROSKI, J. G., ROSEN, C. A. and HASELTINE, W. A. (1984). *Trans*-acting transcriptional activation of the long terminal repeat of human T-lymphotropic viruses in infected cells. *Science*, **225**, 381–5.

UCHIYAMA, T., YODOI, J., SAGAWA, K., TAKATSUKI, K. and UCHINO, H. (1977). Adult T-cell leukemia: Clinical and hematological features of 16 cases. *Blood*, **50**, 481–92.

VARMUS, H. E. (1982). Form and function of retroviral proviruses. *Science*, **216**, 812–20.

VOGT, P. K. (1977). Genetics of RNA tumor viruses. *In Comprehensive Virology*, ed. H. Fraenkel-Conrat and R. Wagner, vol. 9, pp. 341–455. New York, Plenum Press.

WANG-STAAL, R., HAHN, B., MANZARI, V., COLOMBINI, S., FRANCHINI, G., GELMAN, E. P. and GALLO, R. C. (1983). A survey of human leukemias for sequences of a human retrovirus. *Nature*, **302**, 626–8.

WATANABE, T., SEIKI, M. and YOSHIDA, M. (1983). Retrovirus terminology. *Science*, **222**, 1178.

WATANABE, T., SEIKI, M. and YOSHIDA, M. (1984). HTLV type I (US isolate and

ATLV (Japanese isolate) are the same species of human retrovirus. *Virology*, **133**, 238–41.

WEINSTOCK, G. M., AP RHYS, C., BERMAN, M. L., HAMPAR, B., JACKSON, D., SILHAVY, T. J., WEISEMANN, J. and ZWEIG, M. (1983). Open reading frame expression vectors: A general method for antigen production in *Escherichia coli* using protein fusions to β-galactosidase. *Proceedings of the National Academy of Sciences, USA*, **80**, 4432–6.

YAMAGUCHI, K., SEIKI, M., YOSHIDA, M., NISHIZAWA, H., KAWANO, F. and TAKATSUKI, K. (1984). The detection of human T cell leukemia virus proviral DNA and its application for classification and diagnosis of T cell malignancy. *Blood*, **63**, 1235–40.

YOSHIDA, M. (1983). Human leukemia virus associated with adult T-cell leukemia. *Gann (Japanese Journal of Cancer Research)*, **74**, 777–89.

YOSHIDA, M., MIYOSHI, I. and HINUMA, Y. (1982). Isolation and characterization of retrovirus from cell lines of human adult T-cell leukemia and its implication in the disease. *Proceedings of the National Academy of Sciences, USA*, **79**, 2031–5.

YOSHIDA, M., SEIKI, M., HATTORI, S. and WATANABE, T. (1984a). Genome structure of HTLV and its involvement in the development of adult T-cell leukemia (ATL). In *Human T-cell Leukemia–Lymphoma Viruses*, ed. R. C. Gallo, M. E. Essex and L. Gross. New York, Cold Spring Harbor Laboratory, in press.

YOSHIDA, M., SEIKI, M., YAMAGUCHI, K. and TAKATSUKI, K. (1984b). Monoclonal integration of HTLV provirus in all primary tumors of adult T-cell leukemia suggests causative role of HTLV in the disease. *Proceedings of the National Academy of Sciences, USA*, **81**, 2534–7.

HUMAN AND PRIMATE T-LYMPHOTROPIC RETROVIRUSES (HTLV AND PTLV): SUBTYPES, BIOLOGICAL ACTIVITY, AND ROLE IN NEOPLASIAS

L. RATNER, P. S. SARIN, F. WONG-STAAL and R. C. GALLO

Laboratory of Tumor Cell Biology, Developmental Therapeutic Program, Division of Cancer Treatment, National Cancer Institute, Bethesda, Maryland 20205, USA

The isolation and characterization of a human T-cell leukemia virus (HTLV) from patients with T-cell leukemia–lymphoma provided the first evidence of a retroviral etiology for a human disease (Poiesz *et al.*, 1980*a*, 1981; Gallo *et al.*, 1983*b*). HTLV has been isolated by a number of investigators from different laboratories and more than 100 isolates of this virus have been obtained from patients with adult T-cell malignancies or from their healthy family members (Miyoshi *et al.*, 1981*a*; Yoshida, Miyoshi and Hinuma, 1982; Haynes *et al.*, 1983; Popovic *et al.*, 1983*b*; Sarin *et al.*, 1983; Vyth-Drees and de Vries, 1983). Each HTLV-related virus isolated to date falls into one of three main subgroups. HTLV isolates obtained from patients with adult T-cell leukemia-lymphoma from Japan, the Caribbean and United States have similar biochemical and biological properties and belong to HTLV subgroup 1 (HTLV-1). HTLV obtained from one ATLL patient from Zaire shows considerable differences from other HTLV-1 isolates by restriction enzyme mapping, despite its overall high nucleic acid sequence homology. This isolate has been classified as HTLV-1b (Hahn *et al.*, 1984*b*). Two HTLV isolates belonging to subgroup 2 (HTLV-2) have been obtained, one from a patient with T-cell hairy cell leukemia (Kalyanaraman *et al.*, 1982*b*) and another from a patient with acquired immunodeficiency syndrome (AIDS) (Hahn *et al.*, 1984*a*). HTLV-2 isolates show partial nucleic acid sequence (Chen *et al.*, 1983; Gelmann *et al.*, 1984) and structural protein homology (Kalyanaraman *et al.*, 1981; Robert-Guroff *et al.*, 1981) to HTLV-1. More than 50 isolates belonging to a third subgroup (HTLV-3) have been obtained from a number of patients with AIDS and pre-AIDS (Gallo *et al.*, 1984; Popovic *et al.*, 1984; Sargadharan *et al.*, 1984; Schupbach *et al.*, 1984). Recently,

retroviruses related to HTLV have also been obtained from subhuman primates (Old World monkeys (*Papio cynocephalus*), PTLV)
(Guo, Wong-Staal and Gallo, 1984), and seroepidemiological studies indicate that HTLV-related antibodies are present in baboons
(*Hamydryas* sp.), African green monkeys *Cercopithecus aethiops*,
and macaques (*Maccacca suscata*) (Miyoshi *et al.*, 1982; Hunsmann
et al., 1983; Saxinger *et al.*, 1984*a*). Association of these viruses with
development of lymphomas has been demonstrated at least in the
macaque species (Homma *et al.*, 1984). The role that the HTLV
family of viruses plays in the causation of leukemia and immunosuppression, and the mechanism of transformation, are subjects of
current active investigation.

Growth of T-lymphoid cells

A search for growth factors specific for different hematopoietic lineages resulted in the isolation of a T-cell growth factor (TCGF) from
the conditioned medium of peripheral blood lymphocytes stimulated
with phytohemagglutinin (Prival *et al.*, 1974; Morgan, Ruscetti and
Gallo, 1976). TCGF or interleukin 2 (IL-2) has been purified to
homogeneity (Mier and Gallo, 1980; Robb, 1982; Robb and Green,
1983). It is a glycoprotein of 15 000 dalton with some heterogeneity in
carbohydrate content (Gillis, Smith and Watson, 1980; Robb and
Smith, 1981; Welte *et al.*, 1982). The amino-acid sequence for TCGF
has been derived from cDNA and genomic clones (Fujita *et al.*, 1983;
Tanaguchi *et al.*, 1983; Clark *et al.*, 1984; Holbrook *et al.*, 1984). A
signal peptide of 20 amino acids has been identified which is cleaved
prior to secretion of the protein. The mature protein includes three
cysteine residues, two of which form the intramolecular disulfide
bond required for biological activity (Robb, Kutny and Chowdry,
1983). The threonine at residue 3 is one of the glycosylation sites
(Copeland, Smith and Oroszlan, 1984; Robb and Lin, 1984). However, post-transcriptional modifications have no apparent role in the
biological activity of TCGF, since highly active TCGF is expressed in
Escherichia coli (Devos *et al.*, 1983; Rosenberg *et al.*, 1984).

Receptors for TCGF have been identified on lectin or alloantigen-
activated T-cells by the ability of these cells (*a*) to absorb biologically
active TCGF from the medium (Bonnard, Yosaka and Jacobson,
1979; Coutinho *et al.*, 1979; Smith *et al.*, 1979); (*b*) to bind radiolabeled, highly purified TCGF (Robb, Munck and Smith, 1981); and
(*c*) to react with a monoclonal antibody, denoted anti-Tac
(Uchiyama, Broder and Waldmann, 1981; Leonard *et al.*, 1982;

Miyawaki *et al.*, 1982; Greene and Robb, 1984). Receptors for TCGF are present at very low levels or not at all on unstimulated normal T-lymphocytes or a variety of non-T-cells (Greene *et al.*, 1983; Korsmeyer *et al.*, 1983). Some Burkitt's lymphoma cells (Greene and Robb, 1984) express low levels of TCGF receptors, raising the possibility that TCGF may be involved in the differentiation or activation of B-cells. The recent cloning of the gene for the TCGF receptor (Leonard *et al.*, 1984) will allow a more detailed analysis of the structure of this molecule and the factors that control the regulation of its expression.

HTLV-1

Isolation and characterization of HTLV-1

The availability of growth factor for T-cells allowed for the first time the development of a number of T-cell lines from patients with mature T-cell malignancies with the morphological, karyological, biochemical, and antigenic phenotypes of the fresh tumour cells from the patient (Gazdar *et al.*, 1979, 1980; Poiesz *et al.*, 1980*b*). It was from one of the early T-cell lines that type C virus particles were identified by electron microscopy and particle-associated reverse transcriptase activity was also detected (Poiesz *et al.*, 1980*a*, 1981). This virus was named human T-cell leukemia virus or HTLV. A number of virus isolates obtained from patients with T-cell malignancies have been found to be related to the first HTLV isolate and have been grouped into subgroup 1 (HTLV-1). The proviral genome is composed of polyadenylated single-stranded RNA (9 kb) which includes flanking long-terminal repeat (LTR) sequences, *gag*, *pol*, and *env* genes, and an additional region in the 3' portion of the genome, *X* (Seiki *et al.*, 1983). *X* is not a classical oncogene in that it is not closely homologous to a normal cellular sequence.

The LTR of HTLV-1 is somewhat unusual in several respects (Seiki *et al.*, 1983). The U_5 and R regions are 175 and 224 base pairs (bp) long, somewhat larger than that of most other retroviruses but more similar to that of BLV (see Burny *et al.*, this volume). Furthermore, the polyadenylation signal is 250 bp upstream of its expected position, but a hairpin loop structure may be hypothesized which would bring this signal into the proximity of the polyadenylation site.

The *gag* precursor protein is cleaved to a hydrophobic 19 000 dalton protein, (p19) (Robert-Guroff *et al.*, 1981; Kalyanaraman *et al.*, 1983), the 24 000 dalton major structural protein of the virus, (p24) (Kalyanaraman *et al.*, 1981), and a 15 000 dalton nucleic acid binding protein, (p15) (Kalyanaraman *et al.*, 1983). There is no immunological relatedness of these proteins to those isolated from other avian or mammalian retroviruses. However, amino-acid and nucleotide sequence analyses reveal distant structural homology of p24 and p15 of HTLV-1 with BLV p24 and p12 (Oroszlan *et al.*, 1982; Copeland *et al.*, 1983; Seiki *et al.*, 1983). The reverse transcriptase is a 100 000 dalton protein with a divalent cation preference for magnesium over manganese (Rho *et al.*, 1981).

The primary translation product of the *env* gene is predicted by DNA sequence data to be 54 000 daltons (Seiki *et al.*, 1983), which probably undergoes post-translation modification and cleavage. A 61 000–67 000 dalton glycoprotein is found on the surface of HTLV-1 infected cells which reacts with patient sera and this binding is blocked specifically by pre-incubation of the sera with a 46 000 dalton protein in the virus particles (Essex *et al.*, 1983; Schupbach, Sarngadharan and Gallo, 1984). Preliminary amino-acid sequence data of this 61 000–67 000 dalton protein is consistent with that predicted for the *env* gene product (Lee *et al.*, 1984*a*).

Epidemiology of HTLV-1 infection

Seroepidemiological studies to detect HTLV-1 infection in various parts of the world have utilized purified disrupted viral particles or individual viral proteins for antibody assays of a large number of patients and normal individuals (Hinuma *et al.*, 1981; Kalyanaraman *et al.*, 1981, 1982*a*; Robert-Guroff *et al.*, 1982). These studies have demonstrated clustering of HTLV-1 infection in southern Japan (Tajima *et al.*, 1979; Kalyanaraman *et al.*, 1982*a*, Robert-Guroff *et al.*, 1982), the Caribbean (Blattner *et al.*, 1982), parts of South America (Popovic *et al.*, 1983*b*), Southeastern United States (Blattner, Takatsuki and Gallo, 1983; Blayney *et al.*, 1984), and many portions of Africa (Saxinger *et al.*, 1984*b*). The proportion of normal individuals who are seropositive for HTLV-1 is 10–16% in Nagasaki, Kagoshima, and Uwajima in Japan, 2–10% in Jamaica, Capetown, Nigeria, Tunisia, and Ghana, and 1–3% in portions of rural Southeastern United States. These data are to be contrasted with those derived from most other areas of the United States and

Europe which have demonstrated that less than 1% of normal individuals have antibodies to HTLV-1 (Gallo et al., 1983a; Robert-Guroff et al., 1984). The highest prevalence of antibodies (45–48%) to HTLV-1 in normal individuals, however, is found in family members of seropositive patients with lymphoma (Gallo et al., 1983a; Robert-Guroff et al., 1983).

Patients with other hematopoietic malignancies have also been screened for the presence of antibodies to HTLV-1. These studies show a close correlation between the presence of HTLV-1 antibodies and adult T-cell leukemia-lymphoma (ATLL), which is prevalent in Southern Japan, the Caribbean, and to a lesser extent in South America, Southeastern United States and Africa (Gallo, 1984; Robert-Guroff et al., 1984; Saxinger et al., 1984b). Nearly 70% of all individuals with non-Hodgkin's lymphoma seen in Jamaica over a one year period were seropositive. A similar high frequency of HTLV-1 antibodies has been found in Caribbean lymphoma patients who had emigrated to England (Catovsky et al., 1982). Examination of patients with mycosis fungoides and Sezary syndrome has revealed HTLV-1 antibodies in only a few patients (Gallo et al., 1983a). More recent studies of early-stage patients from Denmark have revealed low titers of HTLV-1 antibodies in 15% of patients (Gallo, 1984). Antibodies to HTLV-1 were also found in patients with malignancies of immature T-cells, such as acute lymphoblastic leukemia, at rates similar to that of the normal individuals in the same geographical region, though somewhat higher in individuals who had received numerous blood transfusions (Gallo et al., 1983a). However, B-cell chronic lymphocytic leukemia (CLL) patients in Jamaica who have received no transfusions show a considerably higher antibody prevalence than normal individuals (Gallo, 1984).

In more limited studies, nucleic acid hybridization to cloned HTLV-1 probes have been used to survey individuals for the presence of integrated provirus in their tumor tissue (Manzari et al., 1983b; Wong-Staal et al., 1983; Manzari et al., 1983c). In the large majority of cases, there was a close correlation between the results of surveys of nucleic acids and antibodies. However, a few seronegative individuals were found to be positive for the provirus. Furthermore, a few patients with leukemias other than ATLL were found to have antibodies without integrated provirus in their tumor tissues. This suggests that HTLV probably plays no direct etiological role in such cases, and the HTLV-1 antibodies detected may be the result of a prior infection or recent infection as a result of numerous blood transfusions used during treatment. However, the possibility of an indirect role for HTLV infection cannot be ruled out for many of these cases. This is notable for the seropositive Jamaican B-cell CLL patients in whom integrated provirus was not detected in their malignant B-lymphocytes, but virus could be isolated in several cases from their normal bone marrow T-cells (Gallo, 1984).

Clinical and pathological characteristics of ATLL

ATLL is a distinct clinical syndrome closely associated with infection by HTLV-1 (Takatsuki et al., 1977, 1979, 1982; Catovsky et al., 1982; Blayney et al., 1983). It is described generally as a non-Hodgkin's lymphoma presenting frequently with involvement of the bone marrow, peripheral blood, liver, spleen, and lymph nodes. Characteristic dermal infiltrative lesions are found in one-half of the patients and hypercalcemia in two-thirds. The malignancy is quite aggressive, generally poorly responsive to various chemotherapeutic regimens; the median survival is 3–6 months.

The malignant cells are more mature with convoluted nuclei reminiscent of Sezary cells. Neoplastic infiltrates in various tissues are quite pleomorphic and have been subclassified into a variety of different morphological groups. Immunophenotyping has shown the malignant cells to be T-lymphocytes which are OKT4 positive and T-helper cells, though functional assays indicate that they possess suppressor cell activity (Waldmann et al., 1983; Yamada, 1983). Karyotypic analysis has revealed aneupolidy in most cases, though there is no indication of a consistent chromosomal marker seen in all cases. The most frequent abnormalities include trisomy of chromosomes 7 and 3, deletions of chromosome 7, and translocations involving chromosomes 17 and 14 (Ueshima et al., 1981; Fukuhara et al., 1983; Miyamoto et al., 1983; Whang-Peng et al., 1983, 1984; Nowell et al., 1984; Rowley et al., 1984).

Monoclonality of HTLV-1-induced malignancies

The cloning of the provirus of HTLV-1 has allowed the development of nucleic acid probes to study the molecular aspects of malignancies associated with this virus. Using these probes, one can generally detect one to five integrated proviruses per cell in fresh tumor tissue

(Yoshida *et al.*, 1982; Wong-Staal *et al.*, 1983; Yoshida *et al.*, 1984). Uninvolved fresh tissues from the same patient show no hybridizing sequences (G. Franchini *et al.*, unpublished), and a B-lymphocytic cell line established from one patient is also negative for integrated provirus (Gallo *et al.*, 1982, 1983*b*). These data demonstrate that HTLV-1 is acquired as an exogenous infection. More recently, a few B-lymphocytic cell lines established from patients with typical T-cell ATLL have also been shown to contain integrated viral sequences (Yamamoto *et al.*, 1982a; Gallo, 1984).

Southern blot analysis of DNA using fresh tumor tissues or cell lines has revealed a similar restriction enzyme map for the majority of integrated HTLV-1 proviruses (Wong-Staal *et al.*, 1983). This is demonstrated in Fig. 1 by the conservation of internal 1.05 kb *Bam*HI and 2.4 kb *Pst*I fragments. More comprehensive restriction enzyme maps of proviruses from different individuals cloned in bacteriophage lambda have confirmed this point. Figure 1 also shows that the provirus is integrated in the malignant cells in a mono- or oligo-clonal fashion. This is exemplified by the bands representing flanking sequences as shown for those larger than the 9 kb size of the complete provirus.

Hybrid cells of rodent fibroblasts or lymphocytes fused with HTLV-1 transformed human cells demonstrate the provirus to be present in virtually any human chromosome (Clarke *et al.*, 1984*a*; Seiki *et al.*, 1984), arguing against common integration regions. The failure to find a consistent integration site in this malignancy differs from the situation with many other chronic leukemia viruses, such as avian leukosis virus (ATLV) which gives rise to B-cell lymphomas in chickens in which the provirus is integrated near the c-*myc* oncogene (Fung *et al.*, 1981; Hayward, Neel and Astrin, 1981; Neel *et al.*, 1981; Fung, Crittenden and Kung, 1982; Payne, Bishop and Varmus, 1982), and mouse mammary tumor virus which produces adenocarcinomas in which the provirus is integrated commonly near one of two loci, denoted *int*-1 or *int*-2 (Nusse and Varmus, 1982; Dickson *et al.*, 1984; Dickson and Peters, this volume). Neil *et al.* (1984) have reported a high incidence of clonally integrated copies of feline leukaemia viruses carrying the *myc* oncogene in spontaneous feline lymphoma. Although such viruses could probably integrate at many chromosomal loci and still contribute to oncogenesis, there is no evidence to date that such viruses are involved in ATLV. These considerations suggest the possibility that HTLV may transform lymphocytes by a novel mechanism.

(a)

S.Y.　J.M.　S.D.　T.O.　MT-2　U.K.　C.R.-Sezlll　M2　M.B.　G5
(Jap)　(Jap)　(Jap)　(Jap)　(Jap)　(Is)　(US)　(US)　(Ca)　(US)

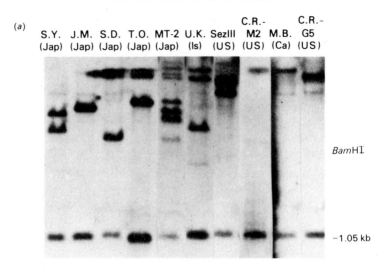

*Bam*HI

−1.05 kb

(b)

S.Y.　J.M.　S.D.　T.O.　MT-2　M.A.　M.I.　C.R.-M2　U.K.　M.J.
(Jap)　(Jap)　(Jap)　(Jap)　(Jap)　(Bra)　(Ca)　(US)　(Is)　(US)

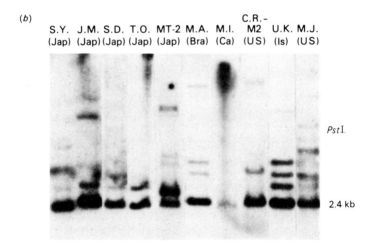

*Pst*I

2.4 kb

Fig. 1. Conservation of the HTLV genomes present in diverse geographical areas. DNA from fresh leukemic cell lines from HTLV-positive patients from different areas of the world was digested with the restriction enzymes *Bam*HI (a) or *Pst*I(b), which cut more than once within HTLV$_{CR}$, and blot hybridized to an HTLV *env–pol* probe. The places of origin are: Jap, Japan; Is, Israel; US, United States, Ca, Caribbean, Bra, Brazil. (From Wong-Staal *et al.*, 1983.)

Transformation by HTLV-1 in vitro

Transformation of lymphocytes

The seroepidemiologic and nucleic acid data described thus far including (a) geographical association of ATLL with areas of antibody prevalence in the population, (b) close correlation of the pres-

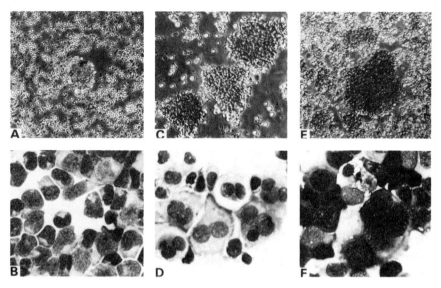

A&B, C5; C&D, C5/MJ; E&F, MJ;

Fig. 2. Growth pattern (A, C, and E) and morphology (B, D, and F) of cultured normal (uninfected) and HTLV-transformed T-cells from cord blood and from HTLV-positive neoplastic T-cells from a patient (MJ) with a cutaneous T-cell lymphoma (mycosis fungoides). (A & B) C5, normal (uninfected) cord blood T-cells cultured with 10% TCGF for 18 days, showing growth as single cells, and the cells have uniform morphology. (C & D) Cord blood T-cells shown in A and B but transformed by an MJ isolate of HTLV. Growth is characterized by large clumps and many bi- and multi-nucleated cells. (E & F) MJ cells in culture form large clumps and have some multinucleated cells (A, C, and E, × 70; B, D, and F, Wright-Giemsa, × 325). (From Popovic *et al.*, 1983*a*.)

ence of antibodies and integrated provirus with the disease, and (*c*) seroepidemiological and molecular biological evidence for horizontal transmission of the virus, imply an etiological role for HTLV-1 in ATLL. The strongest evidence, however, for a causal association is the ability to reproduce certain aspects of the disease *in vitro*. This has been demonstrated by the capability of HTLV-1 to transform normal T-lymphoid cells in culture. Virus can be transmitted either by infection with cell-free virus or by co-cultivating X-irradiated $(6–10 \times 10^3 \text{ rad})$ or mitomycin-treated HTLV-positive T-cell lines with normal T-lymphocytes derived from adult peripheral blood, bone marrow or umbilical cord blood cells (Miyoshi *et al.*, 1981*a*,*b*; Yamamoto *et al.*, 1982*b*; Popovic *et al.*, 1983*a*,*b*). After growth in culture, the resultant cells show the karyotype and HLA phenotype of the recipient cells, but morphologically resemble the neoplastic cells (Fig. 2) and are immortalized in culture. These cells show evidence of HTLV-1 provirus integration and viral RNA and protein expression. All of the other biochemical features of the primary cell lines, some of which will be discussed below are also demonstrated by the cells transformed *in vitro*.

The surface phenotype of the cells transformed *in vitro* shows them to be T-lymphocytes and usually OKT4 positive when using umbilical cord blood as a source of recipient cells (Mann *et al.*, 1983*a,b*; Markham *et al.*, 1983; Popovic *et al.*, 1983*b*). However, when using bone marrow cells as a recipient, the resultant transformed cells may be OKT4 positive, OKT8 positive, or possess neither the OKT4 nor the OKT8 phenotype. This suggests that HTLV-1 may transform a range of different T-cell types, including immature T-cells and thus the preference for OKT4-positive cells is not absolute (Markham *et al.*, 1984).

While HTLV-1 has many of the properties of a chronic leukemia virus including (*a*) nondefectiveness for replication (Popovic *et al.*, 1983*b*; Wong-Staal *et al.*, 1983), (*b*) lack of a cell-derived *onc* gene (Seiki *et al.*, 1983), and (*c*) mono- or oligo-clonal integration (Wong-Staal *et al.*, 1983), the ability of this virus to transform cells *in vitro* is somewhat paradoxical, since this is a property generally ascribed to acute and not chronic leukemia viruses (Bishop, 1978; Weiss *et al.*, 1982).

MOLECULAR MECHANISM OF TRANSFORMATION BY HTLV

The molecular basis for the transforming ability of HTLV has not yet been elucidated. Three lines of research have been followed to gain an understanding of this biological property; (*a*) study of lymphokine production and interaction with malignant cells, especially TCGF; (*b*) the role of a unique portion of the HTLV genome, *X*; and (*c*) the activation of cellular sequences during HTLV infection, particularly oncogenes.

Role of TCGF receptors on HTLV-1-transformed lymphocytes

All of the HTLV-1-transformed cell lines possess receptors for TCGF (Greene and Robb, 1984). The receptors on these malignant cells, however, differ from those on normal cells in several respects. HTLV-1-transformed cells have 5–10-fold more receptors per cell than normal lymphocytes, but the dissociation constant for TCGF is quite similar (Depper *et al.*, 1984). Hut 102 cells, an HTLV-

1-transformed primary cell line, have receptors which are slightly more basic (pI 5.5–6.0) than phytohemagglutinin (PHA)-stimulated lymphocytes (pI 5.4–5.7) (Leonard et al., 1983). Furthermore, the receptor protein migrates on denaturing SDS–polyacrylamide gels as a 50 000 dalton protein rather than the 55 000 dalton protein found for the normal lymphocytes. However, the molecular weight of the precursors for the receptors on Hut 102 cells, 33 000, 35 000, and 37 000 daltons, are quite similar to those of phytohemagglutinin-stimulated lymphocytes. Thus, the structural differences of the mature receptor proteins are most likely due to differences in post-translational processing, e.g. extent of sulfation and sialation (Greene and Robb, 1983).

In addition to the expression of high levels of TCGF receptors on all of the HTLV-1-transformed cell lines, the transformed cells require lower concentrations of TCGF than normal lymphocytes or are TCGF-independent (Popovic et al., 1983a). Furthermore, some of these T-cell lines secrete TCGF into the medium (Gootenberg et al., 1981; Salahuddin et al., 1984). The TCGFs from these malignant T-cell lines differ from that of normal lymphocytes with respect to their mobility on gel filtration and isoelectric focusing columns (Gootenberg, Ruscetti and Gallo, 1982). However, restriction enzyme maps of the TCGF gene using a cloned cDNA probe have revealed no differences between the TCGF genes expressed in PHA-stimulated normal cells or HTLV-1-transformed cells (Clark et al., 1984).

These data, specifically (a) the presence of TCGF receptors, (b) the lower TCGF requirements, and (c) the secretion of TCGF by some of the T-cell lines, suggested the possibility that HTLV-1-transformed cells respond to their own TCGF in an autocrine or paracrine manner and support their own proliferation. Though TCGF was not detectable in the medium of all cell lines growing independently of exogenous TCGF, it still remained possible that levels of growth factor were adequate to promote cell growth which are below the sensitivity of the assay or that intracellular TCGF was effective. These models were essentially ruled out by the failure to find TCGF mRNA in many of the TCGF-independent cell lines (Fig. 3) (Arya, Wong-Staal and Gallo, 1984). The results with HTLV-transformed cell lines in Fig. 3A, lanes 3–14 should be contrasted with the Jurkat immature acute lymphoblastic leukemia cell line (lane 1) which is a good TCGF producer after stimulation with PHA and tetradecanoyl phorbal acetate (TPA). Furthermore, by

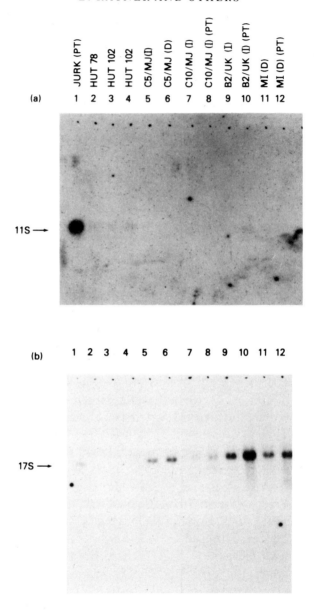

Fig. 3. Expression of (*a*) TCGF gene and (*b*) JD15 gene in HTLV-infected cells. Poly(A)-selected RNA, size-separated by agarose gel electrophoresis, was hybridized with labeled, cloned TCGF DNA. Lanes 1–12 are for RNA from (1) PHA plus TPA-stimulated Jurkat cells, (2) TCGF-independent HUT 78 cells, (3) TCGF-independent HUT 102 cells, (4) another preparation of TCGF-independent HUT 102 cells (5) TCGF-independent C5/MJ cells, (6) TCGF-dependent C5/MJ cells, (7) TCGF-independent C10/MJ cells, (8) TCGF-independent C10/MJ cells treated with TPA–PHA, (9) TCGF-independent B2/UK cells, (10) TCGF-independent B2/UK cells treated with TPA–PHA, (11) TCGF-independent MT-2 cells, (12) TCGF-independent MI cells. Poly(A)-selected RNA was obtained as described

hybridization *in situ* not one in several hundred cells could be detected which expressed TCGF sequences (C. Trainor *et al.*, unpublished). These results eliminate TCGF as playing a role in transformation but do not explain the universal presence of TCGF receptors on these transformed cell lines. It is still conceivable that the TCGF receptor is intrinsically activated in these cells or is activated by another growth factor distinct from TCGF.

Possible biological functions of X

As noted above, the DNA sequence of the complete HTLV-1 provirus demonstrated a region of the genome 3′ to the *env* gene not found in other retroviruses (Seiki *et al.*, 1983). The sequence data show the presence of four open reading frames each beginning with a potential initiator ATG codon predicting protein products of 10, 11, 12, and 27 kilodaltons (kD) though these have not been detected thus far (pX-I, pX-II, pX–III and pX–IV).

A variant of HTLV-1, designated HTLV-1b, isolated from an ATLL patient from Zaire (Hahn *et al.*, 1984*b*) has also allowed a structure–function analysis of the *X* region. The provirus cloned from this patient's tumor tissue hybridizes under high stringency to HTLV-1 probes but restriction enzyme mapping revealed 17 differences among 43 sites mapped. Ten of these restriction enzyme cleavage differences cluster in the 3′ portion of the *env* gene and the 5′ portion of *X*. Nucleotide sequence analysis of portions of the *pol* and *env* genes and *X* have revealed 110 bp changes among 2600 bp examined compared to HTLV-1, with somewhat more diversity in the first open reading frame of *X*, *X*-I, and slightly less in *X*-II than the other portions of the provirus analyzed (our unpublished data). There are no more than three contiguous nucleotide changes compared to HTLV-1, except for an 11-nucleotide deletion including

previously. After denaturation at 65°C in 50% formamide, RNA (10 μg/lane) was electrophoresed in 1% agarose slab gel containing 6% formaldehyde and transferred to a 'Zeta Probe' membrane (BioRad Labs) by electroelution. Hybridization with [^{32}P]-labeled nick-translated cloned TCGF DNA was performed at 37°C for 16 h in a mixture containing 50% formamide, 5 × SSC (SSC = 0.15 mol l^{-1} NaCl + 0.015 mol l^{-1} sodium citrate, pH 7), 0.05 mol l^{-1} sodium phosphate buffer (pH 7), 5 × DM (DM = 0.02% each of bovine serum albumin, polyvinylpyrollidione and Ficoll 400), 200 μgml^{-1} yeast RNA, 20 μgml^{-1} denatured DNA, 0.1% sodium dodecyl sulfate (SDS), and 10% dextran sulfate. The membrane was subsequently washed repeatedly with 1 × SSC, 0.1% SDS at 65°C, air-dried and exposed to a Kodak XAR film using intensifying screens. Where indicated, cells were treated with TPA (10 ngml^{-1}) and PHA (1 μgml^{-1}) for 20 h. The JD15 cDNA probe represents a known constitutively expressed sequence in HTLV-transformed cells, and serves as a control to demonstrate the integrity of the RNA in these preparations.

the potential initiator codon for pX-I. The next ATG codon is near the 3' end of X-I followed after 10 codons by a termination codon. Thus, a functional pX-I protein is unlikely to be made in this transformed cell line, and thus this portion of X probably plays no role in the transformation process.

In comparing the genomes of HTLV-1 and HTLV-2 by Southern blotting analysis and heteroduplex mapping under varying stringency of hybridization, a 1.0 kb sequence within the X region appears to be one of the most conserved portions of the genome (Shaw *et al.*, 1984). DNA sequence analysis of the X region of HTLV-2 shows the 3' 1011 base pairs to be conserved compared to HTLV-1 (77% identity) whereas the 5' region reveals no homology to that of HTLV-1 (Haseltine *et al.*, 1984). The conserved region includes a long open reading frame designated LOR, and a splice acceptor concensus sequence located at the 5' end. No initiation ATG codon is present at the 5' end of this region, and thus it may be translated as part of a fusion protein encoded by a spliced mRNA. This would produce a protein product of at least 38 kD. A protein of 38–42 kD in HTLV-1 infected cells has been identified and preliminary amino-acid sequence data of cyanogen bromide cleavage fragments of this protein are consistent with it being the LOR protein (Lee *et al.*, 1984*b*). These results suggest that LOR may mediate one or more biological properties found in common between these two viruses, such as transformation or T-cell tropism.

A tentative conclusion as to which portions of the X region could be translated might be possible by determining the RNA splice acceptor site(s). A more comprehensive evaluation of the potential protein products of X and their biological function(s) is likely to be derived from transfection experiments. However, it is intriguing that *trans*-acting regulatory factors are present in HTLV-1 and HTLV-2 infected cells which stimulate gene expression directed by the viral LTRs (Sodroski, Rosen and Haseltine, 1984). The LOR protein may represent a mediator in this phenomenon.

Cellular sequences activated in HTLV-transformed cells

Cellular products specifically expressed in HTLV-transformed cells, other than the TCGF receptor, have also been examined to elucidate their role in the transformation process. One approach was to screen a cDNA library from an HTLV-transformed cell line, Hut 102, for sequences expressed differentially compared to an HTLV-negative mature T-lymphocytic cell line, Hut 78 (Manzari *et al.*, 1983*a*). One of the clones (HT3) has been studied in detail. This sequence is represented as a single copy gene in normal and HTLV-transformed cells and is expressed as 2.3 kb mRNA. Expression is also noted in cells not infected with HTLV which produce TCGF, such as PHA-stimulated lymphocytes and the gibbon ape leukemia virus-infected cell line, UCD144. However, HT-3 has no apparent homology to TCGF or the TCGF receptor and the biological function of this gene product is still unclear.

The possible role of oncogenes in ATLL was studied using cloned probes for *sis*, *myc*, *myb*, Ha-*ras*, Ki-*ras*, *abl*, and *src* and no evidence was found for gene amplifications or rearrangements (Gallo, 1984). However, in many of the HTLV-transformed cell lines, *sis* was expressed as a 4.2 kb mRNA in contrast to its lack of expression in a wide range of other hematopoietic tumors and an uninfected T-cell line (Hut 78) (Westin *et al.*, 1982). A *sis* mRNA (4.2 kb) was also detected in several glioblastoma and sarcoma cell lines but not in carcinomas, melanomas, or normal adult tissues (Eva *et al.*, 1982; Slamon *et al.*, 1984).

To analyze the structure and function of these expressed c-*sis* sequences, a cDNA library was constructed from Hut 102 cells (Clarke *et al.*, 1984*b*). Using a v-*sis* probe, three hybridizing clones were isolated, with cDNA inserts of 2.7, 2.7, and 1.8 kb respectively, all representing incomplete transcripts of the gene. Restriction enzyme mapping and DNA sequencing comparisons of the 5′ and 3′ sequences of these three cDNA clones showed that they represented different reverse transcripts of the same mRNA species (our unpublished data with M. Reitz and M. Clarke). Analysis of the first clone, pSM-1, showed the entire v-*sis* homologous sequence to be present in the 5′ portion of the cDNA insert flanked by 182 and about 1300 bp of 5′ and 3′ sequences, respectively (Fig. 4).

Transfection of pSM-1 into NIH3T3 cells resulted in transformed foci of cells which also grew in soft agar and gave rise to tumors in nude mice (Clarke *et al.*, 1984*b*; our unpublished data with E. Westin). Analysis of these c-*sis*-transformed 3T3 cells showed the entire plasmid sequence to be integrated at different loci. Furthermore, these transformed 3T3 cells revealed activation of c-*myc*, analogous to that described for platelet-derived growth factor (PDGF)-stimulated 3T3 cells and PHA-activated lymphocytes (Kelley *et al.*, 1983).

Nucleotide sequence analysis of pSM-1 revealed an ATG codon

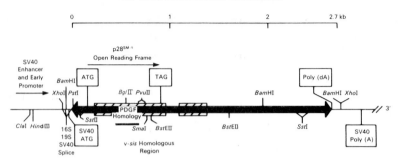

Fig. 4. Organization of the human c-*sis* cDNA clone pSM-1 derived from HUT 102 cell (HTLV-1 infected primary cell line) mRNA. Areas of homology to PDGF and v-*sis* were mapped by Southern blot hybridization and nucleotide sequencing.

64 bp upstream of the v-*sis* homologous region in phase with the long open reading frame (Josephs *et al.*, 1984). The predicted protein product has a hydrophobic amino-terminal sequence which is analogous to the signal peptide characterized for secreted proteins (Blobel, 1980). Most notable was the finding that the nucleotide sequence of the v-*sis* homologous region of the cDNA clone constructed from HTLV-transformed cell RNA is identical with the corresponding exons of the c-*sis* gene of a normal cell. Furthermore, these sequences predict a peptide sequence identical in all 31 amino-acids to one of the peptides of PDGF which have been sequenced (Waterfield *et al.*, 1983). These data strongly suggest that (*a*) the normal c-*sis* sequences are capable of transforming 3T3 cells under certain circumstances, and (*b*) that one peptide chain of human PDGF is encoded by the c-*sis* gene.

Though the c-*sis* cDNA clone is capable of transforming 3T3 cells, PDGF applied externally to those same cells is unable to give rise to transformed foci and tumors in mice (Stiles, 1983). The transforming potential may be because of (*a*) the loss of regulatory sequences in the incomplete c-*sis* sequences, (*b*) higher levels of intracellular PDGF, (*c*) differences in post-translational processing, (*d*) differences in subunit composition, i.e. homodimer versus heterodimer, and/or (*e*) differences in the interaction with cellular receptors.

The role of c-*sis* sequences in HTLV-transformed cells, however, is still not understood. The failure to find evidence of c-*sis* activation in all HTLV-transformed cells would argue that it plays a role in the transformation of some or perhaps no cell lines. However, the unique expression of c-*sis* in these hematopoietic cells suggests an important function for this gene product in this malignancy.

HTLV-2

Though the majority of HTLV isolates belong to the HTLV-1 class of human retroviruses and are quite similar, a second group of retroviruses, HTLV-2, has been detected. Only two isolates of HTLV-2 have been obtained thus far, one each from a patient with a T-cell variant of hairy cell leukemia and from a patient with AIDS. The structural proteins of HTLV-1 and HTLV-2 show only partial immunological cross-reactivity (Kalyanaraman *et al.*, 1982*b*) and the envelope antigens are antigenically distinct (Clapham, Nagy and Weiss, 1984). HTLV-1 DNA probes hybridize to DNA of HTLV-2-infected cells under low hybridization stringency only (Chen *et al.*, 1983; Gelmann *et al.*, 1984). This limited homology allowed the cloning of the HTLV-2 proviruses from both isolates. Restriction enzyme maps of these two clones were indistinguishable from each other, but distinct from that for HTLV-1. DNA sequence comparison of the LTRs from both viral groups showed marked differences throughout their length but conservation of certain structural features including (*a*) large R and U_5 regions; (*b*) presence of conserved 21 bp direct repeat sequences in U_3, and (*c*) identical sequences at the 5' boundary of U_3, the RNA cap site, and the tRNA^{pro} binding site immediately 3' to U_5 (Shimotohno *et al.*, 1984; Sodroski *et al.*, 1984). The DNA sequence of the complete HTLV-2 provirus should allow a more comprehensive comparison with HTLV-1. HTLV-2 is biologically similar to HTLV-1 in its T-cell tropism and its ability to transform T-lymphocytes *in vitro*. However, because HTLV-2 has been isolated on only two occasions, its pathological potential *in vivo* is not evident.

SUBHUMAN PRIMATE HOMOLOGUES OF HTLV (PTLV)

Serological studies have revealed the presence of antibodies against HTLV-1 structural proteins in a number of different Old World primates (Miyoshi *et al.*, 1982; Gallo, 1984). In addition, HTLV-1 proviral sequences have been detected in a T-cell line from a seropositive baboon (Guo *et al.*, 1984). The sequences hybridized under highly stringent conditions to HTLV-1 probes but the provirus has a distinct restriction enzyme map. Homma *et al.* (1984) have shown that PTLV can cause lymphomas in subhuman primates (macaques).

HTLV AND IMMUNOSUPPRESSION

Immunosuppressive or wasting disease may be associated with retro-viral infections in a number of different species (Weiss *et al.*, 1982). This is best exemplified by FeLV, which in some cases causes a T-cell leukemia and in others an immunosuppressive disorder (Essex *et al.*, 1975; Anderson, Jarrett and Laird, 1976; Hardy *et al.*, 1976). There are numerous indications that HTLV may also cause immunosup-pression. First, serological studies of patients hospitalized in infecti-ous disease wards in areas of Japan where HTLV-1 infection is endemic reveal a significantly higher frequency of antibodies towards viral structural proteins than in normal individuals in the same area (M. Essex *et al.*, personal communication). Secondly, a cytotoxic T-cell clone infected with HTLV has been isolated from a patient with mycosis fungoides (Mitsuya *et al.*, 1983, 1984). The cytotoxic activity of these cells is directed only towards cells expressing HTLV antigen(s) and an HLA antigen in common with the cytotoxic clone. Such cytotoxic cells have not been detected in individuals with ATLL, and so these cells may modify the course of the disease. Thirdly, HTLV-1 can infect functional cytotoxic and helper T-cells, and this may result in the abrogation of their function under certain circumstances (our unpublished data with M. Popovic and A. Fauci).

Lastly, another group of HTLV, HTLV-3, appears to be etiologically associated with AIDS (Gallo *et al.*, 1984; Popovic *et al.*, 1984; Sarngadharan *et al.*, 1984; Schupbach *et al.*, 1984). Using a permissive cell line, a cytopathic virus has been cultured from 18 of 21 patients with lymphadenopathy syndrome, 3 of 4 clinically normal mothers of juvenile AIDS, 26 of 72 adult and juvenile AIDS patients, but only one of 22 normal homosexual males and 0 of 115 normal heterosexual patients. Furthermore, the one normal homosexual male from whom the virus was cultured subsequently developed AIDS.

Analysis of this virus has demonstrated proteins similar in size to those of HTLV-1 and HTLV-2. Furthermore, this new viral isolate exhibits partial antigenic cross-reactivity to the other HTLV subgroups, suggesting that the AIDS isolate is a member of the HTLV family of viruses. Antibodies to this virus have been detected in 95% of patients with AIDS, 79% of patients with pre-AIDS, 26% of normal homosexual men presenting for medical evaluation, and 0.6% of patients with unrelated disorders or normal individuals. Therefore, the results are compelling that HTLV-3 has an etiological role in AIDS. Comparison of the structure and function of HTLV-3 to those of HTLV-1 and HTLV-2 should provide clues to the mechanism of immunosuppression.

CONCLUSION

The use of techniques of culture *in vitro* and sensitive reverse transcriptase assays has allowed the detection and isolation of retroviruses associated with various diseases. HTLV-1 is likely to be the etiological agent in ATLL and HTLV-3 probably has a similar relation to AIDS. The disease spectrum of HTLV-2 is unknown because of the rarity of its isolation and the current lack of specific reagents for seroepidemiological studies. HTLV-1 has many of the properties of a chronic leukemia virus, but somewhat paradoxical is its ability to transform T-cells. This suggests that HTLV-1 may transform cells by a novel mechanism. Further studies are required to elucidate the particular biochemical pathways involved with the mechanism of transformation by HTLV-1. The areas of active study in this regard are (*a*) the role of TCGF receptors; (*b*) the role of the X sequence, and (*c*) the activation of cellular sequences such as c-*sis*.

While HTLV-1 has a direct role in the induction of ATLL as evidenced by the finding of virus predominantly in the tumor cells, an indirect role must be considered in other diseases. In B-cell CLL in Jamaica, the frequent finding of antibodies to HTLV-1 and of virus in normal T-lymphocytes but not malignant B-cells suggests such an indirect mechanism. This may be mediated through immunosuppression or via the production of tumorigenic factors by the infected T-cells. These model systems should be expanded to search for the role of viruses, if any, in: (*a*) T-cell malignant disorders such as mycosis fungoides, (*b*) other nonmalignant diseases involving abnormalities of T-cells, such as sarcoidosis and various autoimmune diseases, and (*c*) a wide range of other neoplastic and non-neoplastic conditions.

REFERENCES

ANDERSON, L. J., JARRETT, O. and LAIRD, H. M. (1976). Feline leukemia-virus infection of kittens: Mortality associated with atrophy of the thymus and lymphoid depletion. *Journal of the National Cancer Institute*, **47**, 807–13.

ARYA, S. A., WONG-STAAL, F. and GALLO, R. C. (1984). Expression of T-cell growth factor gene in human T-cell leukemia-lymphoma virus-infected cells. *Science*, **223**, 1086–7.

280　　　　　　　　　　　　L. RATNER AND OTHERS

BISHOP, J. M. (1978). Retroviruses. *Annual Review of Biochemistry*, **47**, 35–88.

BLATTNER, W. A., KALYANARAMAN, V. S., ROBERT-GUROFF, M., LISTER, T. A., GALTON, D. A. G., SARIN, P. S., CRAWFORD, M. H., CATOVSKY, D., GREAVES, M. and GALLO, R. C. (1982). The human type-C retrovirus, HTLV, in blacks from the Caribbean region, and relationship to adult T-cell leukemia/lymphoma. *International Journal of Cancer*, **30**, 257–64.

BLATTNER, W. A., TAKATSUKI, K. and GALLO, R. C. (1983). Human T-cell leukemia/lymphoma virus and adult T-cell leukemia. *Journal of the American Medical Association*, **250**, 1074–81.

BLAYNEY, D. W., BLATTNER, W. A., ROBERT-GUROFF, M., JAFFE, S., FISHER, R. I., BUNN, P. A., PALTON, M. D., RORICK, H. R. and GOLB, L. C. (1984). The human T-cell leukemia/lymphoma virus (HTLV) in the Southeastern United States. *Journal of the American Medical Association*, **250**, 1048–52.

BLAYNEY, D. W., JAFFE, E. S., FISHER, R. I., SCHECHTER, G. P., COSSMAN, J., ROBERT-GUROFF, M., KALYANARAMAN, V. S., BLATTNER, W. A. and GALLO, R. C. (1983). The human T-cell leukemia-lymphoma virus, lymphoma, lytic bone lesions, and hypercalcemia. *Annals of Internal Medicine*, **98**, 144–51.

BLOBEL, G. (1980). Intracellular protein topogenesis. *Proceedings of the National Academy of Sciences, USA*, **77**, 1496–500.

BONNARD, G. D., YOSAKA, D. and JACOBSON, D. (1979). Ligand activated T-cell growth factor-induced proliferation: Absorption of T-cell growth factor by activated T-cells. *Journal of Immunology*, **123**, 2704–8.

CATOVSKY, D., GREAVES, M. F., ROSE, M., GALTON, D. A. G., GOOLDEN, A. W. G., McCLUSKEY, D. R., WHITE, J. M., LAMPERT, I., BOURIKAS, G., IRELAND, R., BROWNELL, A. I., BRIDGES, J. M., BLATTNER, W. A. and GALLO, R. C. (1982). Adult T-cell lymphoma-leukaemia in blacks from the West Indies. *Lancet*, **i**, 639–42.

CHEN, I. S. Y., McLAUGHLIN, J., GASSON, J. C., CLARK, S. C. and GOLDE, D. W. (1983). Molecular characterization of genome of a novel human T-cell leukemia virus. *Nature*, **305**, 502–5.

CLAPHAM, P., NAGY, K. and WEISS, R. A. (1984). Pseudotypes of human T-cell leukemia virus types 1 and 2: Neutralization by patients sera. *Proceedings of the National Academy of Sciences, USA*, **81**, 2806–9.

CLARK, S. C., ARYA, S. K., WONG-STAAL, F., MATSUMATO-KOBAYASHI, M., KAY, R. M., KAUFMAN, R. J., BROWN, E. L., SHOEMAKER, C., COPELAND, T., OROSZLAN, S., SMITH, K., SARNGADHARAN, M. D., LINDEN, S. G. and GALLO, R. C. (1984). Human T-cell growth factor: partial amino acid sequence, cDNA cloning and organization and expression in normal and leukemic cells. *Proceedings of the National Academy of Sciences, USA*, **81**, 2543–7.

CLARKE, M. F., SEIGEL, L. J., MANN, D. L., STRONG, D. M., NASH, W. G., O'BRIEN, S. J. and REITZ, M. S., JR (1984*a*). Independent chromosome assortment of human T-cell leukemia virus (HTLV) induced and host encoded HLA antigen determinants. In preparation.

CLARKE, M. F., WESTIN, E., SCHMIDT, D., JOSEPHS, S. F., RATNER, L., WONG-STAAL, F. and GALLO, R. C. (1984*b*). Transformation of NIH 3T3 cells by a human c-*sis* cDNA clone. *Nature*, **308**, 464–7.

COPELAND, T. D., OROSZLAN, S., KALYANARAMAN, V. S., SARNGADHARAN, M. G. and GALLO, R. C. (1983). Complete amino acid sequence of human T-cell leukemia virus structural protein p15. *FEBS Letters*, **162**, 390–5.

COPELAND, T. D., SMITH, K. A. and OROSZLAN, S. (1984). Characterization of immunoaffinity purified human T-cell growth factor from Jurkat cells. In *Thymic Hormones and Lymphokines '83*, ed. A. L. Goldstein, pp. 165–73. New York, Plenum Publishing Corp.

COUTINHO, A., LARSSON, E.-L., GRONVIK, K. O. and ANDERSON, J. (1979). Studies on T lymphocyte activation. II. The target cells for concanavalin A. *European Journal of Immunology*, **9**, 587–92.

DEPPER, J. M., LEONARD, W., KRONKE, M., PFEFFER, N. J., SVETLIK, P. B., WALDMAN, T. A. and GREENE, W. C. (1984). Augmented T cell growth factor receptor expression in human leukemia/lymphoma virus (HTLV) infection. *Clinical Research*, **32**, 414A.

DEVOS, R., PLAETINCK, G., CHEROUTE, H., SIMONS, G., DEGRAVE, W., TAVERNIER, J., REMOUT, E. and FIERS, W. (1983). Molecular cloning of human interleukin 2 cDNA and its expression in *E. coli. Nucleic Acids Research*, **11**, 4307–23.

DICKSON, C., SMITH, R., BROOKES, S. and PETERS, G. (1984). Tumorigenesis by mouse mammary tumor virus: Proviral activation of a cellular gene in the common integration region *int-2. Cell*, **37**, 520–36.

ESSEX, M., HARDY, W. D., JR, COTTER, S. M., JAKOWSKII, R. M. and SLISKI, A. (1975). Naturally occurring persistent feline oncornavirus infections in the absence of disease. *Infections and Immunity*, **11**, 470–5.

ESSEX, M., MCLANE, M. F., LEE, T. H., FALK, L., HOWE, C. W. S., MULLINS, J. I., CABRADILLA, C. and FRANCIS, D. P. (1983). Antibodies to cell membrane antigens associated with human T-cell leukemia virus in patients with AIDS. *Science*, **220**, 859–62.

EVA, A., ROBBINS, K. C., ANDERSEN, P. R., SRINIVASAN, A., TRONICK, S. L., REDDY, R. P., ELLMORE, N. W., GALEN, A. T., LAUTENBERGER, J. A., PAPAS, T. S., WESTIN, E. H., WONG-STAAL, F., GALLO, R. C. and AARONSON, S. A. (1982). Cellular genes analogous to retroviral *onc* genes are transcribed in human tumour cells. *Nature*, **295**, 116–19.

FUJITA, T., TAKAOKA, C., MATSUI, H. and TANAGUCHI, T. (1983). Structure of the human interleukin 2 gene. *Proceedings of the National Academy of Sciences, USA*, **80**, 7437–41.

FUKUHARA, S., HINUMA, Y., GOTOH, Y.-I. and UCHINO, H. (1983). Chromosome aberrations in T lymphocytes carrying adult T-cell leukemia-associated antigens (ATLA) from healthy adults. *Blood*, **61**, 205–7.

FUNG, Y. K. T., CRITTENDEN, L. B. and KUNG, J. H. (1982). Orientation and position of avian leukosis virus DNA relative to the cellular oncogene c-*myc* in B-lymphoma tumors of highly susceptible $15I_5 \times 7_2$ chickens. *Journal of Virology*, **44**, 742–6.

FUNG, Y. K., FADLY, A. N., CRITTENDEN, L. B. and KUNG, H. J. (1981). On the mechanism of retrovirus-induced avian lymphoid leukosis: Deletion and integration of the provirus. *Proceedings of the National Academy of Sciences, USA*, **78**, 3418–22.

GALLO, R. C. (1984). Human T-cell leukemia–lymphoma virus and T cell malignancies in adults. In *Cancer Surveys*, ed. J. Wyke and R. Weiss, pp. 113–59. Oxford, Oxford University Press.

GALLO, R. C., KALYANARAMAN, V. S., SARNGADHARAN, M. G., SLISKI, A., VONDERHEID, E. C., MAEDA, M., NAKAO, Y., YAMADA, K., ITO, Y., GUTENSOHN, N., MURPHY, S., BUNN, P. A., JR, CATOVSKY, D., GREAVES, M. F., BLAYNEY, D. W., BLATTNER, W., JARRETT, W. F. H., ZUR HAUSEN, H., SELIGMAN, M., BROUET, J. C., HAYNES, B. F., JEGASOTHY, B. V., JAFFE, E., COSSMAN, J., BRODER, S., FISHER, R. I., GOLDE, D. W. and ROBERT-GUROFF, M. (1983*a*). Association of the human type C retrovirus with a subset of adult T-cell cancers. *Cancer Research*, **43**, 3892–9.

GALLO, R. C., MANN, D., BRODER, S., RUSCETTI, F. W., MAEDA, M., KALYAN-ARAMAN, V. S., ROBERT-GUROFF, M. and REITZ, M. S. (1982). Human T-cell leukemia–lymphoma virus (HTLV) is in T- but not B-lymphocytes from a patient

282 L. RATNER AND OTHERS

with cutaneous T-cell lymphoma. *Proceedings of the National Academy of Sciences, USA*, **79**, 4680–3.

GALLO, R. C., POPOVIC, M., LANGE-WANTZIN, G., WONG-STAAL, F. and SARIN, P. S. (1983*b*). Stem cells, leukemia viruses, and leukemia of man. In *Haemopoietic Stem Cells*, ed. Sv.-Aa. Killmann, E. P. Cronkite and C. N. Mullter-Berat, pp. 155–70. Copenhagen, Munksgaard.

GALLO, R. C., SALAHUDDIN, S. Z., POPOVIC, M., SHEARER, G. M., KAPLAN, M., HAYNES, B. F., PALKER, T. J., REDFIELD, R., OLESKE, J., SAFAI, B., WHITE, G., FOSTER, P. and MARKHAM, P. D. (1984). Frequent detection and isolation of cytopathic retroviruses (HTLV-III) from patients with AIDS and at risk for AIDS. *Science*, **224**, 500–3.

GAZDAR, A. F., CARNEY, D. N., BUNN, P. A., RUSSELL, E. K., JAFFE, E. S., SCHECHTER, G. P. and GUCCION, J. G. (1980). Mitogen requirement for the *in vitro* propagation of cutaneous T cell lymphomas. *Blood*, **55**, 409–17.

GAZDAR, A. F., CARNEY, D. N., RUSSELL, E. K., SCHECTER, G. S. and BUNN, P. A. (1979). *In vitro* growth of cutaneous T cell lymphomas. *Cancer Treatment Reports*, **63**, 587–90.

GELMANN, E. P., FRANCHINI, G., MANZARI, V., WONG-STAAL, F. and GALLO, R. C. (1984). Molecular cloning of a unique human T-cell leukemia virus (HTLV-II$_{Mo}$). *Proceedings of the National Academy of Sciences, USA*, **81**, 993–7.

GILLIS, S., SMITH, K. A. and WATSON, J. D. (1980). Biochemical characterization of lymphocyte regulating molecules. II. Purification of a class of rat and human lymphokines. *Journal of Immunology*, **124**, 1954–62.

GOOTENBERG, J. E., RUSCETTI, F. W. and GALLO, R. C. (1982). A biochemical variant of T-cell growth factor produced by a cutaneous T cell lymphoma cell line. *Journal of Immunology*, **129**, 1499–505.

GOOTENBERG, J. E., RUSCETTI, F. W., MIER, J. W., GAZDAR, A. F. and GALLO, R. C. (1981). Human cutaneous T cell lymphoma and leukemia cell lines produce and respond to T cell growth factor. *Journal of Experimental Medicine*, **154**, 1403–18.

GREENE, W. C. and ROBB, R. J. (1984). Receptors for T-cell growth factor: structure, function, and expression on normal and neoplastic cells. In *Contemporary Topics in Molecular Immunology*, eds. S. Gillis and S. P. Inman. New York, Plenum Press, in press.

GREENE, W. C., WALDMANN, T. A., COSSMAN, J., HSU, S.-M., NECHERS, L. M., MARSHALL, S. L., JENSEN, J. A., BAKSHI, A., LEONARD, W. J., DEPPER, J. M., JAFFE, E. S. and KORSMEYER, S. L. (1983). Hairy cell leukemia: A malignant proliferation of B cells which express Tac antigen. In *Normal and Neoplastic Hematopoiesis*, eds. P. Marx and J. Golde, pp. 491–550. New York, Alan R. Liss.

GUO, H.-G., WONG-STAAL, F. and GALLO, R. C. (1984). Novel viral sequences related to human T-cell leukemia virus in T-cells of a seropositive baboon. *Science*, **223**, 1195–7.

HAHN, B. H., POPOVIC, M., KALYANARAMAN, V. S., SHAW, G. M., LoMONICO, A., WEISS, S. H., WONG-STAAL, F. and GALLO, R. C. (1984*a*). Detection and characterization of an HTLV-II provirus in a patient with AIDS. In *UCLA Symposia on Molecular and Cellular Biology, New Series*, vol. 16, ed. M. S. Gottlieb and J. E. Groopman. New York: Alan R. Liss, in press.

HAHN, B. H., SHAW, G. M., POPOVIC, M., LoMONICO, A., GALLO, R. C. and WONG-STAAL, F. (1984*b*). Molecular cloning and analysis of a new varient of human T cell leukemia virus (HTLV-Ib) from an African patient with adult T cell lymphoma. *International Journal of Cancer*, in press.

HARDY, W. D., JR, HEES, P. W., MacEWEN, E. G., McCLELLAND, J. M., ZUCKER-MAN, E. E., ESSEX, M. AND COTTER, S. M. (1976). Biology of feline leukemia virus in the natural environment. *Cancer Research*, **36**, 582–8.

HASELTINE, W. A., SODROSKI, J., PATARCA, R., BRIGGS, D., PERKINS, D. and WONG-STAAL, F. (1984). Structure of 3' terminal region of HTLV-II human T lymphotropic virus: Evidence for a new coding region. *Science*, **225**, 419–24.

HAYNES, B. F., MILLER, S. E., PALKER, T. J., MOORE, J. O., DUNN, P. H., BOLOGNESI, D. P. and METZGAR, R. S. (1983). Identification of human T cell leukemia virus in a Japanese patient with adult T cell leukemia and cutaneous lymphomatous vasculitis. *Proceedings of the National Academy of Sciences, USA*, **80**, 2054–8.

HAYWARD, W. S., NEEL, B. G. and ASTRIN, S. M. (1981). Induction of lymphoid leukosis by avian leukosis virus: Activation of a cellular *'onc'* gene by promotor insertion. *Nature*, **290**, 475–80.

HINUMA, Y., NAGATA, K., MISOKA, M., NAKAI, M., MATSUMOTO, T., KINOSHITA, K. I., SHIRAKAWA, S. and MIYOSHI, I. (1981). Adult T-cell leukemia: Antigen in an ATL cell line and detection of antibodies to the antigen in human sera. *Proceedings of the National Academy of Sciences, USA*, **78**, 6476–80.

HOLBROOK, N. J., SMITH, K. A., FURNACE, A. S., COMEAU, C. M., WISKOCIL, R. L. and CRABTREE, G. R. (1984). T cell growth factor: Complete nucleotide sequence and organization of the gene in normal and malignant cells. *Proceedings of the National Academy of Sciences, USA*, **81**, 1634–8.

HOMMA, T., KANKI, P. J., KING, N. W., HUNT, R. D., O'CONNELL, M. J., LETVIN, N. L., DANIEL, M. D., DESROSIERS, R. C., YANG, C. S. and ESSEX, M. (1984). Lymphoma in macaques: Association with virus of human T-lymphotropic family. *Science*, **225**, 716–18.

HUNSMANN, G., SCHNEIDER, J., SCHMITT, J. and YAMAMOTO, N. (1983). Detection of serum antibodies to adult T-cell leukemia virus in non-human primates and in people from Africa. *International Journal of Cancer*, **32**, 329–32.

JOSEPHS, S., RATNER, L., CLARKE, M. F., WESTIN, F. H., REITZ, M. S. and WONG-STAAL, F. (1984). Transforming potential of human c-*sis* nucleotide sequences encoding platelet-derived growth factor. *Science*, **225**, 636–9.

KALYANARAMAN, V. S., JARVIS-MORAR, M., SARNGHADAHARAN, M. G. and GALLO, R. C. (1983). Immunological characterization of and demonstration of human natural antibodies to the low molecular weight *gag* gene proteins p19 and p15 of human T-cell leukemia lymphoma virus (HTLV). *Virology*, **132**, 61–70.

KALYANARAMAN, V. S., SARNGADHARAN, M. G., NAKAO, Y., ITO, Y., AOKI, T. and GALLO, R. C. (1982*a*). Natural antibodies to the structural core protein (p24) of the human T-cell leukemia (lymphoma) retrovirus found in sera of leukemia patients in Japan. *Proceedings of the National Academy of Sciences, USA*, **79**, 1653–7.

KALYANARAMAN, V. S., SARNGADHARAN, M. G., POIESZ, B. J., RUSCETTI, F. W. and GALLO, R. C. (1981). Immunological properties of type C retrovirus isolated from cultured human T-lymphoma cells and comparison to other mammalian retroviruses. *Journal of Virology*, **38**, 906–15.

KALYANARAMAN, V. S., SARNGADHARAN, M. G., ROBERT-GUROFF, M., MIYOSHI, I., BLAYNEY, D., GOLDE, D. and GALLO, R. C. (1982*b*). A new subtype of human T-cell leukemia virus (HTLV-II) associated with a T-cell variant of hairy cell leukemia. *Science*, **218**, 571–3.

KELLEY, K., COCHRAN, B. H., STILES, C. D. and LEDER, P. (1983). Cell-specific regulation of the c-*myc* gene by lymphocyte mitogens and platelet-derived growth factor. *Cell*, **35**, 603–10.

KORSMEYER, S. J., GREENE, W. C., COSSMAN, J., HSU, S.-M., NECKERS, L. M., MARSHALL, S. L., JENSEN, J. P., BAKHSHI, A., LEONARD, W. J., DEPPER, J. M., JAFFE, E. J. and WALDMANN, T. A. (1983). Rearrangement and expression of immunoglobulin genes and expression of Tac antigen in hairy cell leukemia.

Proceedings of the National Academy of Sciences, USA, **80**, 4522–6.

LEE, T. H., COLIGAN, J. E., HOMMA, T., McLANE, M. F., TACHIBANA, N. and ESSEX, M. (1984*a*). Human T-cell leukemia virus associated membrane antigens; identity of the membrane antigens recognized after virus infection. *Proceedings of the National Academy of Sciences, USA*, **81**, 3856–60.

LEE, T. H., COLIGAN, J. E., SODROSKI, J. A., HASELTINE, W. A., WONG-STAAL, F., GALLO, R. C. and ESSEX, M. (1984*b*). Antigens encoded by the 3'-terminal region of human T-cell leukemia virus: Evidence for a functional gene flanked by the *env* gene and 3' LTR. *Science*, **226**, 57–61.

LEONARD, W. J., DEPPER, J. M., ROBB, R. J., KRONKE, M., SVENTLIK, P. B., PFEFFER, N. J., WALDMANN, T. A. and GREENE, W. C. (1984). Identification of the primary translation product for the human T-cell growth factor receptor in normal and neoplastic cells. *Clinical Research*, **32**, 315A.

LEONARD, W. J., DEPPER, J. M., UCHIYAMA, T., SMITH, K. A., WALDMANN, T. A. and GREENE, W. C. (1982). A monoclonal antibody that appears to recognize the receptor for human T-cell growth factor: Potential characterization of the receptors. *Nature*, **300**, 267–9.

LEONARD, W. J., DEPPER, J. M., WALDMANN, T. A. and GREENE, W. C. (1983). A monoclonal antibody to the human receptor for T cell growth factor. In *Receptors*, ed. M. Greaves, pp. 45–66. London, Chapman & Hall.

MANN, D. L., POPOVIC, M., MURRAY, C., NEULAND, C., STRONG, D. M., SARIN, P., GALLO, R. C. and BLATTNER, W. A. (1983*a*). Cell surface antigen expression in newborn cord blood lymphocytes infected with HTLV. *Journal of Immunology*, **131**, 2021–4.

MANN, D. L., POPOVIC, M., SARIN, P. S., MURRAY, C., REITZ, M. S., STRONG, D. M., HAYNES, B. F., GALLO, R. C. and BLATTNER, W. A. (1983*b*). Cell lines producing human T-cell lymphoma virus show altered HLA expression. *Nature*, **305**, 58–60.

MANZARI, V., AGLIANO, A. M., GALLO, R. C. and WONG-STAAL, F. (1983*c*). A rapid and sensitive assay for proviral sequences of a human retrovirus (HTLV) in leukemic cells. *Leukemia Research*, **7**, 681–6.

MANZARI, V., GALLO, R. C., FRANCHINI, G., WESTIN, E., CECCHERINI-NELLI, L., POPOVIC, M. and WONG-STAAL, F. (1983*a*). Abundant transcription of a cellular gene in T cells infected with human T-cell leukemia–lymphoma virus. *Proceedings of the National Academy of Sciences, USA*, **80**, 11–15.

MANZARI, V., WONG-STAAL, F., FRANCHINI, G., COLOMBINI, S., GELMAN, E. P., OROSZLAN, S., STAAL, S. and GALLO, R. C. (1983*b*). Human T-cell leukemia–lymphoma virus (HTLV): cloning of an integrated defective provirus and flanking cellular sequences. *Proceedings of the National Academy of Sciences, USA*, **80**, 1574–8.

MARKHAM, P. D., SALAHUDDIN, S. Z., KALYANARAMAN, V. S., POPOVIC, M., SARIN, P. S. and GALLO, R. C. (1983). Infection and transformation of fresh human umbilical cord blood cells by multiple sources of human T-cell leukemia–lymphoma virus (HTLV). *International Journal of Cancer*, **31**, 413–20.

MARKHAM, P. D., SALAHUDDIN, S. Z., MACCHI, B., ROBERT-GUROFF, M. and GALLO, R. C. (1984). Transformation of different phenotypic types of human bone marrow T lymphocytes. *International Journal of Cancer*, **33**, 13–17.

MIER, J. W. and GALLO, R. C. (1980). Purification and some characteristics of human T-cell growth factor from phytohemagglutinin stimulated lymphocyte conditioned media. *Proceedings of the National Academy of Sciences, USA*, **77**, 6134–8.

MITSUYA, H., MATIS, L. A., MEGSON, M., BUNN, P. A., MURRAY, C., MANN, D. L., GALLO, R. C. and BRODER, S. (1983). Generation of an HLA-restricted cytotoxic T-cell line reactive against cultured tumor cells from a patient infected with human T-cell leukemia/lymphoma virus. *Journal of Experimental Medicine*, **158**, 994–9.

MITSUYA, H., MEGSON, M., MANN, D. L., MATIS, L. A., COHEN, O. J., GALLO, R. C. and BRODER, S. (1984). Immune T cells reactive against human T-cell leukemia/lymphoma virus. *Lancet*, **i**, 649–52.

MIYAMOTO, K., SATO, J., KITAJIMA, K.-I., TOGAWA, A., SUEMARU, S., SANADA, H. and TANAKA, T. (1983). Adult T cell leukemia. Chromosome analysis of 15 cases. *Cancer*, **52**, 471–8.

MIYAWAKI, T. A., YACHIE, A., UWANDANA, N., OHZEKI, S., NAGAOKI, T. and TANIGUCHI, N. (1982). Functional significance of Tac antigen expressed on activated human T lymphocytes: Tac antigen interacts with T cell growth factor in cellular proliferation. *Journal of Immunology*, **129**, 2474–8.

MIYOSHI, I., KUBONISHI, I., YOSHIMOTO, S., AKAGI, T., OHTSUKI, Y., SHIRAISHI, Y., NAGATO, K. and HINUMA, Y. (1981*a*). Type C virus particles in a cord T-cell line derived by co-cultivating normal human cord leukocytes and human leukemic T-cells. *Nature*, **294**, 770–1.

MIYOSHI, I., YOSHIMOTO, S., FUJISHITA, M., TAGUCHI, H., KUBONISHI, I., NIIYA, K. and MINEZAWA, M. (1982). Natural adult T-cell leukaemia virus infection in Japanese monkeys. *Lancet*, **ii**, 658–9.

MIYOSHI, I., YOSHIMOTO, S., KUBONISHI, I., TAGUCHI, H., SHIRAISHI, Y., OHTSUKI, Y. and AKAGI, T. (1981*b*). Transformation of normal human cord lymphocytes by co-cultivation with a lethally irradiated human T-cell line carrying type C virus particles. *Gann*, **72**, 997–8.

MORGAN, D. A., RUSCETTI, F. W. and GALLO, R. C. (1976). Selective *in vitro* growth of T-lymphocytes from normal human bone marrow. *Science*, **193**, 1007–8.

NEEL, B. G., HAYWARD, W. S., ROBINSON, H. L., FANG, J. and ASTRIN, S. M. (1981). Avian leukosis virus-induced tumors have common proviral integration sites and synthesize discrete new RNAs: Oncogenesis by promotor insertion. *Cell*, **23**, 323–34.

NEIL, J. C., HUGHES, D., MCFARLANE, R., WILKIE, N. M., ONIONS, D. E., LEES, G. and JARRETT, O. (1984). Transduction and rearrangement of the *myc* gene by feline leukaemia virus in naturally occurring T-cell leukaemia. *Nature*, **308**, 814–20.

NOWELL, P. C., FINAN, J. B., CLARK, J., SARIN, P. S. and GALLO, R. C. (1984). Karyotypic differences between primary cultures and cell lines from HTLV-positive leukemia/lymphoma. *Journal of the National Cancer Institute*, **73**, 849–52.

NUSSE, R. and VARMUS, H. E. (1982). Many tumors induced by the mouse mammary tumor virus contain a provirus integrated in the same region of the host genome. *Cell*, **31**, 93–109.

OROSZLAN, S., SARNGADHARAN, M. G., COPELAND, T. D., KALYANARAMAN, V. S., GILDEN, R. V. and GALLO, R. C. (1982). Primary structure analysis of the major internal protein p24 of human type C T-cell leukemia virus. *Proceedings of the National Academy of Sciences, USA*, **79**, 1291–4.

PAYNE, G. S., BISHOP, J. M. and VARMUS, H. E. (1982). Multiple arrangements of viral DNA and an activated host oncogene c-*myc* in bursal lymphomas. *Nature*, **295**, 209–14.

POIESZ, B. J., RUSCETTI, F. W., GAZDAR, A. F., BUNN, P. A., MINNA, J. D. and GALLO, R. C. (1980*a*). Detection and isolation of type C retrovirus particles from fresh and cultured lymphocytes of a patient with cutaneous T-cell lymphoma. *Proceedings of the National Academy of Sciences, USA*, **77**, 7415–19.

POIESZ, B. J., RUSCETTI, F. W., MIER, J. W., WOODS, A. M. and GALLO, R. C. (1980*b*). T-cell lines established from human T-lymphocytic neoplasias by direct response to T-cell growth factor. *Proceedings of the National Academy of Sciences, USA*, **77**, 6815–19.

POIESZ, B. J., RUSCETTI, F. W., REITZ, M. S., KALYANARAMAN, V. S. and GALLO, R. C. (1981). Isolation of a new type-C retrovirus (HTLV) in primary uncultured cells of a patient with Sezary T-cell leukemia. *Nature*, **294**, 268–71.

POPOVIC, M., LANGE-WANTZIN, G., SARIN, P. S., MANN, D. and GALLO, R. C. (1983*a*). Transformation of human umbilical cord blood T cells by human T-cell leukemia/lymphoma virus. *Proceedings of the National Academy of Sciences, USA*, **80**, 5402–6.

POPOVIC, M., SARIN, P. S., ROBERT-GUROFF, M., KALYANARAMAN, V. S., MANN, D., MINOWADA, J. and GALLO, R. C. (1983*b*). Isolation and transmission of human retrovirus (human T-cell leukemia virus). *Science*, **219**, 856–9.

POPOVIC, M., SARNGADAHARAN, M. G., READ, E. and GALLO, R. C. (1984). Detection and isolation and continuous production of cytopathic retroviruses (HTLV-III) from patients with AIDS and pre-AIDS. *Science*, **224**, 497–500.

PRIVAL, J., PARAN, M., GALLO, R. C. and WU, A. M. (1974). Colony-stimulating factors in cultures of human peripheral blood cells. *Journal of the National Cancer Institute*, **53**, 1583–8.

RHO, H. M., POIESZ, B., RUSCETTI, F. W. and GALLO, R. C. (1981). Characterization of the reverse transcriptase from a new retrovirus (HTLV) produced by a human cutaneous T-cell lymphoma cell line. *Virology*, **112**, 355–61.

ROBB, R. J. (1982). T-cell growth factor: Purification to homogeneity and interaction with a cellular pattern. *Federation Proceedings*, **41**, 480.

ROBB, R. J. and GREENE, W. C. (1983). Direct demonstration of the identity of T-cell growth factor binding protein and the Tac antigen. *Journal of Experimental Medicine*, **158**, 1332–7.

ROBB, R. J., KUTNY, R. M. and CHOWDRY, V. (1983). Purification and partial sequence analysis of human T-cell growth factor. *Proceedings of the National Academy of Sciences, USA*, **80**, 5990–4.

ROBB, R. J. and LIN, Y. (1984). T-cell growth factor: Purification, interaction with the cellular receptor, and *in vitro* synthesis. In *Thymic Hormones and Lymphokines '83*, ed. A. L. Goldstein, pp. 247–56. New York, Plenum Publishing Corp.

ROBB, R. J., MUNCK, A. and SMITH, K. A. (1981). T-cell growth factor receptors, quantification, specificity, and biological relevance. *Journal of Experimental Medicine*, **154**, 1455–74.

ROBB, R. J. and SMITH, K. A. (1981). Heterogeneity of human T cell growth factor(s) due to variable glycosylation. *Molecular Immunology*, **18**, 1087–94.

ROBERT-GUROFF, M., KALYANARAMAN, V. S., BLATTNER, W. A., POPOVIC, M., SARNGADHARAN, M. G., MAEDA, M., BLAYNEY, D., CATOVSKY, D., BUNN, P. A., SHIBATA, A., NAKAO, Y., ITO, Y., AOKI, T. and GALLO, R. C. (1983). Evidence for human T cell lymphoma-leukemia virus positive T cell leukemia–lymphoma patients. *Journal of Experimental Medicine*, **157**, 248–58.

ROBERT-GUROFF, M., NAKAO, Y., NOTAKE, K., ITO, Y., SLISKI, A. and GALLO, R. C. (1982). Natural antibodies to human retrovirus HTLV in a cluster of Japanese patients with adult T cell leukemia. *Science*, **215**, 975–8.

ROBERT-GUROFF, M., RUSCETTI, F. W., POSNER, L. E., POIESZ, B. J. and GALLO, R. C. (1981). Detection of the human T-cell lymphoma virus p19 in cells of some patients with cutaneous T-cell lymphoma and leukemia using a monoclonal antibody. *Journal of Experimental Medicine*, **154**, 1957–64.

ROBERT-GUROFF, M., SCHUPBACH, J., BLAYNEY, D. W., KALYANARAMAN, V. S., MERINO, F., SARNGADHARAN, M. G., CLARK, J. W., SAXINGER, W. C., BLATTNER, W. A. and GALLO, R. C. (1984). Seroepidemiologic studies on human T-cell leukemia/lymphoma virus, type I. In *Human T-Cell Leukemia/Lymphoma Virus*, ed. R. C. Gallo, M. Essex and L. Gross, pp. 297–306. New York, Cold Spring Harbor Laboratories.

ROSENBERG, S. A., GRIMM, E. A., McGROGANE, M., DOYLE, M., KAWASAKI C., KOTHO, K. and MARK, D. F. (1984). Biological activity of recombinant human interleukin-2 produced in *Escherichia coli*. *Science*, **223**, 1412–15.

ROWLEY, J. D., HAREN, J. M., WONG-STAAL, F., FRANCHINI, G., GALLO, R. C. and BLATTNER, W. A. (1984). Chromosome pattern in cells from patients positive for human T-cell leukemia/lymphoma virus. In *Human T-Cell Leukemia/Lymphoma Virus*, ed. R. C. Gallo, M. Essex and L. Gross, pp. 85–9. New York, Cold Spring Harbor Laboratories.

SALAHUDDIN, S. Z., MARKHAM, P. D., LINDNER, S. G., GOOTENBERG, J., POPOVIC, M., HEMMI, H., SARIN, P. S. and GALLO, R. C. (1984). Lymphokine production by cultured human T cells transformed by human T-cell leukemia/lymphoma virus. *Science*, **223**, 703–7.

SARIN, P. S., AOKI, T., SHIBATA, A., OHNISHI, Y., AOGAGI, T., MIYAKOSHI, H., EMURA, I., KALYANARAMAN, V. S., ROBERT-GUROFF, M., POPOVIC, M., SARNGADHARAN, M. G., NOWELL, P. C. and GALLO, R. C. (1983). High incidence of human type-C retrovirus (HTLV) in family members of a HTLV-positive Japanese T-cell leukemia patient. *Proceedings of the National Academy of Sciences, USA*, **80**, 2370–4.

SARNGADHARAN, M. G., POPOVIC, M., BRUCH, L., SCHUPBACH, J. and GALLO, R. C. (1984). Antibodies reactive with human T-lymphotropic retroviruses (HTLV-III) in the serum of patients with AIDS. *Science*, **224**, 506–8.

SAXINGER, W. C., LANGE-WANTZIN, G., THOMSEN, K., LAPIN, B., YAKOVLEVA, L., LI, Y.-W., GUO, H.-G., ROBERT-GUROFF, M., BLATTNER, W. A., ITO, Y. and GALLO, R. C. (1984a). Human T-cell leukemia virus: A diverse family of related exogenous retroviruses of human and Old World primates. In Human T-cell Leukemia/Lymphoma Virus, ed. R. C. Gallo, M. Essex and L. Gross, pp. 323–30. New York, Cold Spring Harbor Laboratories.

SAXINGER, W., BLATTNER, W. A., LEVINE, H., CLARK, J., BIGGAR, R., HOH, M., MAGHISSI, J., JACOBS, P., WILSON, L., JACOBSON, P., CROOKES, R., STRONG, M., ANGARI, A. A., DEAN, A. G., NKROMAH, F. K., MOURALI, N. and GALLO, R. C. (1984b). Human T-cell leukemia virus (HTLV) in Africa. *Science*, **225**, 1473–6.

SCHUPBACH, J., POPOVIC, M., GILDEN, R. V., GONDA, M. A., SARNGADHARAN, M. G. and GALLO, R. C. (1984). Serological analysis of a subgroup of human T-lymphotropic retroviruses (HTLV-III) associated with AIDS. *Science*, **224**, 503–5.

SCHUPBACH, J., SARNGADHARAN, M. G. and GALLO, R. C. (1984). Antigens on HTLV-infected cells recognized by leukemia and AIDS sera related to HTLV viral glycoprotein. *Science*, **224**, 607–10.

SEIKI, M., EDDY, R., SHOWS, T. R. and YOSHIDA, M. (1984). Nonspecific integration of the HTLV provirus genome into adult T-cell leukemia cells. *Nature*, **309**, 640–2.

SEIKI, M., HATTORI, S., HIRAYAMA, Y. and YOSHIDA, M. (1983). Human adult T-cell leukemia virus: Complete nucleotide sequence of the provirus genome integrated in leukemia cell DNA. *Proceedings of the National Academy of Sciences, USA*, **80**, 3618–22.

SHAW, G. M., GONDA, M. A., FLICKINGER, G. H., HAHN, B. H., GALLO, R. C. and WONG-STAAL, F. (1984). Genomes of evolutionary divergent members of human T-cell leukemia virus family (HTLV-I and HTLV-II) are highly conserved, especially in pX. *Proceedings of the National Academy of Sciences, USA*, **81**, 4544–8.

SHIMOTOHNO, K., GOLDE, D. W., MIWA, M., SAGIMURA, T. and CHEN, I. S. Y. (1984). Nucleotide sequence analysis of the long term repeat of human T-cell leukemia virus type II. *Proceedings of the National Academy of Sciences, USA*, **81**, 1079–83.

SLAMON, D. J., deKERNION, J. B., VERMA, I. M. and CLINE, M. J. (1984). Expression of cellular oncogenes in human malignancies. *Science*, **224**, 256–62.

SMITH, K. A., GILLIS, S., BAKER, D. E., McKENZIE, D. and RUSCETTI, F. W. (1979). T-cell growth factor-mediated T-cell proliferation. *Annals of the New York Academy of Science*, **332**, 423–32.

SODROSKI, J. G., ROSEN, C. A. and HASELTINE, W. A. (1984). *Trans*-acting transcriptional activation of the long-terminal repeat of human T lymphotropic virus in infected cells. *Science*, **225**, 381–5.

SODROSKI, J., TRUS, M., PERKINS, D., PATARCA, R., WONG-STAAL, F., GELMANN, E., GALLO, R. and HASELTINE, W. A. (1984). Repetitive structure in the long terminal repeat element of a type II human T cell leukemia virus (HTLVII$_{Mo}$). *Proceedings of the National Academy of Sciences, USA*, **81**, 4617–21.

STILES, C. D. (1983). The molecular biology of platelet-derived growth factor. *Cell*, **33**, 653–5.

TAJIMA, K., TOMINAGA, S., KUROISHI, T., SHIMIZU, H. and SUCHI, T. (1979). Geographical features and epidemiological approach to endemic T-cell leukemia/lymphoma in Japan. *Japanese Journal of Clinical Oncology*, **9** (Suppl.), 495–504.

TAKATSUKI, K., UCHIYAMA, J., SAGAWA, K. and YODOI, J. (1977). Adult T-cell leukemia in Japan. In *Topics in Hematology*, ed. S. Seno, F. Takaku and S. Irino, pp. 73–77. Amsterdam–Oxford, Excerpta Medica.

TAKATSUKI, K., UCHIYAMA, T., UESHIMA, Y. and HATTORI, T. (1979). Adult T-cell leukemia: Further clinical observations and cytogenetic and functional studies of leukemic cells. *Japanese Journal of Clinical Oncology*, **9**, 317–24.

TAKATSUKI, K., UCHIYAMA, T., UESHIMA, Y., HATTORI, T., TOIBANA, T., TSUDO, M., WANO, Y. and YODOI, J. (1982). Adult T-cell leukemia: Proposal as a new disease and cytogenetic, phenotypic, and functional studies of leukemic cells. *Gann, Monograph on Cancer Research*, **28**, 13–22.

TANAGUCHI, T., MATSUI, H., FUJITA, T., TAKAOTA, C., KASHIMA, N., YOSHIMOTO, R. and HAMURO, J. (1983). Structure and expression of a cloned complementary DNA for human interleukin 2. *Nature*, **302**, 305–10.

UCHIYAMA, T., BRODER, S. and WALDMANN, T. A. (1981). A monoclonal antibody (anti-Tac) reactive with activated and functionally mature human T-cells. 1. Production of anti-Tac monoclonal antibody and distribution of Tac (+) cells. *Journal of Immunology*, **126**, 1393–7.

UESHIMA, Y., FUKUHARA, S., HATTORI, T., UCHIYAMA, T., TAKATSUKI, K. and UCHINO, H. (1981). Chromosome studies in adult T-cell leukemia in Japan: Significance of trisomy 7. *Blood*, **58**, 420–5.

VYTH-DREES, F. A. and DE VRIES, J. E. (1983). Human T-cell leukemia virus in lymphocytes from a T-cell leukaemia patient originating from Surinam. *Lancet*, **ii**, 993.

WALDMANN, T., BRODER, S., GREENE, W., SARIN, P. S., SAXINGER, C., BLAYNEY, D. W., BLATTNER, W. A., GOLDMAN, C., FROST, K., SHARROW, S., DEPPER, J., LEONARD, W., UCHIYAMA, T. and GALLO, R. C. (1983). A comparison of the function and phenotype of Sezary T-cells with human T-cell leukemia/lymphoma virus (HTLV)-associated adult T-cell leukemia. *Clinical Research*, **31**, 5474–80.

WATERFIELD, M. D., SCRACE, G. T., WHITTLE, N., STROOBANT, P., JOHNSON, A., WASTESON, A., WESTERMARK, B., HELDIN, C.-H., HUANG, J. S. and DEUEL, T. F. (1983). Platelet derived growth factor is structurally related to the putative transforming protein p24 of simian sarcoma virus. *Nature*, **304**, 35–9.

WEISS, R., TEICH, N., VARMUS, H. and COFFIN, J. (ed.) (1982). *RNA Tumor Viruses*. New York, Cold Spring Harbor Laboratories.

WELTE, K., WANG, C. Y., MERTELSMANN, R., VENUTA, S., FELDMAN, S. P. and MOORE, M. A. S. (1982). Purification of human interleukin 2 to apparent homogeneity and its molecular heterogeneity. *Journal of Experimental Medicine*, **156**, 454–64.

WESTIN, E. H., WONG-STAAL, F., GELMANN, E. P., DALLA FAVERA, R., PAPAS, T. S., LAUTENBERGER, J. A., EVA, A., REDDY, E. P., TRONICK, S. R., AARONSON, S. A. and GALLO, R. C. (1982). Expression of cellular homologues of retroviral *onc* genes in human hematopoietic cells. *Proceedings of the National Academy of Sciences, USA*, **79**, 2490–4.

WHANG-PENG, J., BUNN, P. A., KNUTSEN, T., KAO-SHAN, C. S., BRODER, S., JAFFE, E., GELMANN, E., BLATTNER, W. and GALLO, R. C. (1983). Cytogenetic studies in human T-cell lymphoma virus (HTLV) positive leukemia/lymphoma in the USA. *Blood*, **62** (Suppl. 1), 198A.

WHANG-PENG, J., BUNN, P. A., KNUTSEN, T., KAO-SHAN, C. S., BRODER, S., JAFFE, E., GELMANN, E., BLATTNER, W., LOFTERS, W., YOUNG, R. C. and GALLO, R. C. (1984). Cytogenic studies in human T-cell lymphoma virus (HTLV) positive leukemia/lymphoma in the USA. *Journal of the National Cancer Institute*, in press.

WONG-STAAL, F., HAHN, B., MANZARI, V., COLOMBINI, S., FRANCHINI, G., GELMANN, E. P. and GALLO, R. C. (1983). A survey of human leukaemias for sequences of a human retrovirus, HTLV. *Nature*, **302**, 626–8.

YAMADA, Y. (1983). Phenotypic and functional analysis of leukemic cells from 16 patients with adult T-cell leukemia/lymphoma. *Blood*, **61**, 192–9.

YAMAMOTO, N., MATSUMOTO, T., KOYANAGI, T., TANAKA, Y. and HINUMA, Y. (1982*a*). Unique cell lines harboring both Epstein–Barr virus and adult T-cell leukemia virus established from leukemia patients. *Nature*, **249**, 367–9.

YAMAMOTO, N., OKADA, M., KOYANAGI, Y., KANAGI, M. and HINUMA, Y. (1982*b*). Transformation of human leukocytes by cocultivation with an adult T-cell leukemia virus. *Science*, **217**, 737–9.

YOSHIDA, M., MIYOSHI, I. and HINUMA, Y. (1982). Isolation and characterization of retrovirus from cell lines of human adult T-cell leukemia and its implication in the disease. *Proceedings of the National Academy of Sciences, USA*, **79**, 2031–5.

YOSHIDA, M., SEIKI, M., YAMAGUCHI, K. and TAKATSUKI, K. (1984). Monoclonal integration of human T-cell leukemia provirus in all primary tumors of adult T-cell leukemia suggests causative role of human T-cell leukemia virus in the disease. *Proceedings of the National Academy of Sciences, USA*, **81**, 2534–7.

TRANSFORMING *ras* GENES

EUGENIO SANTOS, SARASWATI SUKUMAR, DIONISIO MARTIN-ZANCA, HELMUT ZARBL and MARIANO BARBACID

Developmental Oncology Section,
NCI-Frederick Cancer Research Facility,
Frederick, Maryland 21701, USA

The last few years have been witness to an explosion of new know-ledge on different aspects of malignant transformation. Most of this information has emerged from studies of oncogenes. More than 20 of these transforming genes have been identified, initially in acutely transforming retroviruses (v-*onc* genes) and more recently in tumor cells (Cooper, 1982; Weinberg, 1982; Bishop, 1983). Oncogenes have originated, through different modifications, from normal cellu-lar genes (proto-oncogenes) whose function is assumed to be involved in the regulation of cell proliferation and/or differentiation. Thus, viral oncogenes arose by recombination between the genomes of non-transforming retroviruses and transduced proto-oncogenes (Frankel and Fischinger, 1976; Stehelin *et al.*, 1976). Cellular oncogenes have been shown to become activated by a variety of mechanisms including enhancer activation (Blair *et al.*, 1981; Hayward, Neel and Astrin, 1981), point mutation (Tabin *et al.*, 1982; Reddy *et al.*, 1982; Taparowski *et al.*, 1982), chromosomal translo-cation (Klein, 1983; Rowley, 1983) and gene amplification (Collins and Groudine, 1982; Dalla Favera, Wong-Staal and Gallo, 1982; Schwab *et al.*, 1983).

Initially, cellular proto-oncogenes were identified by their homo-logy to retroviral oncogenes (Frankel and Fischinger, 1976; Stehelin *et al.*, 1976). However, demonstration that these genetic elements could acquire malignant properties without the contribution of exogenous retroviral sequences was first obtained by means of gene transfer experiments (C. Shih *et al.*, 1979). These assays revealed the presence of transforming genes in a significant fraction of human cancers as well as in chemically induced animal tumors (Cooper, 1982). Most of these transforming genes have been shown to be members of a gene family designated *ras*, which was first identified as the transforming principle of several strains of murine sarcoma viruses (Ellis *et al.*, 1981). During the last few years, the genetic structure and mechanism of activation of viral and cellular *ras* oncogenes have been unveiled. This review summarizes experimental data from our laboratory and others on the structure, malignant activation, and role of transforming *ras* genes in human and animal tumors.

DETECTION OF TRANSFORMING *ras* GENES BY GENE TRANSFER EXPERIMENTS

The presence of transforming genes in human and animal tumors has been widely documented by using NIH3T3 cells as recipient cells in gene transfer (transfection) experiments (Cooper, 1982). Results obtained in our laboratory by the transfection of DNA extracted from human tumor cell lines and human tumor biopsies are summarized in Table 1. Eight of 60 human tumor cell lines and 7 of 44 human solid tumors scored as positive in the NIH3T3 transfection assay (Pulciani *et al.*, 1982*a,b*). In all, about 15% of all human samples tested showed the presence of transforming genes (Table 1).

The availability of retroviral DNA probes has made it possible to establish the relationship of these cellular transforming genes to previously known viral oncogenes (Der, Krontiris and Cooper, 1982; Parada *et al.*, 1982; Santos *et al.*, 1982). As shown in Table 1, most of the human transforming genes are members of the *ras* gene family. The H-*ras* and K-*ras* genes are homologous to the oncogenes of the Harvey and Kirsten strains of murine sarcoma virus, respectively (Ellis *et al.*, 1981), whereas the N-*ras* gene is related distantly to both of them and does not possess a retroviral homologue (Shimizu *et al.*, 1983*c*). We found a transforming H-*ras* gene only in the T24 bladder carcinoma cell line; however, N-*ras* and K-*ras* genes were detected in 25% and 60% of positive samples, respectively. Interestingly, K-*ras* oncogenes were found in each of the four lung carcinomas that scored as positive in transfection assays (Pulciani *et al.*, 1982*b*; Santos *et al.*, 1984). Similar observations have been reported by other laboratories (Perucho *et al.*, 1981; Der *et al.*, 1982; McCoy *et al.*, 1983), suggesting a preferential association of this oncogene with human lung carcinoma.

The identification of N-*ras* by gene transfer experiments illustrates that not all transforming genes have yet found their way into retroviruses (Shimizu *et al.*, 1983*c*). In fact, other transforming genes detected by transfection assays do not bear any relationship to known retroviral oncogenes. Such is the case with two different oncogenes

Table 1. *Summary of human oncogenes detected in our laboratory*

Type of tumour	Source of DNA	Oncogene
Carcinomas		
Bladder	T24 cells	H-*ras*
	A1698 cells	K-*ras*
Colon	Solid tumor no. 1665	N-*ras*
	Solid tumor no. 2033	*onc* D[a]
	A2233 cells	K-*ras*
Gall bladder	A1604 cells	K-*ras*
Liver	Solid tumor no. 2193	N-*ras*
	Hep G2 cells	N-*ras*
Lung	A2182 cells	K-*ras*
	A427 cells	K-*ras*
	Solid tumor no. 1615	K-*ras*
	Solid tumor no. LC-10	K-*ras*
Pancreas	Solid tumor no. 1189	K-*ras*
Sarcomas		
Fibrosarcoma	HT-1080 cells	N-*ras*
Rhabdomyosarcoma	Solid tumor no. 1085	K-*ras*
Chemically transformed cells	MNNG-HOS cells	*onc* E[a]

Data presented in this table summarize results from Pulciani *et al.* (1982*a,b*), Santos *et al.* (1984) and unpublished observations from our laboratory; [a]Preliminary characterization of *onc* D and *onc* E indicates that they are not cellular homologs of known retroviral oncogenes.

detected in our laboratory in a colon carcinoma biopsy (solid tumor no. 2033) and in MNNG-HOS cells (Rhim *et al.*, 1975), a cell line derived by 'chemical transformation' of human osteosarcoma cells (Pulciani *et al.*, 1982*b*; Notario *et al.*, 1984). The latter oncogene has also been identified by Blair *et al.* (1982) using a transfection/nude mice combined assay. So far we have been unable to relate either transforming gene to several known oncogenes. Moreover, they appear to be different from other non-*ras* genes previously identified by Cooper and coworkers in human hematopoietic tumors (B-*lym* and T-*lym*) and in mammary carcinoma MCF-7 cells (Lane, Sainten and Cooper, 1981, 1982).

MOLECULAR STRUCTURE OF HUMAN *ras* ONCOGENES

As a result of the combined efforts of a number of laboratories, the molecular structure of the three human *ras* genes is now known in

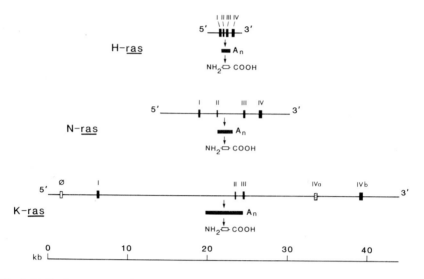

Fig. 1. Molecular structure of the human H-*ras*, N-*ras* and K-*ras* genes. Exons are represented by the vertical boxes. The K-*ras* gene contains two alternative fourth exons (labeled IVa and IVb) as well as an untranslated exon (labeled ϕ) which is located upstream of the initiation codon. Their major mRNA transcripts (sizes of 1.2 kb, 2.2 kb and 5.5 kb for H-*ras*, N-*ras* and K-*ras*, respectively) and their p21 gene products are also represented.

detail (Capon *et al.*, 1983*a,b*; McGrath *et al.*, 1983; Reddy, 1983; Shimizu *et al.*, 1983*a*; Taparowsky *et al.*, 1983). As shown schematically in Fig. 1, these range in genetic complexity from the H-*ras* gene, which is 4.6 kilobases (kb), to the K-*ras* gene, which is 40–45 kb.

The human genome contains at least two loci homologous to H-*ras* (c-Ha-*ras*-1, c-Ha-*ras*-2) and K-*ras* (c-Ki-*ras*-1, c-Ki-*ras*-2) genes (Chang *et al.*, 1982*b*). However, only one of them appears to be functional, the other corresponding to a non-functional pseudogene (McGrath *et al.*, 1983). The use of retroviral and cDNA *ras* probes has allowed the accurate identification of the intron and exon regions of these functional genes. These structural studies have shown that the difference in genetic complexity of human *ras* genes is the result of their different intronic structure. However, H-*ras*, K-*ras* and N-*ras* possess similar codon sequences (Fig. 1). In fact, the splice junctions of all *ras* genes correspond precisely, suggesting that they are derived from a common ancestral gene that contained a minimum of four exons. K-*ras* shows an additional, diverged fourth exon (IVb) that confers the potential to encode for two alternative products (Capon *et al.*, 1983*b*; Shimizu *et al.*, 1983*a*).

As a result of their similar genetic structure, the three known *ras* genes code for highly related proteins of 189 amino acid residues, generally designated p21 (T.Y. Shih *et al.*, 1979). There are three defined domains of structural significance in the gene products of the known *ras* genes (Shimizu *et al.*, 1983a; Taparowsky *et al.*, 1983). A first domain of virtual identity encompasses the first 80 amino acid residues of p21 proteins. The next 80 amino acid residues define a second domain where the structures of the different p21 proteins diverge slightly from each other (85% homology between any pair of H-*ras*, N-*ras* and K-*ras*). Finally, there is a carboxy-terminal variable region of pronounced divergency that varies radically among *ras* genes. This structural organization suggests that the amino-terminal domain has a catalytic or regulatory role that has been conserved in evolution. It is precisely in this region that single amino acid substitutions have been shown to activate the malignant potential of *ras* proteins (see below). The carboxy-terminal variable region may contain the determinants of physiological specificity for each individual member of the *ras* gene family.

The *ras* genes have been shown to be highly conserved during evolution. Less than 5% variation in amino acids has been found between p21 proteins of human and rat origin. More striking is the high degree of relatedness that the recently identified *ras* genes of yeast show with their mammalian counterparts (DeFeo-Jones *et al.*, 1983; Dhar *et al.*, 1984; Papageorge *et al.*, 1984; Powers *et al.*, 1984). Although yeast *ras* genes code for larger products (42 kD), these also conform to the three-domain structure defined in the mammalian p21 proteins. In fact, a homology of more than 90% of significant amino acids is present in the amino-terminal domain between the yeast and mammalian *ras* gene products (Dhar *et al.*, 1984; Powers *et al.*, 1984). These observations strongly suggest a basic role for *ras* genes in eukaryotic cellular proliferation.

CHROMOSOME LOCALIZATION OF HUMAN *ras* GENES

Techniques of chromosome mapping using human/rodent somatic cell hybrids and *in situ* hybridization to human mitotic chromosomes have allowed the assignment of the human *ras* genes to particular chromosomes. N-*ras* has been assigned to the short arm of chromosome 1 (deMartinville *et al.*, 1983a; Ryan *et al.*, 1983), while H-*ras* and K-*ras* have been assigned to the short arms of chromosomes 11 and 12, respectively. H-*ras* has been mapped to

11p15.1–p15.5 (McBride *et al.*, 1982; deMartinville *et al.*, 1983*b*; Jahnwar *et al.*, 1983; O'Brien *et al.*, 1983), and K-*ras* to 12p12.05–pter (O'Brien *et al.*, 1983; Ryan *et al.*, 1983; Sakaguchi *et al.*, 1983, 1984). The non-functional human *ras* homologues have also been mapped: c-Ha-*ras*-2 to chromosome X (Rowley, 1983) and c-Ki-*ras*-1 to 6p23-12 (Sakaguchi *et al.*, 1984). Jahnwar *et al.* (1983), using *in situ* hybridization, have identified K-*ras*-related sequences in the long arm of chromosome 12 and the short arm of chromosome 3. These sequences may represent additional members of the human *ras* gene family; however, this possibility remains to be tested.

A duplication mechanism has been proposed for H-*ras* and K-*ras* loci based on the observation that the short arms of human chromosomes 11 (encoding D-LDH A and H-*ras*) and 12 (encoding D-LDH B and K-*ras*) share at least two pairs of genes that probably evolved from common ancestral genes (Ohno, 1973; Doolittle, 1981). If the duplication–divergence schedule allows for fine tuning of developmental processes, it may follow that each member of the *ras* gene family performs distinct but related functions.

MECHANISM OF MALIGNANT ACTIVATION OF *ras* ONCOGENES

Comparative analysis of the nucleotide sequences of *ras* oncogenes with their normal cellular counterparts isolated from normal human cells has allowed the identification of the critical change that confers on the *ras* oncogenes their transforming properties. In all cases a single point mutation responsible for a single amino acid change in the p21 protein product is responsible for the malignant activation of *ras* oncogenes. Two 'hot spots' for activation have been detected so far: codon 12, which is located in the first exon, and codon 61, which is located in the second exon (Table 2).

The activated H-*ras* oncogene detected in T24 bladder carcinoma cells differs from its normal counterpart in a single base change $(G \rightarrow T)$ at codon 12 that results in the substitution of valine for the normal glycine residue (Reddy *et al.*, 1982; Tabin *et al.*, 1982; Taparowsky *et al.*, 1982). Substitution of aspartic acid at this position also results in malignant activation (Santos *et al.*, 1983). Indeed, computer-assisted models predict similar conformational changes of the p21 products regardless of the residue replacing glycine: a flexible

Table 2. *Mutations responsible for malignant activation of human ras genes*

Gene	Codon	Human cells	Sequence	Amino acid	References
H-*ras*	12	Normal	GGC	Glycine	Tabin *et al.* (1982), Reddy *et al.* (1982), Taparowsky *et al.*(1982)
		T24	GTC	Valine	*Ibid.*
		134–51[a]	GAC	Aspartic acid	Santos *et al.* (1983)
	61	Normal	CAG	Glutamine	Capon *et al.* (1983*a*)
		Hs242	CTG	Leucine	Yuasa *et al.* (1983)
N-*ras*	12	Normal	GGT	Glycine	Taparowsky *et al.* (1983)
		PA1	GAT	Asparatic acid	Tainsky *et al.* (1984)
	61	Normal	CAA	Glutamine	Taparowsky *et al.* (1983)
		Sk-N-SH	AAA	Lysine	*Ibid.*
K-*ras*	12	Normal	GGT	Glycine	McGrath *et al.* (1983), Santos *et al.* (1984)
		Calu 1	TGT	Cysteine	Shimizu *et al.* (1983*a*)
		SW480	GTT	Valine	Capon *et al.* (1983*b*)
		PR371	TGT	Cysteine	Nakano *et al.* (1984*b*)
		A2182	CGT	Arginine	Santos *et al.* (1984)
		A1698[b]	CGT	Arginine	*Ibid.*
		LC-10[b]	CGT	Arginine	*Ibid.*
	61	Normal	CAA	Glutamine	McGrath *et al.* (1983)
		PR310	CAT	Histidine	Nakano *et al.* (1984*a*)

[a] NIH3T3 cells transformed by a spontaneously activated human H-*ras* oncogene.
[b] Determined by *Sac*I polymorphism.

hinge region that allows the amino terminus of normal p21 to fold into the core of the molecule disappears in the transforming p21. As a consequence, mutant p21 molecules may acquire a more rigid tertiary structure (Pincus *et al.*, 1983; Santos *et al.*, 1983). Transforming H-*ras* genes can also result from mutations at codon 61, leading to the substitution of the normal glutamine residue by leucine (Yuasa *et al.*, 1983). Whether the structural consequences of mutations at this codon are similar to those discussed above remains to be determined.

Activated N-*ras* oncogenes detected in HT-1080 fibrosarcoma cells (Marshall, Hall and Weiss, 1982; Hall *et al.*, 1983) and SK-N-SH neuroblastoma cells (Shimizu *et al.*, 1983*b*) show the substitution of lysine at position 61 for the normal glutamine (Taparowsky *et al.*,

1983). Recently an N-*ras* oncogene mutated in codon 12 has been detected in late passage human tetratocarcinoma cells (Tainsky *et al.*, 1984). In the case of K-*ras* oncogenes, amino acid substitutions at positions 12 and 61 have also been detected (Table 2). Cysteine (Capon *et al.*, 1983*b*; Shimizu *et al.*, 1983*a*; Nakano *et al.*, 1984*b*), valine (Capon *et al.*, 1983*b*) and arginine (Santos *et al.*, 1984) have been found replacing the normal glycine at position 12, and also in one case the normal glutamine at position 61 has been found changed to histidine (Nakano *et al.*, 1984*a*).

Increased expression of non-mutant *ras* genes may also lead to malignant transformation. Placing the human H-*ras* proto-oncogene under the regulatory control of a retroviral LTR confers on this proto-oncogene transforming properties in the NIH3T3 transfection assay (Chang *et al.*, 1982*a*) and the ability to rescue from senescence ('immortalization') primary embryo cells of rodent origin (Spandidos and Wilkie, 1984). The high levels of expression mediated by the interaction of the proto-oncogene with the retroviral LTR are responsible for both of these biological activities. In related experiments, we have shown that the normal H-*ras* proto-oncogene, on its own, can also transform NIH3T3 cells, provided that multiple copies become stably integrated in the recipient transfected cells (Santos *et al.*, 1983). Linkage of LTR sequences to the mutant H-*ras* oncogene derived from T24 bladder carcinoma cells also results in increased tumorigenic properties. Whereas *ras* oncogenes require the cooperation of other oncogenes to transform primary rat embryo cells (Land, Parada and Weinberg, 1983), these chimeric *ras* oncogenes transform these primary cells in a single step (Spandidos and Wilkie, 1984).

Increased expression of *ras* genes has also been implicated in the development of naturally occurring malignancies. Amplification of the K-*ras* locus has been reported in a mouse adrenocortical tumor cell line (Schwab *et al.*, 1983) and in a human lung carcinoma biopsy (our published observations). Elevated expression of apparently non-mutated *ras* genes in premalignant as well as malignant tissues has been reported by several authors (Horan-Hand *et al.*, 1984; Kerr and Spandidos, 1984; Slamon *et al.*, 1984; Thor *et al.*, 1984). However, similar findings have been obtained in regenerating rat liver (Goyette *et al.*, 1983, 1984). Thus, assessment of the role of normal (non-mutated) *ras* proto-oncogenes in malignancies must await careful and detailed investigation of the crucial threshold of *ras* expression that separates normal from neoplastic growth.

ACTIVATION OF *ras* ONCOGENES AND TUMOR DEVELOPMENT: HUMAN STUDIES

The central aim of researchers in the field of cellular transforming genes is to establish unequivocally the role of these genes in the initiation and/or maintenance of tumor development. One factor hindering these studies is the technical difficulty in detecting mutant *ras* alleles. To overcome this, we have taken advantage of the specific nature of mutations that activate *ras* oncogenes. Some of these mutations introduce restriction endonuclease polymorphisms that can be used as molecular markers to identify transforming *ras* genes in human DNAs. Some of these polymorphisms are listed in Table 3.

In the case of the K-*ras* proto-oncogene, which so far has been the most frequently activated in human tumors, none of the nucleotides of its critical twelfth codon, GGT, is part of a sequence specifically recognized by known restriction endonucleases (McGrath *et al.*, 1983; Santos *et al.*, 1984). However, substitution of the first deoxyguanosine by a deoxycytidine creates the sequence GAGCTC which is specifically recognized by *Sac*I (Santos *et al.*, 1984). The first exon of the K-*ras* locus is conveniently located within a 14 kb *Sac*I DNA fragment that can be easily identified by Southern blot analysis. The creation of this polymorphic *Sac*I cleavage site splits the first exon sequences between two *Sac*I fragments of 5.8 kb and 8.2 kb that can be easily distinguished from the single 14 kb *Sac*I fragment characteristic of the normal allele (Fig. 2*a*).

To test whether this molecular marker could be used to identify transforming K-*ras* genes, we analyzed eight human tumor DNA samples in which we had previously demonstrated the presence of activated K-*ras* oncogenes by gene transfer experiments (Pulciani *et al.*, 1982*b*). DNAs from these samples were digested with *Sac*I and submitted to Southern blot analysis (Fig. 3). Six of these tumor DNAs showed the wild-type 14 kb *Sac*I DNA fragment; however, two of them (A1698 bladder carcinoma cell line and A2182 lung carcinoma cell line) exhibited 8.2 kb and 5.8 kb *Sac*I DNA fragments indicative of a GGT→ CGT mutation at codon 12. Molecular cloning of the K-*ras* oncogene of A2182 cells allowed us to determine the nucleotide sequence of its first exon, to demonstrate directly that the observed polymorphism was due to this specific point mutation (Santos *et al.*, 1984). The role of the G→ C transversion in the malignant activation of the A2182 K-*ras* gene was demonstrated by constructing a chimeric molecule containing a 4.1 kb *Eco*RI–*Xba*I

Table 3. *Molecular polymorphisms created by point mutations in codons 12 and 61 of human ras oncogenes*

Gene	Codon	Mutation[a]	Polymorphism
H-*ras*	12	$G^{34} \to N$	Lost *Hpa*II/*Msp*I sites
		$G^{35} \to N$	Lost *Hpa*II/*Msp*I sites
		$G^{34} \to C$	New *Sac*I site
	61	$C^{181} \to N$	Lost *Bst*NI site
		$A^{182} \to G,C$	Lost *Bst*NI site
K-*ras*	12	$G^{34} \to C$	New *Sac*I site
		$G^{35} \to C$	New *Fnu*4HI site
	61	$C^{181} \to G$	New *Hph*I, *Mbo*II sites
		$A^{182} \to C$	New *Asu*I, *Ava*II sites
		$A^{182} \to G$	New *Taq*I site
		$A^{182} \to T$	New *Xba*I site
N-*ras*	12	$G^{34} \to C$	New *Hgi*AI site
		$G^{35} \to C$	New *Pvu*II site
	61	$G^{181} \to G$	New *Mbo*II site
		$A^{182} \to C$	New *Asu*I, *Ava*II sites

[a] Not all of these mutations have yet been detected in human transforming *ras* genes.

DNA fragment that contains the mutant first exon of K-*ras* and a 4.6 kb *Xba*I–*Bam*HI DNA fragment containing the second, third and fourth exons of the normal human H-*ras* gene (Fig. 2*c*). This hybrid *ras* gene showed transforming activity in transfection assays, thus demonstrating that the G→ C transversion observed in codon 12 is responsible for the transforming properties of the A2182 K-*ras* oncogene (Santos *et al.*, 1984).

This molecular polymorphism has been used for rapid screening of human tumor DNAs for the presence of activated K-*ras* oncogenes. Of a total of 40 tumor biopsies analyzed, one sample, designated LC-10, was shown to possess the diagnostic *Sac*I polymorphism (Santos *et al.*, 1984). LC-10 was a lung carcinoma removed from a 66-year-old man at the Istituto di Tumori in Milan and sent to us by Drs M. Pierotti and G. Della Porta. The patient was a heavy smoker and did not receive chemotherapy before surgery. The tumor was classified at the Istituto di Tumori as a moderately differentiated squamous cell carcinoma. It was located in the upper right lobe, measured 7 cm × 7 cm, and infiltrated the pleura as well as the thoracic wall. No metastases were detected in this patient. Transfection

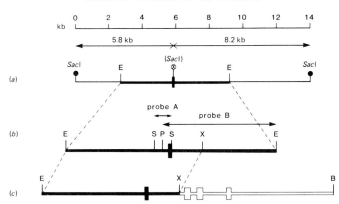

Fig. 2. (*a*) *Sac*I polymorphism created by a G→C mutation in codon 12 of the first exon (■) of the human K-*ras* gene. Wild-type (●) and polymorphic (⊗) *Sac*I cleavage sites. (*b*) DNA probes used to identify this *Sac*I polymorphism in human DNA. Probe A contains 374 base pairs (bp) of upstream sequences and 96 bp of the first exon. Probe B contains 176 bp of upstream sequences, the entire exon, and 3.1 kb of downstream intron sequences. (*c*) Chimeric *ras* oncogene containing the first exon (black box) of the A2182 K-*ras* oncogene and the second, third and fourth exons (open boxes) of the normal human H-*ras* gene. E, *Eco*RI; P, *Pst*I; S, *Sau*3AI; X, *Xba*I. Only those restriction endonuclease sites needed to define the probes are indicated.

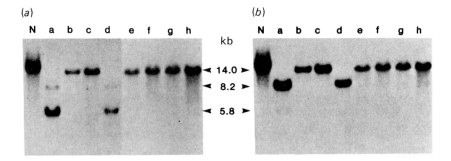

Fig. 3. Screening for *Sac*I polymorphisms in the K-*ras* locus of human tumor cells containing transforming K-*ras* genes. DNAs digested with *Sac*I were hybridized to the indicated probes (Fig. 2). (*a*) probe A; (*b*) probe B. Human DNAs listed included: N, normal human foreskin cells; a, A1698 bladder carcinoma cell line; b, A2233 colon carcinoma cell line; c, A1604 gall bladder carcinoma cell line; d, A2182 lung carcinoma cell line; e, A427 lung carcinoma cell line; f, lung carcinoma biopsy no. 1615; g, pancreatic carcinoma biopsy no. 1189; h, embryonal rhabdomyosarcoma biopsy no. 1085.

studies done in parallel showed that LC-10 DNA had transforming activity. Transformed NIH3T3 cells were cloned in agar, and their DNAs were submitted to hybridization analysis. All LC-10-derived transformants tested contained the expected K-*ras* 8.2 kb and 5.8 kb *Sac*I DNA fragments that are diagnostic of the activating G→C mutation (Fig. 4).

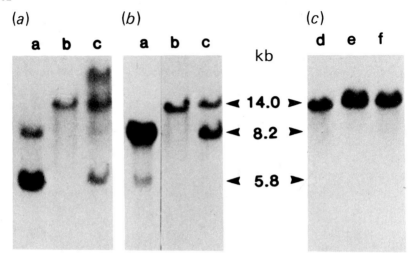

Fig. 4. Somatic point mutation responsible for the malignant activation of the K-*ras* locus in tumor but not in normal tissue of a patient with squamous cell lung carcinoma (LC-10). DNAs digested with *Sac*I were hybridized to probe A (panel *a*), probe B (panel *b*) or probes A and B (panel *c*). a, NIH3T3 cells transformed by LC-10 tumor DNA; b, normal human foreskin cells; c, squamous cell lung carcinoma LC-10; d, normal bronchia; e, normal parenchyma; and f, blood lymphocytes of the same patient.

Examination of DNAs isolated from normal tissue of the same patient, including bronchial and parenchymal cells and blood lymphocytes, showed that none of them contained the G→C mutation detected in tumor tissue (Fig. 4*c*). Moreover, DNA extracted from these normal tissues did not show transforming activity in NIH3T3 transfection assays (Santos *et al.*, 1984). These results indicate that the mutational event responsible for the malignant activation of the K-*ras* gene in the LC-10 lung carcinoma is specifically associated with tumor development. Moreover, these findings unequivocally establish the somatic nature of the activating point mutation, implying that external agents should be responsible for or causative of it.

ARE *ras* ONCOGENES DOMINANT?

T24 bladder carcinoma cells have been shown to harbor the mutant H-*ras* oncogene but not its normal allele (Feinberg *et al.*, 1983). Similarly, as shown in Fig. 3, *Sac*I analysis has shown that A1698 and A2182 human tumor DNAs contain only the transforming K-*ras* allele (Santos *et al.*, 1984). These results suggest that malignant activation of *ras* loci in human tumor cells may occur through the same mutational event in both alleles or that the normal allele is lost in these cells. In a related series of experiments, Capon *et al.* (1983*b*) have shown that the human tumor cell line SW 480 expresses only the mutant allele, while another human tumor cell line, Calu 1, expresses both the mutant and normal alleles but the latter is heavily under-represented. These results, taken together, suggest that *ras* on-cogenes may not be fully dominant in the tumor cells in which they become activated. Such a mechanism would be in agreement with reports from somatic cell hybridization studies showing that the tumorigenic phenotype behaves as a recessive trait in hybrids result-ing from the fusion of normal and transformed cells (Harris *et al.*, 1969; Stanbridge *et al.*, 1982).

The picture presented by these studies is obscured by results ob-tained with tumor biopsies. As seen in Fig. 4, LC-10 DNA possesses a significant amount of the normal K-*ras* allele. Therefore, before a definitive conclusion can be drawn regarding the dominance of *ras* oncogenes, it will be necessary to ascertain whether the normal allele present in tumor DNA is contributed by the tumor cells or by the accompanying stroma. Recent experiments have clearly demon-strated the dominance of mutated *ras* genes in yeast (Kataoka *et al.*, 1984). Thus, it is likely that the loss of the normal allele may reflect a requirement for the establishment of these tumor cells as cell lines.

ACTIVATION OF *ras* GENES AND TUMOR DEVELOPMENT: CHEMICAL CARCINOGENESIS STUDIES

A complete understanding of the precise role that *ras* oncogenes play in the development of neoplasia requires the use of appropriate model systems. The suitability of one such model, the induction of mammary carcinomas in rats by a single dose of nitrosomethylurea (NMU) (Gullino, Pettigrew and Grantham, 1975; McCormick *et al.*, 1981), has been demonstrated recently in our laboratory (Sukumar *et al.*, 1983). NMU, a direct-acting carcinogen with a very short half-life (about 20 minutes when injected intravenously), was injected into animals at the onset of sexual maturity. The organ specificity dis-played by this carcinogen suggests that its action is closely linked to a special stage in growth and/or differentiation of certain cells within the mammary gland.

More than 85% of all tumor DNAs tested but not DNAs isolated

```
                      HindIII
     +1                 ▼                      +25                                    +50
     met thr glu tyr lys leu val val val gly ala GLY gly val gly lys ser ala leu
H-ras-1  ATG ACA GAA TAC AAG CTT GTG GTG GTG GGC GCT GGA GGC GTG GGA AAG AGT GCC CTG

         met thr glu tyr lys leu val val val gly ala GLU gly val gly lys ser ala leu
NMU-H-ras ATG ACA GAA TAC AAG CTT GTG GTG GTG GGC GCT GAA GGC GTG GGA AAG AGT GCC CTG

                   PvuII
                     ▼           +75                               +100
         thr ile gln leu ile gln asn his phe val asp glu tyr asp pro thr ile glu
H-ras-1  ACC ATC CAG CTG ATC CAG AAC CAT TTT GTG GAC GAG TAT GAT CCC ACT ATA GAG

         thr ile gln leu ile gln asn his phe val asp glu tyr asp pro thr ile glu
NMU-H-ras ACC ATC CAG CTG ATC CAG AAC CAT TTT GTG GAC GAG TAT GAT CCC ACT ATA GAG
```

Fig. 5. Comparative analysis of the nucleotide sequence of the first exon of the rat H-ras-1 locus and its transforming allele, the NMU-H-ras oncogene. The deduced amino acid sequence is also indicated. The activating G→ A mutation at nucleotide +35 is underlined. As a consequence, the NMU-H-ras oncogene will direct the incorporation of glutamic acid instead of the normal glycine as the twelfth amino acid residue of its gene product (stippled area).

from normal breasts, contained transforming genes as determined by the NIH3T3 transfection assay. Characterization of the transforming principle in these tumors consistently revealed the presence of an H-ras oncogene (Sukumar et al., 1983). These results suggest the possibility that the reproducible activation of the H-ras locus may reflect a specific role of this gene in the growth and/or differentiation of the target cells in the mammary gland.

Molecular analysis of the transforming H-ras gene present in an NMU-induced mammary carcinoma revealed that its malignant activation was due to the same mutational mechanism previously identified in transforming ras genes of human tumors (Sukumar et al., 1983; Notario et al., 1984). Substitution of the second deoxyguanosine of the twelfth codon by a deoxyadenosine resulted in the incorporation of glutamic acid instead of the normal glycine as the twelfth amino acid residue of the p21 protein (Fig. 5). This mutant deoxyguanosine is part of the sequence GAGG (residues +35 to +38) which is specifically recognized by the restriction endonuclease MnlI. Mutations in this particular nucleotide or in either of the two deoxyguanosines responsible for the coding properties of codon 13 (residues +37 and +38) will eliminate this cleavage site (Fig. 6). As a consequence, a molecular polymorphism with which to identify rat H-ras oncogenes will be created by point mutations in any of these three deoxyguanosines (substitution of the deoxyadenosine residue will not have any biological significance). Preliminary observations indicate that the H-ras oncogenes present in NMU-induced mammary carcinomas always exhibit this MnlI polymorphism (unpublished observations). The tremendous specificity observed strongly

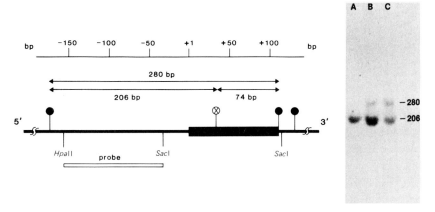

Fig. 6. Detection of H-*ras* oncogenes in NMU-induced mammary carcinomas of Buf/N rats by molecular polymorphisms. Wild-type (●) and polymorphic (⊗) *Mnl*I cleavage sites either within or in the vicinity of the first exon (■, nucleotides +1 to +111) of the Buf/N rat H-*ras*-1 locus are indicated. Southern blot analysis of *Mnl*I-cleaved normal rat DNA yields a fragment of 206 bp that can be identified with the indicated 120 bp *Hpa*II–*Sac*I ^{32}P-labeled DNA probe. DNAs containing mutations within nucleotides +35 to +38 (GAGG) exhibit a 280 bp DNA fragment because of the elimination of the polymorphic *Mnl*I cleavage site (⊗). DNAs tested include those isolated from a normal Buf/N breast (A) and representative NMU-induced Buf/N mammary carcinomas (B, C).

argues against the possibility that malignant activation of *ras* genes may be a consequence of the genetic disarray of neoplastic cells. Instead, it provides support for the concept that malignant activation of *ras* genes is an important event in tumor development.

Further support for this concept comes from similar studies in other carcinogen-induced animal tumor systems. Balmain and Pragnell (1983) have reported the reproducible activation of the H-*ras* locus in DMBA-induced mouse skin carcinomas. More importantly, this oncogene could be identified in premalignant papillomas, suggesting an early role of *ras* activation in tumorigenesis (Balmain *et al.*, 1984). Guerrero *et al.* (1984) have recently reported the specific activation of *ras* genes in mouse thymomas. Consistently, N-*ras* oncogenes were present in NMU-induced tumors, whereas K-*ras* oncogenes were detected when the thymomas were induced by X-rays. Finally, we have recently shown the activation of a *ras* oncogene in guinea-pig fetal cells mutated either *in vivo* or *in vitro* with four different chemical carcinogens (Sukumar *et al.*, 1984).

The wealth of animal tumor models showing the reproducible activation of specific *ras* genes supports the idea that these oncogenes play an important role in tumorigenesis. Furthermore, the similarity between the mechanism of activation of human and animal *ras* oncogenes suggests that at least certain chemically induced animal tumors may proceed to malignancy through the same pathways as human neoplasia. If so, these pathways can now be conveniently dissected and studied at the molecular level.

EXPRESSION OF p21 PROTEINS IN *E. coli*

Biochemically the main activity associated with p21 is guanine nucleotide binding (Scolnick, Papageorge and Shih, 1979; Shih *et al.*, 1980, Furth *et al.*, 1982; Papageorge, Lowy and Scolnick, 1982), although an associated autophosphorylating activity has been reported (Shih *et al.*, 1982). The p21 gene product appears to be localized to the inner face of the plasma membrane (Willingham *et al.*, 1980, Furth *et al.*, 1982) and undergoes post-translational acylation (Sefton *et al.*, 1982). Recently, normal as well as mutated p21 proteins have been efficiently expressed in *E. coli* (Lautenberger *et al.*, 1983; Feramisco *et al.*, 1984; Lacal *et al.*, 1984; McGrath *et al.*, 1984; Stein, Robinson and Scolnick, 1984). These bacterially synthesized proteins exhibit the same autophosphorylating and nucleotide-binding properties as their mammalian counterparts. More importantly, it has been recently reported that p21 proteins possess an intrinsic GTPase activity, which is reduced 4-fold to 10-fold in p21 proteins carrying a mutated amino-acid in position 12 (Gibbs *et al.*, 1984; McGrath *et al.*, 1984; Sweet *et al.*, 1984). These findings represent the first biochemical evidence of a functional difference between the gene product of a *ras* proto-oncogene and that of its transforming allele. Finally, microinjection of bacterially synthesized p21 proteins into rodent cell lines has provided direct evidence for the transforming activity of these proteins (Feramisco *et al.*, 1984; Stacey & Kung, 1984).

Computer analysis of the predicted amino acid sequences of normal and transforming p21 proteins predicts the existence of significant structural differences (Pincus *et al.*, 1983; Santos *et al.*, 1983). These observations raise the possibility that monoclonal antibodies may be elicited against structural domains specific for transforming p21 proteins. Existing monoclonal antibodies obtained from retrovirus-induced tumor-bearing rats do not exhibit such specificity (Furth *et al.*, 1982). The availability of purified normal and transforming p21 proteins will allow the design of appropriate immunization protocols to obtain monoclonal antibodies that may be capable of specifically recognizing transforming proteins. Such reagents should have a significant diagnostic value.

CONCLUDING REMARKS

Transforming *ras* genes identified in human and animal tumors have been characterized molecularly. As a consequence, the molecular basis for their malignant activation has been established. Point mutations leading to a single amino acid substitution in two critical domains of the *ras* gene product are sufficient to generate a transforming p21 protein. Although less compelling, there is also good evidence suggesting that *ras* genes can acquire malignant properties by quantitative mechanisms. Malignant transformation has been achieved *in vitro* by using either non-mutant *ras* genes placed under the regulatory control of retroviral enhancer sequences, or transfection assays in which multiple (>30) copies of a *ras* proto-oncogene become stably integrated in the recipient NIH3T3 cell. Thus, it is very likely that the amplified K-*ras* oncogenes detected in some human tumor cells may play a role in tumorigenesis by means of increasing expression of normal p21 proteins rather than by directing the synthesis of a structurally altered product. Yet, little is known regarding the mechanisms by which *ras* gene products induce malignant transformation and even less regarding their role in normal physiological processes. Availability of large amounts of purified p21 proteins and the possibility of genetic manipulation of *ras* genes in lower eukaryotic organisms should provide the necessary means to answer these fundamental questions.

At present, the pivotal question regarding transforming *ras* genes is whether they have a causative role in tumorigenesis. The apparent simplicity with which *ras* genes acquire transforming properties has become some sort of conceptual barrier to admitting that they play a fundamental role in the multi-step process of malignancy. To date, the most compelling evidence regarding a causative role of *ras* oncogenes in neoplasia comes from studies on chemical carcinogenesis. Transforming *ras* genes have been found in a high percentage (over 70%) of chemically induced tumors in a variety of animal model systems. Moreover, in each model system only a specific oncogene (so far any of the three known members of the *ras* family) becomes reproducibly activated. Recent results from our laboratory have extended the specificity of oncogene activation in NMU-induced mammary tumors to the nucleotide level. This tremendous specificity *in vivo* can only be explained if *ras* oncogenes play a major role in the initiation and/or maintenance of carcinogenesis. However, further efforts are necessary to establish the exact role of *ras* gene activation as well as to define other genetic and epigenetic steps needed to complete the neoplastic process.

Research sponsored by the National Cancer Institute, DHHS, under contract No. NO1-CO-23909 with Litton Bionetics, Inc. The contents of this publication do not necessarily reflect the views or policies of the Department of Health and Human Services, nor does mention of trade names, commercial products, or organizations imply endorsement by the US Government.

REFERENCES

BALMAIN, A. and PRAGNELL, I. B. (1983). Mouse skin carcinomas induced *in vivo* by chemical carcinogens have a transforming Harvey-*ras* oncogene. *Nature, London*, **303**, 72.

BALMAIN, A., RAMSDEN, M., BOWDEN, G. T. and SMITH, J. (1984). Activation of the mouse cellular Harvey-*ras* gene in chemically induced benign skin papillomas. *London, Nature*, **307**, 658.

BISHOP, J. M. (1983). Cellular oncogenes and retroviruses. *Annual Review of Biochemistry*, **52**, 301.

BLAIR, D. G., COOPER, C. S., OSKARSSON, M. K., EADER, L. A. and VANDE WOUDE, G. F. (1982). New method for detecting cellular transforming genes. *Science*, **218**, 1122.

BLAIR, D. G., OSKARSSON, M. K., WOOD, T. G., McCLEMENTS, W. L., FISHINGER, P. J. and VANDE WOUDE, G. F. (1981). Activation of the transforming potential of a normal cell sequence: a molecular model for oncogenesis. *Science*, **212**, 941.

CAPON, D. J., ELLSON, Y., LEVINSON, A., SEEBURG, P. H. and GOEDDEL, D. V. (1983*a*). Complete nucleotide sequences of the T24 bladder carcinoma oncogene and its normal homologue. *Nature, London*, **302**, 33.

CAPON, D. J., SEEBURG, P. H., McGRATH, J. P., HAYFLICK, J. S., EDMAN, U., LEVINSON, A. D. and GOEDDEL, D. V. (1983*b*). Activation of Ki-*ras2* gene in human colon and lung carcinomas by two different point mutations. *Nature, London*, **304**, 507.

CHANG, E. H., FURTH, M. E., SCOLNICK, E. M. and LOWY, D. R. (1982*a*). Tumorigenic transformation of mammalian cells induced by a normal gene homologous to the oncogene of Harvey murine sarcoma virus. *Nature, London*, **297**, 479.

CHANG, E. H., GONDA, M. A., ELLIS, R. W., SCOLNICK, E. M. and LOWY, D. R. (1982*b*). Human genome contains four genes homologous to transforming genes of Harvey and Kirsten murine sarcoma viruses. *Proceedings of the National Academy of Sciences, USA*, **79**, 4848.

COLLINS, S. and GROUDINE, M. (1982). Amplification of endogenous *myc*-related DNA sequences in a human myeloid leukemia cell line. *Nature, London*, **298**, 679.

COOPER, G. M. (1982). Cellular transforming genes. *Science*, **217**, 801.

DALLA FAVERA, R., WONG-STAAL, F. and GALLO, R. C. (1982). Oncogene amplification in promyelocytic leukemia cell line HL-60 and primary leukemic cells of the same patient. *Nature, London*, **299**, 61.

DeFeo-Jones, D., Scolnick, E., Koller, R. and Dhar, R. (1983). *ras*-related gene sequences identified and isolated from *Saccharomyces cerevisiae*. *Nature, London*, **306**, 707.

deMartinville, B., Cunningham, J. M., Murray, M. J. and Francke, U. (1983a). The N-*ras* oncogene assigned to the short arm of human chromosome 1. *Nucleic Acids Research*, **11**, 5267.

deMartinville, B., Giacalone, J., Shih, C., Weinberg, R. A. and Francke, U. (1983b). Oncogene from human EJ bladder carcinoma is located on the short arm of chromosome 11. *Science*, **219**, 498.

Der, C., Krontiris, T. and Cooper, G. (1982). Transforming genes of human bladder and lung carcinoma cell lines are homologous to the *ras* genes of Harvey and Kirsten sarcoma viruses. *Proceedings of the National Academy of Sciences, USA*, **79**, 3637.

Dhar, R., Nieto, A., Koller, R., DeFeo-Jones, D., Robinson, P., Temeles, G. and Scolnick, E. M. (1984). Nucleotide sequence of two H-*ras*-related genes isolated from the yeast *Saccharomyces cerevisiae*. *Nucleic Acids Research*, **12**, 3611.

Doolittle, R. F. (1981). Similar amino acid sequences: chance or common ancestry? *Science*, **214**, 149.

Ellis, R. W., DeFeo, D., Shih, T., Gonda, M., Young, H. A., Tsuchida, N., Lowy, D. and Scolnick, E. M. (1981). The p21 *src* genes of Harvey and Kirsten sarcoma viruses originate from divergent members of a family of normal vertebrate genes. *Nature, London*, **292**, 506.

Feinberg, A. P., Vogelstein, B., Droller, M. J., Baylin, S. B. and Nelkin, B. D. (1983). Mutations affecting the 12th amino acid of the c-Ha-*ras* oncogene product occur infrequently in human cancer. *Science*, **220**, 1175.

Frankel, A. E. and Fischinger, P. J. (1976). Nucleotide sequences in mouse DNA and RNA specific for Moloney sarcoma viruses. *Proceedings of the National Academy of Sciences, USA*, **73**, 3705.

Furth, M. E., Davis, L. J., Fleurdelys, B. and Scolnick, E. M. (1982). Monoclonal antibodies to the p21 products of the transforming gene of Harvey murine sarcoma virus and of a cellular *ras* gene family. *Journal of Virology*, **243**, 294.

Guerrero, I., Calzada, P., Mayer, A. and Pellicer, A. (1984). A molecular approach to leukemogenesis: mouse lymphomas contain an activated c-*ras* oncogene. *Proceedings of the National Academy of Sciences, USA*, **81**, 202.

Gullino, P. M., Pettigrew, H. H. and Grantham, F. H. (1975). *N*-Nitrosomethylurea as mammary gland carcinogen in rats. *Journal of the National Cancer Institute*, **54**, 401.

Hall, A., Marshall, C., Spurr, N. and Weiss, R. (1983). Identification of the transforming gene in two human sarcoma cell lines as a new member of the *ras* gene family located on chromosome 1. *Nature, London*, **303**, 396.

Harris, H., Miller, O. J., Klein, G., Worst, P. and Tachibaka, T. (1969). Suppression of malignancy by cell fusion. *Nature, London*, **223**, 363.

Hayward, W. S., Neel, B. G. and Astrin, S. M. (1981). Activation of a cellular *onc* gene by promoter insertion in ALV-induced lymphoid leukosis. *Nature, London*, **290**, 475.

Jahnwar, S. C., Neel, B. G., Hayward, W. S. and Chaganti, R. S. K. (1983). Localization of c-*ras* oncogene family on human germ-line chromosomes. *Proceedings of the National Academy of Sciences, USA*, **80**, 4794.

Kerr, I. B. and Spandidos, D. A. (1984). Elevated expression of the human *ras* oncogene family in premalignant and malignant tumours of the colorectum. *British Journal of Cancer*, **49**, 681–8.

KLEIN, G. (1983). The role of gene dosage and genetic transpositions in carcinogenesis. *Nature, London*, **294**, 313.

LACAL, J. C., SANTOS, E., NOTARIO, V., BARBACID, M., YAMAZAKI, S., KUNG, H., SEAMANS, C., McANDRE, S. and CROWL, R. (1984). Expression of normal and transforming H-*ras* genes in *E. coli* and purification of their encoded p21 proteins. *Proceedings of the National Academy of Sciences, USA*, **81**, 5305.

LANE, M. A., SAINTEN, A. and COOPER, G. (1981). Activation of related transforming genes in mouse and human mammary carcinomas. *Proceedings of the National Academy of Sciences, USA*, **78**, 5185.

LANE, M. A., SAINTEN, A. and COOPER, G. (1982). Stage-specific transforming genes of human and mouse B- and T-lymphocyte neoplasms. *Cell*, **28**, 875.

LAUTENBERGER, J. A., VESH, L., SHIH, T. Y. and PAPIS, T. S. (1983). High level expression in *E. coli* of enzymatically active Harvey murine sarcoma virus p21 *ras* protein. *Science*, **221**, 858.

McBRIDE, O. W., SWAN, D. C., SANTOS, E., BARBACID, M., TRONICK, R. and AARONSON, S. A. (1982). Localization of the normal allele of T24 human bladder carcinoma oncogene to chromosome 11. *Nature, London*, **300**, 773.

McCORMICK, D. L., ADAMOWSKI, C. B., FIKS, A. and MOON, R. C. (1981). Life-time dose–response relationships for mammary tumor induction by a single administration of *N*-methyl-*N*-nitrosourea. *Cancer Research*, **41**, 1690.

McCOY, M. S., TOOLE, J. J., CUNNINGHAM, J. M., CHANG, E. H., LOWY, D. R. and WEINBERG, R. A. (1983). Characterization of a human colon/lung carcinoma oncogene. *Nature, London*, **302**, 79.

McGRATH, J. P., CAPON, D. J., SMITH, D. H., CHEN, E. Y., SEEBURG, P. H., GOEDDEL, D. V. and LEVINSON, A. D. (1983). Structure and organization of the human Ki-*ras* proto-oncogene and a related processed pseudogene. *Nature, London*, **304**, 510.

MARSHALL, C. J., HALL, A. and WEISS, R. (1982). A transforming gene present in human sarcoma cell lines. *Nature, London*, **299**, 171.

NAKANO, H., NEVILLE, C., YAMAMOTO, F., GARCIA, J. L., FOGH, J. and PERUCHO, M. (1984*a*). Structure and mechanism of activation of the c-K-*ras* oncogene from two human lung tumors. In *Cancer Cells 2*, Oncogenes and Viral Oncogenesis, ed. G. Vande Woude *et al.*, pp. 447–54. New York, Cold Spring Harbor Laboratories.

NAKANO, H., YAMAMOTO, F., NEVILLE, C., EVANS, D., MIZUNO, T. and PERUCHO, M. (1984*b*). Isolation of transforming sequences of two human lung carcinomas: structural and functional analysis of the activated c-K-*ras* oncogenes. *Proceedings of the National Academy of Sciences, USA*, **81**, 71.

NOTARIO, V., SUKUMAR, S., SANTOS, E. and BARBACID, M. (1984). A common mechanism for the malignant activation of *ras* oncogenes in human neoplasia and in chemically induced animal tumors. In *Cancer Cells 2*, Oncogenes and Viral Oncogenesis, ed. G. Vande Woude *et al.*, pp. 425–32. New York, Cold Spring Harbor Laboratories.

O'BRIEN, S. J., NASH, W. G., GOODWIN, J. L., LOWY, D. R. and CHANG, E. H. (1983). Dispersion of the *ras* family of transforming genes to four different chromosomes in man. *Nature, London*, **302**, 839.

OHNO, S. (1973). Ancient linkage groups and frozen accidents. *Nature, London*, **244**, 259.

PAPAGEORGE, A. G., DeFEO-JONES, D., ROBINSON, P., TEMELES, G. and SCOLNICK, E. M. (1984). *Saccharomyces cerevisiae* synthesizes proteins related to the p21 gene product of *ras* genes found in mammals. *Molecular and Cellular Biology*, **4**, 23.

PAPAGEORGE, A. G., LOWY, D. and SCOLNICK, E. M. (1982). Comparative biochemical properties of p21 *ras* molecules coded for by viral and cellular *ras* genes. *Journal of Virology*, **44**, 1509–19.

PARADA, L. F., TABIN, C. J., SHIH, C. and WEINBERG, R. A. (1982). Human EJ bladder carcinoma oncogene is homologue of Harvey sarcoma virus *ras* gene. *Nature, London*, **297**, 474.

PERUCHO, M., GOLDFARB, M., SHIMIZU, K., LAMA, C., FOGH, J. and WIGLER, M. (1981). Human-tumor-derived cell lines contain common and different transforming genes. *Cell*, **27**, 467.

PINCUS, M. R., VAN RENSWOUDE, J., HARFORD, J. B., CHANG, E. H., CARTY, R. P. and KLAUSNER, R. D. (1983). Prediction of the three-dimensional structure of the transforming region of the EJ/T24 human bladder oncogene and its normal cellular homologue. *Proceedings of the National Academy of Sciences, USA*, **80**, 5253.

POWERS, S., KATAOKA, T., FASANO, O., GOLDFARB, M., STRATHERN, J., BROACH, J. and WIGLER, M. (1984). Genes in *S. cerevisiae* encoding proteins with domains homologous to the mammalian *ras* proteins. *Cell*, **36**, 607.

PULCIANI, S., SANTOS, E., LAUVER, A. V., LONG, L. K., ROBBINS, K. C. and BARBACID, M. (1982a). Oncogenes in human tumor cell lines: molecular cloning of a transforming gene from human bladder carcinoma cells. *Proceedings of the National Academy of Sciences, USA*, **79**, 2845.

PULCIANI, S., SANTOS, E., LAUVER, A. V., LONG, L. K., AARONSON, S. A. and BARBACID, M. (1982b). Oncogenes in solid human tumors. *Nature, London*, **300**, 539.

REDDY, E. P. (1983). Nucleotide sequence analysis of the T24 human bladder carcinoma oncogene. *Science*, **220**, 1061.

REDDY, E. P., REYNOLDS, R. K., SANTOS, E. and BARBACID, M. (1982). A point mutation is responsible for the acquisition of transforming properties by the T24 human bladder carcinoma oncogene. *Nature, London*, **300**, 149.

RHIM, J. S., PARK, D. K., ARNSTEIN, P., HUEBNER, R. J., WEISBURGER, E. K. and NELSON-REES, W. A. (1975). Transformation of human cells in culture by *N*-methyl-*N'*-nitro-*N*-nitrosoguanidine. *Nature, London*, **256**, 751.

ROWLEY, J. D. (1983). Human oncogene locations and chromosome aberrations. *Nature, London*, **301**, 290.

RYAN, J. P., BARKER, E., SHIMIZU, K., WIGLER, M. and RUDDLE, F. H. (1983). Chromosomal assignments of a family of human oncogenes. *Proceedings of the National Academy of Sciences, USA*, **80**, 4460.

SAKAGUCHI, A. Y., NAYLOR, S. L., SHOWS, T. B., TOOLE, J. J., McCOY, M. and WEINBERG, R. A. (1983). Human c-Ki-*ras*2 proto-oncogene on chromosome 12. *Science*, **219**, 1081.

SAKAGUCHI, A. Y., ZABEL, B. V., GRZESCHIK, K. H., LAW, M. L., ELLIS, R. W., SCOLNICK, E. M. and NAYLOR, S. L. (1984). Regional localization of two human cellular Kirsten *ras* genes on chromosomes 6 and 12. *Molecular and Cellular Biology*, **4**, 989.

SANTOS, E., MARTIN-ZANCA, D., REDDY, E. P., PIEROTTI, M. A., DELLA PORTA, G. and BARBACID, M. (1984). Malignant activation of a K-*ras* oncogene in lung carcinoma but not in normal tissue of the same patient. *Science*, **223**, 661.

SANTOS, E., REDDY, E. P., PULCIANI, S., FELDMAN, R. J. and BARBACID, M. (1983). Spontaneous activation of a human proto-oncogene. *Proceedings of the National Academy of Sciences, USA*, **80**, 4679.

SANTOS, E., TRONICK, S. R., AARONSON, S. A., PULCIANI, S. and BARBACID, M. (1982). T24 human bladder carcinoma oncogene is an activated form of the normal human homologue of BALB- and Harvey-MSV transforming genes. *Nature, London*, **298**, 343.

Schwab, M., Alitalo, K., Varmus, H. E., Bishop, J. M. and George, D. (1983). A cellular oncogene (c-Ki-*ras*) is amplified, overexpressed, and located within karyotypic abnormalities in mouse adrenocortical tumor cells. *Nature, London*, **303**, 497.

Scolnick, E. M., Papageorge, A. G. and Shih, T. Y. (1979). Guanine nucleotide-binding activity as an assay for *src* protein of rat-derived murine sarcoma virus. *Proceedings of the National Academy of Sciences, USA*, **76**, 5355–9.

Shih, C., Shilo, B., Goldfarb, M. P., Dannenberg, A. and Weinberg, R. A. (1979). Passage of phenotypes of chemically transformed cells via transfection of DNA and chromatin. *Proceedings of the National Academy of Sciences, USA*, **76**, 5714.

Shih, T. Y., Papageorge, A. G., Stokes, P. E., Weeks, M. O. and Scolnick, E. M. (1980). Guanine nucleotide-binding and autophosphorylating activities associated with the p21^{src} protein of Harvey murine sarcoma virus. *Nature, London*, **287**, 686–91.

Shih, T. Y., Weeks, M. O., Young, H. O. and Scolnick, E. M. (1979). Identification of a sarcoma virus-coded phosphoprotein in nonproducer cells transformed by Kirsten or Harvey murine sarcoma virus. *Virology*, **96**, 64.

Shimizu, K., Birnbaum, D., Ruley, M. A., Fasano, O., Suard, Y., Edlund, L., Taparowsky, E., Goldfarb, M. and Wigler, M. (1983*a*). Structure of the Ki-*ras* gene of the human lung carcinoma cell line Calu-1. *Nature, London*, **304**, 497.

Shimizu, K., Goldfarb, M., Perucho, M. and Wigler, M. (1983*b*). Isolation and preliminary characterization of the transforming gene of a human neuroblastoma cell line. *Proceedings of the National Academy of Sciences, USA*, **80**, 383.

Shimizu, K., Goldfarb, M., Suard, Y., Perucho, M., Li, Y., Kamata, T., Feramisco, J., Stavnezer, E., Fogh, J. and Wigler, M. (1983*c*). Three human transforming genes are related to the viral *ras* oncogenes. *Proceedings of the National Academy of Sciences, USA*, **80**, 2112.

Spandidos, D. A. and Wilkie, N. M. (1984). Malignant transformation of early passage rodent cells by a single mutated human oncogene. *Nature, London*, **310**, 469–75.

Stanbridge, E. J., Der, C. J., Doerson, C. J., Nishimi, R. Y., Peehl, D. M., Weissman, B. E. and Wilkinson, J. E. (1982). Human cell hybrids: analysis of transformation and tumorigenicity. *Science*, **215**, 252.

Stehelin, D., Varmus, H. E., Bishop, J. M. and Vogt, P. K. (1976). DNA related to the transforming gene(s) of avian sarcoma viruses is present in normal avian DNA. *Nature, London*, **260**, 170.

Stein, R. B., Robinson, P. S. and Scolnick, E. M. (1984). Photoaffinity labeling with GTP of viral p21 *ras* protein expressed in *Escherichia coli*. *Journal of Virology*, **50**, 343.

Sukumar, S., Notario, V., Martin-Zanca, D. and Barbacid, M. (1983). Induction of mammary carcinomas in rats by nitroso-methyl-urea involves the malignant activation of the H-*ras*-1 locus by single point mutations. *Nature, London*, **306**, 658.

Sukumar, S., Pulciani, S., Doniger, J., DiPaolo, J. A., Evans, C., Zbar, B. and Barbacid, M. (1984). A transforming *ras* gene in tumorigenic guinea pig cell lines initiated by diverse chemical carcinogens. *Science*, **223**, 1197.

Tabin, C. J., Bradley, S., Bargman, C., Weinberg, R., Papageorge, A., Scolnick, E., Dhar, R., Lowy, D. and Chang, E. (1982). Mechanism of activation of human oncogene. *Nature, London*, **300**, 143.

Taparowsky, E., Suard, Y., Fasano, O., Shimizu, K., Goldfarb, M. and Wigler, M. (1982). Activation of the T24 bladder carcinoma transforming gene is linked to a single amino acid change. *Nature, London*, **300**, 762.

TAPAROWSKY, E., SHIMIZU, K., GOLDFARB, M. and WIGLER, M. (1983). Structure and activation of the human N-*ras* gene. *Cell*, **34**, 581.

WEINBERG, R. A. (1982). Fewer and fewer oncogenes. *Cell*, **30**, 3.

WILLINGHAM, M. C., PASTAN, I., SHIH, T. Y. and SCOLNICK, E. M. (1980). Localization of the *src* gene product of the Harvey strain of MSV to plasma membrane of transformed cells by electron microscopic immunocytochemistry. *Cell*, **19**, 1005–14.

YUASA, Y., STRIVASTAVA, S. K., DUNN, C. Y., RHIM, J. S., REDDY, E. P. and AARONSON, S. A. (1983). Acquisition of transforming properties by alternative point mutations within c-*bas*/*has* human proto-oncogene. *Nature, London*, **303**, 775.

References added in proof

FERAMISCO, J. R., GROSS, M., KOMATA, T., ROSENBERG, M. and SWEET, R. W. (1984). Microinjection of the oncogene form of the human H-*ras* (T24) protein results in rapid proliferation of quiescent cells. *Cell*, **38**, 109.

GIBBS, J. B., SIGAL, I. S., POE, M. and SCOLNICK, E. M. (1984). Intrinsic GTPase activity distinguishes normal and oncogenic *ras* p21 molecules. *Proceedings of the National Academy of Sciences, USA*, **81**, 5704.

GOYETTE, M., PETROPOULOS, C. J., SHANK, P. R. and FAUSTO, N. (1983). Expression of a cellular oncogene during liver regeneration. *Science*, **219**, 510.

GOYETTE, M., PETROPOULOS, C. J., SHANK, P. R. and FAUSTO, N. (1984). Regulated transcription of c-Ki-*ras* and c-*myc* during compensataory growth of rat liver. *Molecular and Cellular Biology*, **4**, 1493.

HORAN-HAND, P., THOR, D., WUNDERLICH, D., MURARO, R., CARUSO, A. and SCHLOM, J. (1984). Monoclonal antibodies of predefined specificity detect activated *ras* gene expression in human mammary and colon carcinomas. *Proceedings of the National Academy Sciences, USA*, **81**, 5227.

KATAOKA, T., POWERS, S., McGILL, C., FASANO, O., STRATHERN, J., BROACH, J. and WIGLER, M. (1984). Genetic analysis of yeast *ras*1 and *ras*2 genes. *Cell*, **37**, 437.

LAND, H., PARADA, L. F. and WEINBERG, R. A. (1983). Tumourigenic conversion of primary embryo fibroblasts requires at least two cooperating oncogenes. *Nature*, **304**, 596–602.

McGRATH, J. P., CAPON, D., GOEDDEL, D. and LEVINSON, A. (1984). Comparative biochemical properties of normal and activated human *ras* p21 protein. *Nature*, **310**, 644.

SEFTON, B. M., TROWBRIDGE, I. S. and COOPER, J. A. (1982). The transforming proteins of Rous sarcoma virus, Harvey sarcoma virus and Abelson sarcoma virus contain tightly bound lipid. *Cell*, **31**, 465.

SLAMON, D. J., deKERNION, J. B., VERMA, I. and CLINE, M. J. (1984). Expression of cellular oncogenes in human malignancies. *Science*, **244**, 256.

STACEY, D. W. and KUNG, H. F. (1984). Transformation of NIH3T3 cells by microinjection of Ha-*ras* p21 protein. *Nature*, **310**, 508.

SWEET, R. W., YOKOHAMA, S., KOMATA, T., FERMISCO, J. R., ROSENBERG, M. and GROSS, M. (1984). The product of *ras* is a GTPase and the T24 oncogenic mutant is deficient in this activity. *Nature*, **311**, 273.

TAINSKY, M. A., COOPER, C. S., GIOVANELLA, B. C. and VANDE WOUDE, G. F. (1984). An activated *ras*[N] gene: Detected in late but not early passage human PA1 teratocarcinoma cells. *Science*, **225**, 643.

THOR, A., HORAN-HAND, P., WUNDERLICH, D., CARUSO, A., MURARO R. and SCHLOM, J. (1984). Monoclonal antibodies define differential *ras* gene expression in malignant and benign colonic diseases. *Nature*, **311**, 562.

INDEX